Introduction to Flight Testing

Aerospace Series

Introduction to Flight Testing

James W. Gregory
The Ohio State University

Tianshu Liu
Western Michigan University

This edition first published 2021
© 2021 John Wiley & Sons Ltd.

The right of James W. Gregory and Tianshu Liu to be identified as the authors of this work has been asserted in accordance with law.

Registered Office
John Wiley & Sons, Inc., 111 River Street, Hoboken, NJ 07030, USA

Editorial Office
111 River Street, Hoboken, NJ 07030, USA

For details of our global editorial offices, customer services, and more information about Wiley products visit us at www.wiley.com.

Wiley also publishes its books in a variety of electronic formats and by print-on-demand. Some content that appears in standard print versions of this book may not be available in other formats.

Library of Congress Cataloging-in-Publication Data

Names: Gregory, James W., author. | Liu, T. (Tianshu), author.
Title: Introduction to flight testing / James W. Gregory, The Ohio State
    University; Tianshu Liu, Western Michigan University.
Description: First edition. | Hoboken, NJ : Wiley, 2021. | Includes
    bibliographical references and index.
Identifiers: LCCN 2020048350 (print) | LCCN 2020048351 (ebook) | ISBN
    9781118949825 (hardback) | ISBN 9781118949795 (adobe pdf) | ISBN
    9781118949801 (epub)
Subjects: LCSH: Airplanes–Flight testing–Textbooks.
Classification: LCC TL671.7 .G74 2021 (print) | LCC TL671.7 (ebook) | DDC
    629.134/53–dc23
LC record available at https://lccn.loc.gov/2020048350
LC ebook record available at https://lccn.loc.gov/2020048351

Cover Design: Wiley
Cover Image: © NNehring/Getty Images

Set in 9.5/12.5pt STIXTwoText by SPi Global, Chennai, India

10 9 8 7 6 5 4 3 2 1

Dedicated to
Deb, Alina, and Maggie – J. W. G.
Ruomei and Ranya – T. L.

Contents

About the Authors

James W. Gregory is professor and chair of the Department of Mechanical and Aerospace Engineering at The Ohio State University (OSU). He received his Bachelor of Aerospace Engineering from Georgia Tech in 1999 and his PhD in Aeronautics and Astronautics from Purdue University in 2005. He has been a faculty member at OSU since 2008 and served as Director of OSU's Aerospace Research Center from 2017 to 2020. In 2017, he led a team of research staff and students to set FAI/NAA-sanctioned world records for speed and distance for an autonomous drone. He teaches classes at OSU on *Flight Test Engineering* and *Introduction to Aerospace Engineering*. Prof. Gregory also recorded a series of video lectures on the *Science of Flight*, produced by the Great Courses. He is an instrument-rated commercial pilot and holds a remote pilot certificate.

 Tianshu Liu is a professor in Department of Mechanical and Aerospace Engineering at Western Michigan University (WMU). He received a PhD in Aeronautics and Astronautics from Purdue University in 1996. He was a research scientist at NASA Langley Research Center from 1999 to 2004. His research focuses on experimental aerodynamics and fluid mechanics, particularly on global measurement techniques for various physical quantities such as pressure, temperature, heat flux, skin friction, velocity, aeroelastic deformation, and aerodynamic force. He teaches classes in aerodynamics and flight testing at WMU.

Series Preface

The field of aerospace is multidisciplinary and wide-ranging, covering a large variety of platforms, disciplines, and domains, not merely in engineering but in many related supporting activities. These combine to enable the aerospace industry to produce innovative and technologically advanced vehicles. The wealth of knowledge and experience that has been gained by expert practitioners in the various aerospace fields needs to be passed onto others working in the industry and also researchers, teachers, and the student body in universities.

The *Aerospace Series* aims to be a practical, topical, and relevant series of books aimed at people working in the aerospace industry, including engineering professionals and operators, engineers in academia, and allied professions such as commercial and legal executives. The range of topics is intended to be wide-ranging, covering design and development, manufacture, operation and support of aircraft, as well as topics such as infrastructure operations and current advances in research and technology.

Flight testing is a vital part of the certification and validation phase of all new aircraft and is performed to determine or verify the performance and handling qualities. Although the flight characteristics are predicted in the design and development stages of new aircraft programs, the real-world capabilities are not known until the aircraft is flown and tested. Most aircraft flight testing programs are focused on meeting airworthiness certification requirements and demonstrate all aspects of the flight vehicle's performance and handling characteristics to ensure flight safety.

This book, *Introduction to Flight Testing*, is aimed at advanced-level undergraduate students, graduate students, and practicing engineers who are looking for an introduction to the field of flight testing. With a focus on light aircraft and UAVs, the book covers the engineering fundamentals of flight, including the flight environment, aircraft performance and stability and control, combined with the piloting, sensors, and digital data acquisition and analysis required to perform flight tests. This book is a very welcome addition to the Wiley Aerospace Series.

October 2020   *Peter Belobaba, Jonathan Cooper and Allan Seabridge*

Preface

The goal of this book is to provide an accessible introduction to the fascinating and intriguing world of aircraft flight testing. This unique discipline directly straddles the domains of engineering and piloting, requiring knowledge of both the theory and practice of flight. Our target audience is advanced-level undergraduate students, beginning graduate students, and practicing engineers who are looking for an introduction to the field of flight testing. Flight testing professionals (engineers, pilots, managers, etc.) may also find this to be a helpful resource if they wish to solidify their understanding of the fundamentals beyond what is provided in most other flight testing resources. We have attempted to write this book in an engaging, conversational style that invites the reader into understanding the fundamental principles.

Both authors teach a senior-year technical elective course at our home universities on the topic of flight test engineering. Within this context, we have found that students best learn the material when they actively engage with flight testing practice. Experiencing flight in an aircraft is the best way to develop a tacit understanding of the principles of flight, to augment and deepen the intellectual knowledge of engineering practice that students receive in the classroom. In working with our senior-year engineering students, we have developed the following learning objectives for our courses, as well as for this book. Our aim is that readers of this book will:

- Have an appreciation for the purpose, scope, and magnitude of historical and modern flight test programs in the commercial and military aircraft sectors.
- Understand the theoretical foundations of the flight environment, aircraft performance, and stability and control as it applies to flight testing.
- Be familiar with aircraft cockpit instrumentation, supplemental sensors for flight testing, and digital data acquisition techniques.
- Be able to plan a flight test to evaluate the performance or handling qualities of a general aviation or unmanned aircraft.
- Have the ability to coordinate with an experienced pilot to successfully conduct the flight test.
- Have the knowledge and background needed to perform postflight analysis and data reduction.
- Be able to professionally and succinctly communicate the findings of a flight test program via oral and written communication.
- Have a general familiarity with piloting, aviation weather, and flight planning.

Thus, this book is best approached in conjunction with flight in an actual aircraft. No specially instrumented aircraft are needed in order to do most of the flight tests in this book. No exceptional training is needed in order to fly the basic maneuvers described in this book. In fact, all of the procedures described herein are routine maneuvers that are encountered often in piloting practice. An interested reader can simply head to their nearest airport and work with a qualified, professional pilot (*e.g.*, a certified flight instructor) to conduct the flights described here. In our educational contexts, we collaborate with flight instructors in the aviation programs at The Ohio State University and Western Michigan University to conduct flight tests for our students, but any flight instructor would be capable of performing these basic maneuvers.

We must be sure to emphasize that flying an aircraft involves elevated risk compared to other routine activities in daily life. It is critical for the pilot in command to always maintain positive control of the aircraft and to maintain flight within the performance envelope of the aircraft. All operating limitations of the Pilot's Operating Handbook, as well as all regulatory limitations and best practices for safety of flight, must be observed. Flights should be conducted with a minimum crew of two, where the pilot is solely focused on safe operation of the aircraft. Since precise flying is important for acquiring quality data, the pilot should be experienced – a pilot with a commercial license is likely a safe minimum standard for piloting credentials. The second crew member – the flight test engineer – should be dedicated to acquiring flight data and not have any responsibilities related to ensuring safe operation of flight. The flight test engineer is essentially a passenger for these flights, and all piloting authority and responsibility for the flight rest with the pilot in command. Chapter 6 of this book describes the principles of flight test safety and risk management, which form an essential foundation for the flight test profession. Fly safely!

While there are several other resources on flight testing already available, we saw a specific need for this textbook. Some of the existing resources are targeted toward flight testing professionals and may not be as accessible to the general student. Other resources have become dated, with the relatively recent rewrite of airworthiness certification standards for normal category airplanes (Title 14 of the U.S. Code of Federal Regulations, Part 23). In writing this book, we have sought to provide a modern and accessible resource for flight test educators and students, with several unique features that we hope will set it apart as a helpful and leading resource. Our primary audience is engineering students, with the goal of drawing connections between engineering practice and flight testing experience. We have provided guidelines on how to conduct each flight test, which will guide the reader in the flight test planning process. We have also included unique chapters on digital data acquisition and analysis techniques, uncertainty analysis, and unmanned aircraft flight testing. These are all modern topics that are not covered in the flight testing literature, but are now critical topics. And, with the proliferation of smartphones (repurposed as digital data acquisition devices in manned flight testing) and drones, the modern principles of flight testing are more accessible than ever.

The focus of our book is predominantly on light aircraft (small general aviation airplanes) and small unmanned aircraft. We have homed in on this subset of aviation since these aircraft are generally accessible to the public. While our focus is on light aircraft, the principles described here are equally applicable to all regimes of flight testing. This book provides an introduction, while other resources can be consulted for more advanced topics. The discussion here has been tailored to academic classroom instruction to convey the main principles of flight testing, rather than as a "field manual" for definitive best practices in all situations for flight testing. Having said that, we have made a reasonable effort to align the guidance provided here with accepted best practices. Also, we have decided to omit discussions of spin flight testing and flutter flight testing. These are significant and important topics in flight testing practice, but these are hazardous flight tests. We wish to encourage the reader to engage only with the safer dimensions of flight testing as an entry point.

Throughout this book, we'll predominantly use English units. This is primarily because aviation practice in North America has mostly converged on English units. For example, most air traffic control organizations around the world assign altitude in units of feet and airspeed in units of knots. While SI units are generally preferable in science and engineering environments, we'll generally work with the aviation standard. We view this choice as an educational opportunity for the reader to become acquainted with and proficient in multiple unit systems. There is clear pedagogical value in learning how to quickly convert between and track various units – we hope that students and professional engineers alike will become comfortable with all units and how to convert between them. Appendix B includes a range of unit conversion factors and discussion on best practices for handling units in aviation and engineering practice.

Furthermore, we have avoided embedding implied units into equations. This practice can be convenient for some cases when input and output units for a formula are well established and clearly documented – this can facilitate situations where rapid computations are needed without encumbering the analysis with unit conversions. However, in many cases this practice leads to confusion or ambiguity since the input and output units are seldom clearly documented or agreed upon. Another disadvantage of embedded units is that constants must be embedded in the formula, which have no basis on the physics. This can be confusing to a student who is exposed to theory for the first time. Finally, embedded units force the reader into one specific unit system. Our approach with equations that are unit-agnostic will allow the reader to use either English or SI consistent units as desired. Thus, the assumption throughout this text (unless otherwise specified) is that equations are based on standard, consistent units.

To the reader – thank you for picking up this book. We are passionate about flight testing and are eager to share our deep interest in this domain with you. We hope that this book will be rewarding, enriching, and fascinating.

September 6, 2020

James W. Gregory
Columbus, Ohio

Tianshu Liu
Kalamazoo, Michigan

Acknowledgements

The authors are grateful for a wide array of colleagues, collaborators, and students who have shaped our thinking and provided input and feedback on this book.

JG would like to thank Profs. Stacy Weislogel and Gerald M. Gregorek, who as pioneers of educational flight testing in the 1970s were inspirations for this work. Prof. Hubert C. "Skip" Smith was also generous with ideas and resources along the way. The flight education department and colleagues in Aviation at The Ohio State University have been extremely helpful in providing tactical support over the years – D. Gelter, D. Hammon, B. Mann, S. Morgan, S. Pruchnicki, C. Roby, B. Strzempkowski, and S. Young. Special thanks go to Profs. Jeffrey Bons and Cliff Whitfield, who helped review significant portions of a near-final version of the manuscript. (Any remaining errors or inaccuracies are solely the responsibility of the authors). Portions of this book were written in 2014–15 while JG was on sabbatical at the Technion in Israel; the support of Ohio State University and the Fulbright foundation is gratefully acknowledged.

JG also wishes to extend special thanks to Dr. Matt McCrink, who assisted with many of the flight tests presented in this book and coauthored the final chapter on UAV flight testing. He has been an instrumental sounding board and key partner throughout this project.

TL would like to thank M. Schulte, M. Mandziuk, S. Yurk, M. Grashik, S. Woodiga, P. Wewengkang, D. M. Salazar, and WMU's College of Aviation.

Numerous colleagues, including M. Abdulrahim, J. Baughn, K. Colvin, C. Cotting, K. Garman, B. Gray, C. Hall, J. Jacob, J. Kidd, K. Kolsti, J. Langelaan, B. Martos, N. Sarigul-Klijn, R. Smith, J.P. Stewart, A. Suplisson, A. Tucker, J. Valasek, C. Walker, O. Yakimenko, and M. Yukish, helped influence the presentation of ideas in this book. The Flight Test Education Workshop, hosted by the USAF Test Pilot School, was a particularly helpful resource for materials and connections. A. Bertagnolli of Continental Aerospace Technologies (Continental Motors) and J. J. Frigge of Hartzell Propeller generously provided data and resources for the text. C. Daniloff, K. King, R. Heidersbach, H. Henley, K. King, N. Lachendro, H. Rice, H. Sakaue, B. Stirm, and R. Winiecki gracefully allowed their images or likenesses to be published in this book.

JG and TL also wish to thank their doctoral advisor, Prof. J. P. Sullivan of Purdue University, for his encouragement for us to pursue and finish this project.

Finally, JG and TL especially thank their families for their patience and support over the many years it took to complete this project.

About the Companion Website

The companion website for this book is at

https://www.wiley.com/go/flighttesting

Scan this QR code to visit the companion website.

1

Introduction

Flight testing is seemingly the stuff of legends, with tales of derring-do and bravery, spearheaded by great pilots such as Yeager, Armstrong, Glenn, and others. But what exactly is flight testing all about? What is being tested, and why? What's the difference between a test pilot and a flight test engineer? Is flight testing an inherently dangerous or risky activity?

With this book, we hope to show that flight testing is both exciting and accessible – we hope to make flight testing understandable and achievable by the typical undergraduate aerospace engineering student. The basic principles of flight testing can be explored in any aircraft, all the while remaining safely well within the standard operating envelope of an aircraft. This book will introduce students to the principles that experienced flight test engineers work with as they evaluate new aircraft systems.

Flight testing is all about determining or verifying the performance and handling qualities of an aircraft. These flight characteristics may be predicted in the design and development stages of a new aircraft program, but we never really know the exact capabilities until the full system is flown and tested. Most aircraft flight testing programs are focused on airworthiness certification, which is the rigorous demonstration of all facets of the flight vehicle's performance and handling characteristics in order to ensure safety of flight.

We also wish to highlight that most flight testing should not incur the levels of risk and danger that we associate with the great test pilots of the 20th century. Their bravery was indeed laudable, since they ventured into flight that no human had done before, such as breaking the "sound barrier" or being the first person to walk on the Moon. But, if done correctly, flight testing should be a methodical process where risks are managed at an acceptable level, where human life and property are not exposed to undue risk. Even more hazardous flight testing such as flutter boundary determination or spin recovery should be done in a methodical, well-controlled manner that mitigates risk. In fact, most flight testing, at least to an experienced professional, can be almost mundane (Corda 2017).

Nor is flight testing an individualistic activity where an intrepid pilot relies solely on their superlative piloting skills to push the aircraft to its limits, as suggested by the caricature in Figure 1.1. Quite the contrary, flight testing is a team effort with many individuals carefully contributing to the overall success of a flight testing program (see Figure 1.2). There is, of course, a pilot involved whose job it is to fly the aircraft as precisely and accurately as possible to put the aircraft through the necessary maneuvers to extract the needed performance or handling data. If an aircraft can carry more than just the pilot, then there is almost always a flight test engineer on board. The flight test engineer is responsible for preparing the plan for the flight test and for acquiring the data in flight while the pilot puts the aircraft through the required maneuvers. Beyond the role of the flight test engineer, there are many others involved including those who monitor systems and downlinked data on the ground, data analysts who post-process and interpret

Introduction to Flight Testing, First Edition. James W. Gregory and Tianshu Liu.
© 2021 John Wiley & Sons Ltd. Published 2021 by John Wiley & Sons Ltd.
Companion website: https://www.wiley.com/go/flighttesting

Figure 1.1 The caricature view of flight test is of an individualistic, cowboy-like, rugged test pilot who single-handedly defies danger. Here, Joe Walker playfully boards the Bell X-1A in a moment of levity. Source: NASA.

Figure 1.2 A more realistic view of the people behind flight testing – a team effort is required to promote safety and professionalism of flight. Depicted here is the team of NACA scientists and engineers who supported the XS-1 flight test program. Source: NASA.

the data after the test is complete, and program managers who set the strategic direction for the program and make budgetary decisions.

Flight testing is a critical endeavor in the overall design cycle of a new aircraft system. The main objective is to prove out the assumptions that are inherent to every design process and to discover any hidden anomalies in the performance of the aircraft system. Aircraft design typically proceeds by drawing upon historical data to estimate the performance of a new aircraft concept, but there is always uncertainty in those design estimates. The initial stages of design have very crude estimates made for a wide range of parameters and theories applied to the design. Over time, the design team reduces the uncertainty in the design by refining the analysis with improved design tools and higher-fidelity (more expensive!) analysis, wind tunnel testing, and ground testing of functional systems and even the entire aircraft. But, then the moment of truth always comes, where it is time for first flight of the aircraft. It is at this point that the flight test team documents the true performance of the airplane. If differences arise between actual and predicted performance, minor tweaks to the design may be needed (*e.g.*, the addition of vortex generators on the wings). Also, the insight gleaned from flight testing is documented and fed back into the design process for future aircraft.

This chapter will provide a brief overview of the flight testing endeavor through a historical anecdote that illustrates the key outcomes of flight testing, how flight testing is actually done, and the roles of all involved. Following this, we'll take a look at the various kinds of flight testing that are done, with a particular emphasis on airworthiness certification of an aircraft, which is the main objective of many flight testing programs. We'll then conclude this chapter with an overview of the rest of the book, including our objectives in writing this book and what we hope the reader will glean from this text.

1.1   Case Study: Supersonic Flight in the Bell XS-1

A great way to learn about the essential elements of a successful flight test program is to look at a historical case study. We'll consider the push by the Army Air Forces (AAF) in 1947 to fly an aircraft faster than the speed of sound. Along the way, we'll pick up some insight into how flight testing is done and some of the values and principles of the flight test community.

At the time, many scientists and engineers did not think that supersonic flight could be achieved. They observed significant increases in drag as the flight speed increased. On top of that, there were significant loss-of-control incidents where pilots found that their aircraft could not be pulled out of a high-speed dive. These highly publicized incidents led some to conclude that the so-called "sound barrier" could not be broken. We now know, however, that this barrier only amounted to a lack of insight into the physics of shock–boundary layer interaction, shock-induced separation, and the transonic drag rise, along with a lack of high-thrust propulsion sources to power through the high drag. Scientific advancements in theoretical analysis, experimental testing, and flight testing, along with engineering advancements in propulsion and airframe design, ultimately opened the door to supersonic flight.

In a program kept out of public sight, the U.S. Army Air Forces, the National Advisory Committee for Aeronautics (NACA, the predecessor to NASA), and the Bell aircraft company collaborated on a program to develop the Bell XS-1 with the specific intent of "breaking the sound barrier" to supersonic flight. (Note that the "S" in XS-1 stands for "supersonic"; this letter was dropped early in the flight testing program, leaving us with the commonly known X-1 notation.) The XS-1 (see Figure 1.3) was a fixed-wing aircraft with a gross weight of 12,250 lb, measured 30-ft

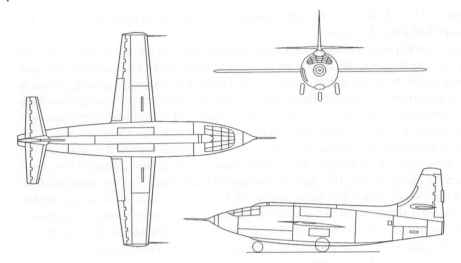

Figure 1.3 Three-view drawing of the Bell XS-1. Source: NASA, X-1/XS-1 3-View line art. Available at http://www.dfrc.nasa.gov/Gallery/Graphics/X-1/index.html.

11-in. long, had a straight (unswept) wing with an aspect ratio of 6.0 and a span of 28 ft, and an all-moving horizontal tail (a detail that we'll soon see was important!). The XS-1 was powered by a four-chamber liquid-fueled rocket engine producing 6000 lb of thrust. The overarching narrative of the program is well documented in numerous historical and popular sources (*e.g.*, see Young 1997; Gorn 2001; Peebles 2014; Hallion 1972; Hallion and Gorn 2003; or Wolfe 1979), but we'll pick up the story in the latter stages of the flight test program at Muroc Army Airfield, positioned on the expansive Rogers Dry Lake bed that is today the home of Edwards Air Force Base and NASA Armstrong Flight Research Center.

The XS-1 had an aggressive flight test schedule, with not too many check-out flights before going for the performance goal of supersonic flight. The extent of the test program was actually a matter of contentious debate between the AAF, the NACA, and Bell. In the end, Bell dropped out of the mix for contractual and financial reasons, and the NACA and AAF proceeded to collaborate on the flight test program. But the continued collaboration was not without tension. The AAF leaders and pilots continually pushed for an aggressive flight test program, making significant steps with each flight. The NACA, on the other hand, advocated for a much slower, methodical pace where substantial data would be recorded with each flight and carefully analyzed before proceeding on to the next boundary. In the end, the AAF vision predominantly prevailed, although there was a reasonable suite of instrumentation on board the aircraft. The XS-1 was outfitted with a six-channel telemeter, where NACA downlinked data on airspeed, altitude, elevator position, normal acceleration, stabilizer position, aileron position, and elevator stick force, along with strain gauges to record airloads and vibrations (Gorn 2001, p. 195). On the ground, the NACA crew had five trucks to support the data acquisition system – one to supply power, one for telemetry data, and three for radar. The radar system was manually directed through an optical sight, but if visual of the aircraft was lost, the radar system could be switched to automatic direction finding (Gorn 2001, pp. 187–188).

To lead the flying of the aircraft toward the perceived "sound barrier," the AAF needed a pilot with precision flying capabilities, someone who was unflappable under pressure, and someone with scientific understanding of the principles involved. The Army turned to Captain Charles E. "Chuck" Yeager – a young, 24-year-old P-51 ace from World War II – for the honor and responsibility of being primary pilot. According to Colonel Albert Boyd who selected him, Yeager had impeccable instinctive piloting skills and could work through the nuance of the aircraft's response to figure out exactly how it was performing (Young 1997, p. 41). Not only could he fly with amazing skill, but the engineering team on the ground loved him for his postflight debriefs. Yeager was able to return from a flight and relate in uncanny detail exactly how the aircraft responded to his precise control inputs, all in a vernacular that the engineering staff could immediately appreciate (Peebles 2014, p. 29). But it wasn't just Yeager doing all of the work – he had a team around him. Backing him up and flying an FP-80 chase plane was First Lieutenant Robert A. "Bob" Hoover, who was also well known as an exceptional pilot. Captain Jackie L. "Jack" Ridley, an AAF test pilot and engineer with an MS degree from Caltech, was the engineer in charge of the project. Others involved included Major Robert L. "Bob" Cardenas, pilot of the B-29 Superfortress carrier aircraft and officer in charge, and Lieutenant Edward L. "Ed" Swindell, flight engineer for the B-29. Backing up these AAF officers was Richard "Dick" Frost, a Bell engineer and test pilot who already had flight experience in the XS-1 and got Yeager up to speed on the intricacies of the aircraft. This cast of characters is depicted in Figure 1.4.

Beyond this core group of military professionals was a team of NACA scientists and engineers led by Walt Williams (see Figure 1.2). This team was focused predominantly on understanding the flight physics in this exploratory program, providing deep technical insight and support to the Air Force crew. Yet, this objective was inherently at odds with the AAF's stated desire to push to supersonic flight as quickly and safely as possible. This tension was aptly described by Williams: "We were enthusiastic, there is little question. The Air Force group – Yeager, Ridley – were very, very enthusiastic. We were just beginning to know each other, just beginning to work together. There had to be a balance between complete enthusiasm and the hard, cold facts. We knew that if this program should fail, the whole research airplane program would be set back. So, our problem became one of maintaining the necessary balance between enthusiasm and eagerness to get

Figure 1.4 The Air Materiel Command XS-1 flight test team, composed of (from left to right): Ed Swindell (B-29 Flight Engineer), Bob Hoover (XS-1 Backup and Chase Pilot), Bob Cardenas (Officer-in-charge and B-29 Pilot), Chuck Yeager (XS-1 Pilot), Dick Frost (Bell Engineer), and Jack Ridley (Project Engineer). Source: U.S. Air Force.

the job completed with a scientific approach that would assure success of the program. That was accomplished" (Gorn 2001, pp. 194–195).

In the run-up to the first supersonic flight, the team carefully pushed forward. On Yeager's first powered flight on August 29, 1947, he accelerated up to Mach 0.85, exceeding the planned test point of Mach 0.8. This negated NACA's need to acquire telemetered data in the Mach 0.8–0.85 range, leading to further tension between Yeager and Williams. In Yeager's words, "They [the NACA engineers and technicians] were there as advisers, with high-speed wind tunnel experience, and were performing the data reduction collected on the X[S]-1 flights, so they tried to dictate the speed in our flight plans. Ridley, Frost, and I always wanted to go faster than they did. They would recommend a Mach number, then the three of us would sit down and decide whether or not we wanted to stick with their recommendation. They were so conservative that it would've taken me six months to get to the [sound] barrier" (Young 1997, p. 51 – quoted from Yeager and Janos (1985), p. 122).

Yeager was admonished by Colonel Boyd to cooperate more carefully with the NACA technical specialists and to follow the test plan. This led to careful preflight briefings that, while Yeager considered to be tedious, were essential to flight safety and accomplishment of the test objectives. At each briefing, Williams would review the lessons learned from the previous flight and detail the objectives of the upcoming mission (Gorn 2001, pp. 195–196).

As Yeager flew at progressively higher flight speeds, he noticed significant changes in the trim condition of the aircraft. At certain Mach numbers, the trim condition would change nose-up, and at other Mach numbers it would trend toward nose-down, all accompanied by buffeting at various flight conditions. For example, on one flight at Mach 0.88 and 40,000 ft, Yeager was unable to put the aircraft into a light stall (even with the stick full aft), due to the lack of control authority. Then, on October 10, 1947, Yeager piloted another mission in a series of powered flights to ever-higher Mach numbers to test the response of the aircraft in this untested regime. After accelerating up to an indicated Mach number of 0.94 at an altitude of 40,000 ft, Yeager found that he had lost virtually all pitch control! He moved the control stick full fore and aft, yet obtained very little pitch response. Fortunately, the XS-1 was still stable at this flight condition, if not controllable. At this point, Yeager cut off the engines and came back for a landing on the expansive Rogers lakebed (Young 1997).

All of these various anomalies were due to compressibility effects, which were only poorly understood at the time. As the aircraft exceeded the critical Mach number, shock waves would form at various locations on the aircraft body. Furthermore, these shock waves could move substantially, with only a minor adjustment in freestream Mach number. Since there is a significant pressure gradient across a shock wave, this could result in dramatic changes in the forces and moments produced on control surfaces, and the strong pressure gradient across the shock would often lead to boundary layer separation. Thus, if a shock happened to be present at a hinge line for the elevator, the shock-induced boundary layer separation would create a thick unsteady wake flow over the elevator, causing the dynamic pressure on this control surface to drop dramatically and the elevator to lose effectiveness. With some foresight, researchers at NACA and designers at Bell anticipated this eventuality and designed the XS-1 to enable pitch control by moving the incidence angle of the entire horizontal tail (rather than inducing pitch changes using the elevator alone). So, as Yeager and Ridley discussed the phenomena occurring on October 10 and earlier, Ridley encouraged Yeager to adjust the horizontal tail angle of incidence to achieve pitch control, instead of using the elevator.

The plan for the next flight was to go for it – Yeager's intent was to fly supersonic. However, with the technical uncertainty associated with loss of elevator control and shock-induced buffeting, the NACA engineering team admonished Yeager to not exceed Mach 0.96 unless he was completely certain that he could do so safely. Beyond the NACA team, however, Jack Ridley was the one whom Yeager trusted the most. Ridley thought Yeager would be just fine controlling pitch with the moving horizontal tail, actuating it in increments of a quarter or a third of a degree to achieve pitch control without using the elevator. Ridley explained, "It may not be much, and it may feel ragged to you up there, but it will keep you flying" (Young 1997, p. 56). Yeager trusted Ridley implicitly – much more so than the NACA team. He later recounted, "I trusted Jack with my life. He was the only person on earth who could have kept me from flying the X [S]-1" (Young 1997, p. 56).

So, on the morning of October 14, 1947, Yeager set out with his team to fly faster than Mach 1. With Cardenas at the controls of the B-29, the Superfortress carried the Bell XS-1 up to altitude. On the way up, Bob Hoover and Dick Frost joined up in formation in their FP-80s. Hoover positioned himself in the "high chase" position: 10 mi ahead of the B-29 at an altitude of 40,000 ft, to give Yeager an aiming point as he climbed and accelerated in the XS-1. Frost joined up slightly to the right of and behind the B-29 in order to observe the rocket firing and drop of the XS-1.

When everyone was ready, Cardenas put the B-29 in a slight dive and started a countdown: "10-9-8-7-6-5-3-2-1" (yes, he skipped "4"!, as he often skipped a number on these flights) and pulled the release mechanism at 10:26 a.m. an altitude of 20,000 ft and an airspeed of 250 knots. This airspeed was slightly lower than planned, causing the XS-1 to nearly stall. Yeager pitched the nose down to regain airspeed and then lit all four burners to rapidly accelerate upward. As he breezed past the high-chase FP-80, Hoover was able to snap the world-famous photo of Yeager's flight (Figure 1.5) as the XS-1 continued going faster and higher. Yeager then shut down two of the rocket chambers in order to keep the vehicle's acceleration in check. Accelerating through Mach 0.83, 0.88, and 0.92, he tested the aircraft's response to horizontal stabilizer control. With the small increments of a quarter or a third of a degree that Ridley recommended, Yeager was able to maintain effective control of the aircraft. Then, as Yeager recounted in his postflight report: "At 42,000' in approximately level flight, a third cylinder was turned on. Acceleration was rapid and speed increased to .98 $Mach_i$. The needle of the machmeter fluctuated at this reading momentarily, then passed off the scale. Assuming that the off-scale reading remained linear, it is estimated that 1.05 $Mach_i$ was attained at this time. Approximately 30% of fuel and lox remained when this speed was reached and the motor was turned off" (Young 1997, p. 75).

Figure 1.5 Yeager accelerates in the Bell XS-1 on his way to breaking the "sound barrier" on October 14 1947. Source: NASA.

Yeager had done it! As mentioned in his postflight report, his Machmeter indications were a bit unusual. In fact, during the flight he radioed: "Ridley! Make another note. There's something wrong with this Machmeter. It's gone screwey!" (Young 1997, p. 73). That radio transmission heralded the dawn of a new era in aviation to supersonic speeds and well beyond. After maintaining supersonic flight for about 15 seconds, he shut down the rocket motors, performed a 1-g stall, and descended for a landing on Rogers dry lakebed.

Postflight analysis of the data, including corrections of the Machmeter reading for installation error, revealed that Yeager had reached a maximum flight Mach number of 1.06. A reproduction of this data is shown in Figure 1.6, where the initial jump in total and static pressures heralded the formation of a shock wave in front of the probe tip, causing a loss of total pressure. This is the characteristic "Mach jump" experienced by every Machmeter as the aircraft accelerates to supersonic speeds.

There are a number of interesting and revealing features of this story that can tell us something about flight testing. First, we see that this endeavor was anything but an individual effort. There was a large team with many players involved – pilots, engineers, managers, analysts, range safety officers, and so on. In this particular case, the flight test program was a collaboration between two separate organizations – the AAF was leading the program execution, and were supported by NACA's technical experts. Even though there was tension between these two groups, they were able to rise above those difficulties to work together in an effective manner to achieve the test objectives.

The source of the tension was inherently due to different test objectives – the AAF crew was tasked with breaking the sound barrier as quickly and safely as possible, while the NACA team was focused on developing a scientific understanding of transonic and supersonic flight, requiring a slower and more methodical approach. Flight test programs sometimes have such competing objectives in mind, which requires deft coordination and program management in order to ensure

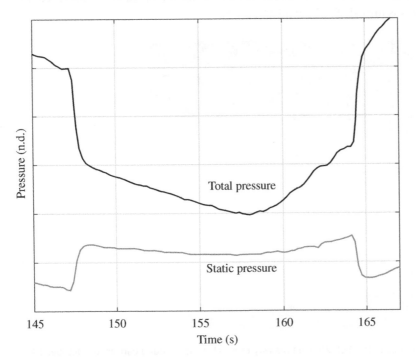

Figure 1.6 Plot of the total and static pressure for the first supersonic flight of the XS-1 on October 14 1947. Source: Data from NASA.

safety of flight and accomplishment of the test objectives. There is always a tension between programmatic needs, budget, and safety.

Another hallmark of successful flight testing is the careful probing of the edges of the flight envelope. Notice how the team approached the uncertain conditions associated with loss of control and buffeting. They gingerly pushed the Mach limits higher and higher, with the hope that any loss-of-control situation could be quickly recovered from. Despite the accelerated nature of the test program, the team took the time to carefully analyze the data and debrief after each flight. This was essential for gleaning insight from each test condition and informing the next step in the flight test program. They took an incremental buildup approach – starting from low-risk flights with known characteristics and carefully advancing to higher-risk flights, where the flight characteristics were unknown and potentially hazardous.

Also note how the aircraft was instrumented beyond what a normal production aircraft would have been. In fact, the record-setting XS-1 (the first airframe built) was only lightly instrumented compared to its sister ship, the second airframe off the production line, which was targeted for a much more detailed exploration of supersonic flight by the NACA team. This instrumentation is critical for understanding exactly what is happening during flight and preserving a record for post-flight analysis. The analytical work was done by a large team of engineers, technicians, and, in that day, human "computers" who did many of the detailed computations of the data (see Figure 1.2).

After some initial renegade flying by Yeager, the flight test team settled into a rhythm of carefully planned and executed flights. Before each flight they carefully planned the objectives and specific maneuvers to fly on the next mission. The injunction was that the flight must proceed exactly as planned, with specific plans for various contingencies and anomalies. This culture of flight testing is absolutely essential for the safety and professionalism of the process. One common phrase captures this mentality of flight testing: "plan the flight, and fly the plan."

This initial foray into exploring the flight testing program of the XS-1 illustrates many of the hallmarks of flight test programs. We'll next discuss some of the different kinds of flight testing being done today. Clearly, not every flight test program is as ambitious or adventurous as the XS-1 program, but a common objective is to answer the remaining unknown questions that are always present in an aircraft development program, even after rigorous design work backed up by wind tunnel testing and computational studies.

1.2 Types of Flight Testing

There are several different kinds of flight testing, driven by the objective of a particular program. These motivations include scientific research, development of new technologies or experimental capabilities, evaluation of operational performance, or airworthiness certification of new aircraft for commercial use. Other kinds of flight tests include production flight test (first flight of a new airframe of an already certified type, to verify compliance with design performance standards), systems flight test (new systems installed, new external stores on a fighter aircraft that must be tested for separation, new avionics systems), and post-maintenance test flight. Here, we'll focus our attention on flight testing for scientific research, assessment of experimental technologies, developmental test and evaluation, operational test and evaluation, and airworthiness certification programs. Other perspectives on the different kinds of flight testing are provided by Kimberlin (2003), Ward et al. (2006), or Corda (2017).

1.2.1 Scientific Research

In many instances, the highest-quality scientific research can only be done in actual flight. Even though wind tunnels are commonly available, results from these facilities are always limited in some way – facility effects such as streamwise pressure gradients in the test section, wall boundary layer effects, test section blockage, turbulence intensity level, constraints on model size, lack of Mach and/or Reynolds scaling, etc. are always present (see Tavoularis 2005 or Barlow et al. 1999 for a discussion of wind tunnels and their limitations). Similarly, computational fluid dynamics simulations are inherently limited in their ability to model viscous, unsteady separated flows, particularly when the model – such as a full aircraft – is large (see Cummings et al. 2015 for the limitations on computational aerodynamics). Grid resolution, turbulence modeling strategies, and time-accurate solutions will always need validation of some kind. Thus, the ultimate proof of scientific principles associated with flight is to actually conduct experiments in flight.

The range of scientific experiments that can be studied via flight testing can be very broad and conducted by government labs, universities, and industry. University flight test efforts have included Purdue University's development of pressure-sensitive paint (PSP) for in-flight measurements of chordwise surface pressure distribution on an aircraft wing (Figure 1.7). The advantage of PSP is that there is minimal flow intrusiveness, compared to the traditional pressure belts mounted on top of the wing, which are banded and flexible tubes. Furthermore, it is much simpler to instrument the aircraft with PSP, since no tubing has to be run into the fuselage and connected to pressure transducers. In fact, the production Beechjet 400 shown in Figure 1.7 was returned to normal flight under its regular airworthiness certification immediately following flight testing (Lachendro 2000).

Another leading flight test program for scientific research is the University of Notre Dame's Airborne Aero-optics flight research program (Jumper et al. 2015). Researchers at Notre Dame, led by Prof. Eric Jumper and Prof. Stanislav Gordeyev, study approaches for correcting optical aberrations to laser beams propagating through unsteady shear flows and turbulence. Their active correction schemes allow them to focus a laser beam emitted from one aircraft on the fuselage of a target aircraft such as the Dassault Falcon 10 shown in Figure 1.8. These concepts are used for applications ranging from optical air-to-air communications to directed energy for military applications.

The US government is also active with scientific research enabled by flight testing programs. One notable example is NASA's F-18 high alpha research vehicle (HARV). The goal of the first phase of this program was to understand vortex formation, trajectory, and breakdown on the F-18 operated at

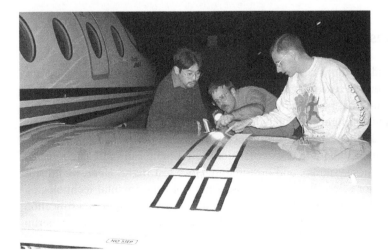

Figure 1.7 Inspection of pressure-sensitive paint on Purdue University's Beechjet 400 following a flight test in 1999 (depicted left to right are Hirotaka Sakaue, Brian Stirm, and Jim Gregory). Source: Photo courtesy of Nate Lachendro.

Figure 1.8 Notre Dame's Dassault Falcon 10. Source: U.S. Air Force.

Figure 1.9 Smoke and tuft flow visualization on the NASA F-18 High Alpha Research Vehicle at an angle of attack of 20°. Source: NASA.

high angle of attack. The specially instrumented F-18 had tufts (short pieces of yarn) taped to the top of the wing, smoke tracer particles released from orifices near the nose, dye flow visualization, and hundreds of pressure taps. These various techniques were used to study local flow separation and vortex trajectories. In-flight measurements, shown in Figure 1.9, clearly documented the formation of vortices on the leading-edge extension (LEX) of the F-18 at high angle of attack, the trajectory of these vortices, and the specific location of vortex breakdown. The vortex breakdown phenomenon, when occurring in the vicinity of the aft tail, led to significant tail buffeting and issues with fatigue (see Fisher et al. 1990).

1.2.2 Experimental Flight Test

Now, we turn our attention from basic scientific and engineering studies to development and test of new vehicle concepts. NASA Armstrong Flight Research Center (formerly known as NASA Dryden) has led the way over the years with this type of flight research (for a good historical overview of NASA's many flight research programs, see Gorn 2001 or Hallion and Gorn 2003). This type of flight testing is all about pushing the boundaries of what is possible, through development and demonstration of new flight technologies. Beyond the Bell XS-1 discussed earlier, there are numerous flight research programs that the US Government has conducted (Miller 2001; Jenkins et al. 2003). These cutting-edge aircraft are generally classified as X-planes, with the goal of proving out new technologies or advanced concepts (see Figure 1.10). The Bell XS-1 was the first aircraft in this distinguished lineup, which includes over 60 aircraft (and counting!). Many of these X-planes led to successful production flight vehicles after a period of focused flight testing (see Miller 2001; Jenkins et al. 2003; Corda 2017).

One interesting example is the X-wing project, which had the goal of improving the forward flight speed of helicopters. This interesting vehicle, the Sikorsky S-72 shown in Figure 1.11, is a hybrid between a fixed wing aircraft and a traditional rotorcraft. It could take off vertically like a traditional helicopter, but then its rigid rotors could be stopped mid-flight as the aircraft transitioned from vertical flight to forward flight. Instead of articulating the rotor blades as a traditional helicopter does, the S-72 used compressed air blown from the edges of the blades to achieve lift control (called circulation control – see Reader and Wilkerson 1977 for details). This innovative aircraft from the

Figure 1.10 Early X-planes, including the Douglas X-3 Stiletto (center) and (clockwise, from lower left) Bell X-1A, Douglas D-558-1 Skystreak, Convair XF-92A, Bell X-5, Douglas D-558-2 Skyrocket, and the Northrop X-4 Bantam. Source: NASA.

Figure 1.11 Sikorsky S-72 X-wing testbed aircraft. Source: NASA.

early 1980s has paved the way for high-speed helicopters today, such as the Sikorsky S-97 Raider or the Airbus RACER program.

Vehicle flight testing programs are also pushing into the domain of unmanned aircraft systems (UAS), commonly known as drones. For example, The Ohio State University developed and flight tested the Avanti UAS, which is a 70-lb jet capable of autonomous, unmanned, high-speed flight (Figure 1.12). This flight vehicle featured dual-redundant radio control links and a third independent satellite communications link, to provide robust beyond-line-of-sight flight. Flight research with this vehicle assessed the robustness of the control links, along with adaptive control laws for real-time in-flight system identification (see Warwick 2017; McCrink and Gregory 2021;

Figure 1.12 The Ohio State University's Avanti jet unmanned aircraft system. Source: Photo courtesy of Kamilah King.

or Chapter 16 for details). In the midst of the flight testing program, the Ohio State team set official world records for speed (147 mph) and out-and-back distance (28 mi) of an autonomous unmanned aerial vehicle (UAV), as certified by the Fédération Aéronautique Internationale (FAI) and the National Aeronautic Association (NAA).

1.2.3 Developmental Test and Evaluation

Within the US military, a significant amount of time and energy are invested in development test and evaluation (DT&E) flight testing. This aspect of flight testing involves a careful assessment of how an aircraft flies, including evaluation of aircraft performance, stability, and handling qualities. DT&E also includes performance assessment of new weapons, new software, and new airframes. These tests are centered on assessment of compliance with performance standards and focus on identifying anomalies in new systems. Test pilots (see Figure 1.13) push the performance limits of the system and are often involved in test planning very early in the design cycle. For example, if a new weapon system is designed for an aircraft, the developmental test pilot will evaluate the separation characteristics, compatibility of the new weapon with the aircraft system across a wide range of flight conditions, and evaluation of flutter flight characteristics. This testing and evaluation are done through a gradual build-up approach that minimizes (but does not eliminate) risk.

1.2.4 Operational Test and Evaluation

Operational test and evaluation (OT&E) involves assessment of an air vehicle's performance under representative operational conditions. This often includes operation on different runways under different conditions (*e.g.*, rain, sleet, snow, etc.) or at high-density altitude (high elevation, hot day). Operational testing also involves determination of crosswind limits on landing and taxiing operations. Aircraft manufacturers will also assess aircraft system robustness and reliability under a wide range of extreme weather conditions, including heat, cold, and icing.

Figure 1.13 Maj Rachael Winiecki, a developmental test pilot with the 461st Flight Test Squadron at Edwards Air Force Base, and the first F-35 female test pilot. Also shown is Airman 1st Class Heather Rice, a crew chief with the 412th Aircraft Maintenance Squadron. Source: U.S. Air Force.

1.2.5 Airworthiness Certification

Airworthiness certification is the process by which an aircraft is demonstrated to conform to approved design principles and that it is in a condition for safe operation. But what constitutes safe flight? This generally involves an insignificant hazard to people or property on the ground and minimal hazard to the occupants of the aircraft. Typically, a government's civil aviation authority, such as the Federal Aviation Administration (FAA) in the United States, grants an airworthiness certificate to an applicant submitting reports that document airworthiness for a new aircraft type. This process can be lengthy, involving flight testing to document aircraft performance and demonstrate compliance with safety standards.

In the United States, the regulatory authority for the FAA to certify the airworthiness of light aircraft is Title 14 of the Code of Federal Regulations ("Aeronautics and Space"), Chapter I ("Federal Aviation Administration, Department of Transportation"), Subchapter C ("Aircraft"), Part 23 ("Airworthiness Standards: Normal Category Airplanes") – we'll refer to this as 14 CFR §23 or simply part 23 (U.S. Code of Federal Regulations 2021). Part 23 covers the certification standards for general aviation aircraft, which have a maximum takeoff weight of 19,000 lb or less and carry 19 or fewer passengers. Since the scope of this book focuses on light aircraft, Part 23 is most relevant for our purposes. The subpart that is most relevant for flight testing is Subpart B (14 CFR §23.2100 through §23.2165), which defines the requirements for flight testing of aircraft for airworthiness certification.

Aircraft certified under Part 23 are grouped into different certification and performance levels based on number of passengers that can be carried and flight speed (14 CFR §23.2005), which are summarized in Table 1.1. Each level indicates a higher hazard, and a correspondingly higher bar is set to mitigate the risks associated with those hazards. Aircraft at the higher certification levels and higher performance levels will have higher standards to meet for certification.

Part 23 details the standards of safe flight that must be met for an aircraft to be certified as airworthy by the FAA, organized into broad categories of performance metrics and flight characteristics. Performance metrics include defining limits on the aircraft weight and center of gravity position, the stall speed of the aircraft under various operating conditions, takeoff performance, climb performance, glide performance, and landing distance required. The flight characteristics for certification include demonstration that the airplane is controllable and maneuverable; that the airplane can be trimmed in flight; that it has static and dynamic longitudinal, lateral, and directional stability; that the aircraft has controllable stall characteristics in all maneuvers and that

Table 1.1 Airworthiness certification levels defined by part 23.

	Airplane certification levels		Airplane performance levels
Level 1	Maximum seating configuration of 0–1 passengers	Low speed	Airplanes with a V_{NO} and $V_{MO} \leq 250$ KCAS (and $M_{MO} \leq 0.6$)
Level 2	Maximum seating configuration of 2–6 passengers		
Level 3	Maximum seating configuration of 7–9 passengers	High speed	Airplanes with a V_{NO} or $V_{MO} > 250$ KCAS (and $M_{MO} > 0.6$)
Level 4	Maximum seating configuration of 10–19 passengers		

V_{NO} = maximum structural cruising speed, V_{MO} = maximum operating limit speed, M_{MO} = maximum operating Mach number, and KCAS represents the units for knots calibrated airspeed.
Source: Based on FAA (2011).

sufficient stall warning is provided; that spins are recoverable; that the airplane has controllable ground handling characteristics; and that vibration and buffeting do not interfere with control of the airplane or cause excessive fatigue. If certification is requested for flight into known icing conditions, then the aircraft performance and handling characteristics must be shown to the same level of safety even in icing conditions. This textbook provides an introduction to the underlying principles for some of these flight tests; more detailed information is available from Kimberlin (2003) or FAA Advisory Circulars (2003, 2011).

While the regulatory framework and overall safety criteria are defined in Part 23, the regulations are intentionally sparse on details on *how* to actually demonstrate compliance for certification. Instead, means of compliance (§23.2010) can be determined by the applicant, subject to approval by the FAA. Typically, the means of compliance is established by a consensus standard. A type certificate applicant for a new light aircraft could demonstrate compliance with a consensus-based industry standard, which has been approved by the FAA. This compliance mechanism is a dynamic and flexible approach (compared to explicitly defining the compliance mechanisms in part 23), since consensus-forming bodies can quickly respond to new technologies and develop consensus standards. One key example of such a body is ASTM International. The ASTM convenes a number of committees, which are populated by representatives from various industry groups, and also includes government (FAA) representatives. The key ASTM committee that covers certification standards for light aircraft is the F44 committee on General Aviation Aircraft and specifically the F44.20 subcommittee on Flight. At the time of writing this book, ASTM F44.20 had published standard specifications for flight test demonstration of aircraft weight and center of gravity, operating limitations, aircraft handling characteristics, performance, and low-speed flight characteristics (ASTM 2017, 2018a, 2018b, 2019a, 2019b). Historical guidance from the FAA is also available for means of compliance with 14 CFR part 23 through nonregulatory advisory circulars (FAA 2003, 2011).

It's important to also be familiar with the historical approaches to airworthiness certification, since there are many aircraft flying today that were certified under older versions of the regulations. Predating certification of general aviation aircraft under part 23, certification was granted under the Civil Air Regulations (from the late 1930s until 1965). Kimberlin (2003, chapter 1) provides a good synopsis of these older regulations and how antique aircraft are still flying under airworthiness certificates granted under the older regulations.

For decades, certification of light general aviation aircraft followed regimented flight testing protocols that were explicitly defined in part 23. Over the years, the part grew more complex as additional safety measures and compliance protocols were codified. The resulting regulation was a rigid document that could not easily accommodate new technologies. For example, part 23 was strictly written to document how a type applicant must demonstrate the performance of internal combustion engines and the associated fuel system. This strict delineation of a compliance pathway was fine when all general aviation aircraft were powered by internal combustion engines running off Avgas. However, there are new propulsion system concepts emerging such as electric motors driven by fuel cells, batteries, or hybrid battery-generator systems, but these could not be certified under the former regimented structure of part 23. Type certificate applicants would have had to demonstrate an equivalent level of safety and obtain waivers, but there was no established and agreed-upon process for doing so. Thus, certification of new technologies such as electric propulsion would have been costly, with an uncertain outcome.

The current certification framework was developed in response to these challenges, leading to a complete rewrite of part 23 in 2016. With the rewrite of part 23, the FAA removed historical designations of various certification categories for airplanes. While these categories no longer exist for new aircraft certifications, any aircraft certified under the old part 23 will retain its category designation. These categories are normal, utility, acrobatic, and commuter. The commuter category is the designation for the largest general aviation aircraft, with a maximum takeoff weight of 19,000 lb, a passenger seating capacity of up to 19, and multiple engines. The normal, utility, and acrobatic categories all have a much lower weight limit of 12,500 lb and a seating capacity of up to 9. Normal category airplanes are approved for normal (routine) flying, stalls (but not "whip stalls"), and routine commercial maneuvers (less than 60° bank). Airplanes certified for utility category are approved for limited aerobatic maneuvers, which may include spins and commercial maneuvers at higher bank angles (up to 90°). Acrobatic category airplanes are approved for acrobatic maneuvers, which is basically any maneuver that a pilot can fly, and found to be safe in the flight testing program. For the normal, utility, and acrobatic categories, a given airplane could be certified for one, two, or all three, with varying operating limitations corresponding to each. Given that there are many aircraft routinely flying today that are well over 60 years old, one can anticipate that these legacy certification categories will persist for quite some time as historical and current aircraft continue flying.

1.3 Objectives and Organization of this Book

Our objective for this book is to provide the reader with an introduction to the exciting world of flight testing of light aircraft and UAS. Within the broad theme of that overarching objective, we specifically seek to:

(1) Provide a solid foundation for the reasons why flight testing is done the way it is. This involves a clear and concise establishment of the theoretical principles. Each equation that is presented here is backed up by physical explanations of the phenomena involved.

(2) Offer aerospace engineering students the context for connecting engineering theory with practice through guided flights in an aircraft. This provides the student with a visceral, empirical way of connecting their theoretical knowledge of flight with practical knowledge. The goal is for the student to develop a tacit understanding of flight beyond the explicit knowledge gleaned in traditional classrooms.

(3) Introduce the concepts and practice of digital data acquisition and signal processing, which is the underpinning of complex industrial and governmental flight test programs. These concepts are typically not taught in the undergraduate aerospace curriculum, but are important for knowing how to acquire and analyze flight test data using advanced, micro-scale sensors and digital data acquisition systems.

(4) Provide an overview of many of the foundational flight test topics encountered in performance flight testing. Individual chapters address each topic in turn, starting with the theoretical basis for that aspect of aircraft performance and moving on to flight test methods for acquiring and analyzing data for each performance metric.

This text is partitioned into two main segments – the first half of the book (Chapters 1–6) deals with preliminary content and fundamental principles, while the second half (Chapters 7–16) covers a series of flight test topics in detail. The flight tests covered here focus predominantly on the performance and stability characteristics of an aircraft. We predominantly focus on light general aviation aircraft and UAVs, since these are accessible to most students, and optimal learning takes place when a student can experience flight testing firsthand. The material is designed to be accessible such that a student can go with a qualified pilot in nearly any general aviation aircraft and acquire meaningful flight test data. Dedicated flight test instrumentation, modifications to the aircraft, or expensive hardware is not required. Thus, many of the flight test methods presented here may be simplified relative to what is done in industry.

Figure 1.14 Ohio State University students Greg Rhodes and Jennifer Haines following turn performance flight testing in a Piper PA-28R at the Ohio State University Airport. Source: Courtesy of Greg Rhodes and Jennifer Haines.

This textbook should not be regarded as a definitive or even advisory source on how to conduct flight testing. Instead, this book should be considered a general introduction to the ideas, scientific principles, theoretical foundations, and some of the best practices associated with flight testing. We provide a mix of aircraft performance theory with flight testing methods. Our goal is to invite the student or practitioner into understanding the physical fundamentals underlying flight testing – this will enable the reader to more fully appreciate why flight testing is done the way it is, to spot errors or problems in theory or procedures, and to know how to adapt established practices to unanticipated circumstances or new vehicle concepts. So, our aim is to provide a general overview and introduction to flight testing: a general idea of the nature of the field and a sound theoretical basis for what is done. We hope that this book will be a good first step as preparation for entry into the flight testing domain, where more detailed methods can be picked up on the way.

Official publications, standards, and advisory documents from the relevant civil aviation authority must be regarded as the definitive source for guidance on how to safely conduct flight testing and how to provide sufficient information to comply with the certification requirements. In the United States, this documentation is primarily found in 14 CFR 23, FAA Advisory Circular 23-8C, and any consensus standards accepted by the FAA (such as standards produced by ASTM International's F44 committee on General Aviation Aircraft). Other helpful sources of procedural and practical information are found in Hamlin (1946), Smith (1981), Stoliker et al. (1996), Stinton (1998), Kimberlin (2003), Ward et al. (2006, 2007), McCormick (2011), Mondt (2014), Corda (2017), and the publicly available flight test guides from the governmental flight test organizations (Herrington et al. 1966; USAF TPS 1986; USN TPS 1977, 1997; Gallagher et al. 1992; Stoliker 1995; Olson 2003). More advanced details on system identification for aircraft are available from Klein and Morelli (2006), Tischler and Remple (2012), or Jategaonkar (2015).

Flight testing is a fascinating, exhilarating field of aerospace engineering. It's incredibly rewarding to connect theory with practice, and we hope that the thoughts we provide here will draw students into a deeper understanding of flight through the intertwined approaches of theory and flying in flight test. And we hope to inspire the next generation of flight test professionals (Figure 1.14) to pursue this fascinating line of work. Hang on for a wild ride!

Nomenclature

M_{MO} maximum operating Mach number
V_{MO} maximum operating limit speed
V_{NO} maximum structural cruising speed

Acronyms and Abbreviations

AAF   Army Air Forces
CFR   Code of Federal Regulations
CG   center of gravity
DT&E   developmental test and evaluation
FAA   Federal Aviation Administration
HARV   high alpha research vehicle
LEX   leading-edge extension
NACA   National Advisory Committee for Aeronautics
NASA   National Aeronautics and Space Administration
OT&E   operational test and evaluation
PSP   pressure-sensitive paint
UAV   unmanned aerial vehicle

References

ASTM Committee F44 on General Aviation Aircraft, Subcommittee F44.20 on Flight. (2017). *Standard Specification for Weights and Centers of Gravity of Aircraft*. F3082/F3082M-17, approved 15 October 2017, West Conshohocken, PA: ASTM International. doi: https://doi.org/10.1520/F3082_F3082M-17.

ASTM Committee F44 on General Aviation Aircraft, Subcommittee F44.20 on Flight. (2018a). *Standard Specification for Performance of Aircraft*. F3179/F3179M-18, approved 1 May 2018, West Conshohocken, PA: ASTM International. doi: https://doi.org/10.1520/F3179_F3179M-18.

ASTM Committee F44 on General Aviation Aircraft, Subcommittee F44.20 on Flight. (2018b). *Standard Specification for Aircraft Handling Characteristics*. F3173/F3173M-18, approved 1 December 2018, West Conshohocken, PA: ASTM International. doi: https://doi.org/10.1520/F3173_F3173M-18.

ASTM Committee F44 on General Aviation Aircraft, Subcommittee F44.20 on Flight. (2019a). *Standard Specification for Establishing Operating Limitations and Information for Aeroplanes*. F3174/F3174M-19, approved 1 May 2019, West Conshohocken, PA: ASTM International. doi: https://doi.org/10.1520/F3174_F3174M-19.

ASTM Committee F44 on General Aviation Aircraft, Subcommittee F44.20 on Flight. (2019b). *Standard Specification for Low-Speed Flight Characteristics of Aircraft*. F3180/F3180M-19, approved 1 May 2019, West Conshohocken, PA: ASTM International. doi: https://doi.org/10.1520/F3180_F3180M-19.

Barlow, J., B., Rae, W. H., and Pope, A., 1999, *Low-Speed Wind Tunnel Testing*, 3, New York: Wiley.

Corda, S. (2017). *Introduction to Aerospace Engineering with a Flight Test Perspective*. Chichester, West Sussex, UK: Wiley.

Cummings, R.M., Mason, W.H., Morton, S.A., and McDaniel, D.R. (2015). *Applied Computational Aerodynamics*. Cambridge, UK: Cambridge University Press.

Federal Aviation Administration (2003). *Small Airplane Certification Compliance Program, Advisory Circular 23-15A*. Washington, DC: U.S. Department of Transportation.

Federal Aviation Administration (2011). *Flight Test Guide for Certification of Part 23 Airplanes, Advisory Circular 23-8C*. Washington, DC: U.S. Department of Transportation.

Fisher, D.F., Del Frate, J.H., and Richwine, D.M. (1990). In-flight flow visualization characteristics of the NASA F-18 high alpha research vehicle at high angles of attack. *NASA Technical Memorandum 4193* http://hdl.handle.net/2060/19910010742.

Gallagher, G.L., Higgins, L.B., Khinoo, L.A., and Pierce, P.W. (1992). *Fixed Wing Performance*. USNTPS-FTM-No. 108,. Patuxent River, MD: Naval Air Warfare Center.

Gorn, M.H. (2001). *Expanding the Envelope: Flight Research at NACA and NASA*. Lexington, KY: University Press of Kentucky.

Hallion, R.P. (1972). *Supersonic Flight; The Story of the Bell X-1 and Douglas D-558*. New York: Macmillan.

Hallion, R.P. and Gorn, M.H. (2003). *On the Frontier: Experimental Flight at NASA Dryden*. Washington, DC: Smithsonian Books.

Hamlin, B. (1946). *Flight Testing Conventional and Jet-Propelled Airplanes*. New York: Macmillan Company.

Herrington, R. M., Shoemacher, P. E., Bartlett, E. P., and Dunlap, E. W. (1966). *Flight Test Engineering Handbook*, USAF Technical Report 6273, Edwards AFB, CA: US Air Force Flight Test Center. Defense Technical Information Center Accession Number AD0636392, https://apps.dtic.mil/docs/citations/AD0636392.

Jategaonkar, R.V. (2015). Chapter 2. In: *Flight Vehicle System Identification: A Time-Domain Methodology*, 2e. Reston, VA: American Institute of Aeronautics and Astronautics.

Jenkins, D. R., Landis, T., and Miller, J. (2003). *American X-Vehicles: An Inventory – X-1 to X-50*. Monographs in Aerospace History No. 31, NASA SP-2003-4531.

Jumper, E.J., Gordeyev, S., Davalieri, D. et al. (2015). *Airborne Aero-Optics Laboratory – Transonic (AAOL-T)*. AIAA 2015-0657,. Kissimmee, FL: American Institute of Aeronautics and Astronautics, 53rd Aerospace Sciences Meeting.

Kimberlin, R.D. (2003). *Flight Testing of Fixed-Wing Aircraft*. Reston, VA: American Institute of Aeronautics and Astronautics.

Klein, V. and Morelli, E.A. (2006). Chapter 9. In: *Aircraft System Identification: Theory and Practice*. Reston, VA: American Institute of Aeronautics and Astronautics.

Lachendro, N. (2000). Flight testing of pressure sensitive paint using a phase based laser scanning system. MS thesis, West Lafayette, IN: School of Aeronautics and Astronautics, Purdue University.

McCormick, B.W. (2011). *Introduction to Flight Testing and Applied Aerodynamics*. Reston, VA: American Institute of Aeronautics and Astronautics.

McCrink, M. H. and Gregory, J. W. (2021). Design and development of a high-speed UAS for beyond visual line-of-sight operations. *Journal of Intelligent & Robotic Systems* 101: 31. https://doi.org/10.1007/s10846-020-01300-2.

Miller, J. (2001). *The X-planes – X-1 to X-45*, 3e. Hinckley, UK: Midland Publishing.

Mondt, M.J. (2014). *The Tao of Flight Test: Principles to Live By*. Boone, IA: J. I. Lord.

Olson, W.M. (2003). *Aircraft Performance Flight Testing, AFFTC-TIH-99-01*. Edwards AFB, CA: Air Force Flight Test Center.

Peebles, C. (2014). *Probing the Sky: Selected NACA Research Airplanes and their Contributions to Flight*. Washington, DC: National Aeronautics and Space Administration.

Reader, K. R. and Wilkerson, J. B. (1977). *Circulation Control Applied to a High Speed Helicopter Rotor*. Report 77-0024, David W. Taylor Naval Ship Research and Development Center, Bethesda, MD: DTIC accession number ADA146674.

Smith, H.C. (1981). *Introduction to Aircraft Flight Test Engineering*. Basin, WY: Aviation Maintenance Publishers.

Stinton, D. (1998). *Flying Qualities and Flight Testing of the Airplane*. Reston, VA: American Institute of Aeronautics and Astronautics.

Stoliker, F.N. (1995). *Introduction to Flight Test Engineering, AGARD Flight Test Techniques Series, AGARD-AG-300*, vol. 14. Neuilly-sur-Seine, France: Advisory Group for Aerospace Research and Development.

Stoliker, F., Hoey, B., and Armstrong, J. (1996). *Flight Testing at Edwards: Flight Test Engineers' Stories 1946–1975*. Lancaster, CA: Flight Test Historical Foundation.

Tavoularis, S. (2005). *Measurement in Fluid Mechanics*. Cambridge, UK: Cambridge University Press.

Tischler, M.B. and Remple, R.K. (2012). *Aircraft and Rotorcraft System Identification*, 2e. Reston, VA: American Institute of Aeronautics and Astronautics.

U.S. Code of Federal Regulations. (2021). *Airworthiness Standards: Normal Category Airplanes*. Title 14, Chapter I, Subchapter C, Part 23 (14 CFR §23). http://www.ecfr.gov (accessed 01 January 2021).

U.S. Air Force Test Pilot School (1986). *Performance Phase Textbook*, vol. 1, USAF-TPS-CUR-86-01. Edwards AFB, CA: US Air Force Test Pilot School.

U.S. Naval Test Pilot School. (1977). *Fixed Wing Performance: Theory and Flight Test Techniques*. USNTPS-FTM-No. 104, Patuxent River, MD: Naval Air Test Center.

U.S. Naval Test Pilot School. (1997). *Fixed Wing Stability and Control: Theory and Flight Test Techniques*. USNTPS-FTM-No. 103, Patuxent River, MD: Naval Air Warfare Center.

Ward, D., Strganac, T., and Niewoehner, R. (2006). *Introduction to Flight Test Engineering*, 3e, vol. 1. Dubuque, IA: Kendall/Hunt.

Ward, D., Strganac, T., and Niewoehner, R. (2007). *Introduction to Flight Test Engineering*, vol. 2. Dubuque, IA: Kendall/Hunt.

Warwick, G. (2017). Ohio State pushes speed envelope in UAS Research. *Aviation Week and Space Technology* 18 (2017): 44.

Wolfe, T. (1979). *The Right Stuff*. New York: Farrar, Straus and Giroux.

Yeager, C. and Janos, L. (1985). *Yeager: An Autobiography*. Toronto: Bantam Books.

Young, J.O. (1997). *Meeting the Challenge of Supersonic Flight*. Edwards AFB, CA: Air Force Flight Test Center.

2

The Flight Environment: Standard Atmosphere

In this chapter, we will discuss the properties of the environment for flight testing – Earth's atmosphere. It is critical to understand the nature of the atmosphere, since aircraft performance depends significantly on the properties of air. For example, the lift produced by the aircraft is proportional to the air density, and the amount of power produced by an internal combustion engine also varies with density. For these two reasons, aircraft performance decreases as density decreases. We will see in this chapter that density decreases with altitude, so key aircraft performance metrics such as takeoff distance, rate of climb, acceleration, etc. all degrade with altitude. Since aircraft performance depends significantly on the local properties of air, we need some way to factor out altitude effects. We also need to be able to predict the performance of an aircraft as a function of altitude, once its baseline performance is known. Thus, we need an agreed-upon standard definition of the properties of the atmosphere – this is the standard atmosphere. Definition of the standard atmosphere allows us to evaluate and compare aircraft performance in a consistent manner, no matter what the altitude is.

The important atmospheric parameters are the atmospheric temperature, pressure, density, and viscosity, which depend on the distance from the earth surface, geographic location, and time. In order to describe the atmosphere in a universal way, a standard atmosphere model has been developed, where the atmospheric parameters are determined as the univariate functions of altitude from sea level. Temperature exhibits strong variations with time of year, geographic location, and altitude. And, on a daily basis, temperature depends on current weather conditions in a stochastic manner. It is impossible to develop a first-principles model that will capture all of these parameters that influence the temperature profile; thus, the standard temperature profile is determined from an average of a large ensemble of atmospheric measurements. The variation of pressure with altitude, however, is rigorously described by some basic physical principles – we will derive these here. In fact, pressure is so intricately and reliably linked to altitude that aircraft altimeters measure pressure and convert the measurement to an indicated altitude through the definition of the standard atmosphere. Density is related to the estimated value of temperature and the derived value of pressure via the ideal gas law. Finally, we will provide a relationship that determines the viscosity of air as a function of temperature. Based on these developments, we will define a standard atmosphere that can be expressed in tabular form, or equations coded for computational analysis. This chapter will start with a physical description of the atmosphere and then present a detailed development of the standard atmosphere. Most of the development of the standard atmosphere presented in this chapter will rely on SI units, since this was the unit system used to define the standard atmosphere and the boundaries of atmospheric regions. The input and output of the standard atmosphere can be easily converted from SI units to English units as needed.

Introduction to Flight Testing, First Edition. James W. Gregory and Tianshu Liu.
© 2021 John Wiley & Sons Ltd. Published 2021 by John Wiley & Sons Ltd.
Companion website: https://www.wiley.com/go/flighttesting

2.1   Earth's Atmosphere

Earth's atmosphere is an envelope of air surrounding the planet Earth, where dry air consists of 78.08% nitrogen, 20.95% oxygen, 0.93% argon, 0.031% carbon dioxide, and small amounts of other gases (NOAA et al. 1976). In addition, air contains a small amount of water vapor (about 1%). The entire atmosphere has an air mass of about 5.15×10^{18} kg (1.13×10^{19} lb), and three quarters of the total air mass are contained within a layer of about 11 km (~36,000 ft) from the Earth's surface.

There is a general stratification of Earth's atmosphere, which leads to the definition of distinct regions of the atmosphere: the troposphere (0–11 km), stratosphere (11–50 km), mesosphere (50–85 km), and thermosphere (85–600 km). The atmosphere becomes thinner as the altitude increases, and there is no clear boundary between the atmosphere and outer space. However, the Kármán line has been defined at 100 km and is often used as the border between the atmosphere and outer space. Atmospheric effects become noticeable during atmospheric reentry of spacecraft at an altitude of around 120 km. Aircraft propelled by internal combustion engines and propellers are generally limited to operating in the troposphere, while jet-propelled aircraft routinely operate in the stratosphere.

Figure 2.1 illustrates the bottom three layers of Earth's atmosphere, which is where all atmospheric flight vehicles conduct flight. The delineation between the various regions of the atmosphere is based on historical measurements of temperature profiles, which lead to distinct regions with different temperature lapse rates. In the troposphere (the layer of the atmosphere nearest the surface), the air temperature generally decreases linearly with the altitude. This temperature reduction with altitude is due to the increasing distance from Earth and a concomitant reduction in heating from Earth's surface. Weather phenomena are directly dependent on this temperature reduction with altitude, causing most storms and other weather phenomena to develop and reside within the troposphere. The dividing boundary between the troposphere and the next layer (the stratosphere) is called the tropopause, at 11 km. Within the lower portion of the

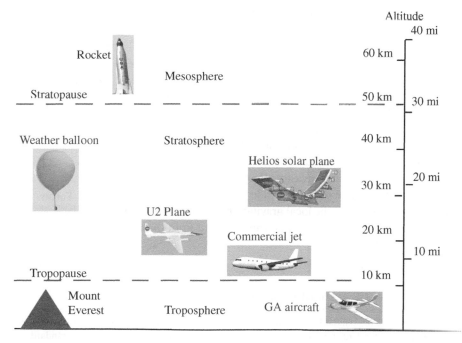

Figure 2.1   The layers of Earth's atmosphere.

stratosphere (11–20 km), the air temperature remains constant; it then increases with altitude in the upper stratosphere (20–50 km), due to absorption of the sun's ultraviolet radiation by ozone in this region of the atmosphere.

In contrast with the temperature–altitude profile, the variation of pressure with altitude is highly repeatable and deterministic. Air pressure continually decreases with altitude from Earth's surface all the way to the edge of the atmosphere. The primary reason for this is the action of Earth's gravitational acceleration on air, causing a given mass of air to exert a force on the air below it. Air at a given altitude must support the weight of all of the air mass above it, and it balances this force by pressure. As altitude increases, there is less air mass above that altitude, so there is less force (weight) acting on the air at that point and the pressure decreases. Thus, pressure decreases as altitude increases. We will discuss this physical mechanism in greater detail in Section 2.2, when we derive an expression for the variation of pressure with altitude.

2.2 Standard Atmosphere Model

A standardized model of the atmosphere allows scientists, engineers, and pilots in the flight testing community to have a commonly agreed-upon definition of the properties of the atmosphere. The definition of the standard atmosphere includes the variation of gravitational acceleration, temperature, pressure, density, and viscosity as a function of altitude. There is actually more than one definition of the standard atmosphere: the U.S. Standard Atmosphere (NOAA et al. 1976) and the International Civil Aviation Organization (ICAO) Standard Atmosphere (ICAO 1993). Thankfully, the two definitions are identical at lower altitudes where aircraft fly – the only differences are in the upper stratosphere and beyond. Our discussion here will generally follow the development of the U.S. Standard Atmosphere (NOAA et al. 1976).[1]

2.2.1 Hydrostatics

The development of the standard atmosphere directly results from the hydrostatic equation, which is derived here based on a control volume analysis. Figure 2.2 illustrates an arbitrary control volume, measuring $dx \times dy \times dh_G$, and the forces acting upon it (here, h_G is the geometric altitude, or height above mean sea level (MSL)). The forces due to pressure acting on all of the side walls balance one another out in this static equilibrium condition, and we will consider only the forces acting in the vertical direction. The force acting upward on the bottom surface of the control volume is the pressure, p, times the cross-sectional area $dx\,dy$. Similarly, on the top surface, we have a force of $(p + dp)dx\,dy$ acting downward. (Here, the differential pressure dp accounts for pressure changes in the vertical direction.) Finally, we have the weight of the air inside the control volume acting downward, $W = mg$, where g is the local gravitational acceleration and the mass of the air inside the control volume can be found from the product of density and the volume,

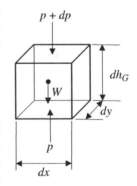

Figure 2.2 Forces acting on a hydrostatic control volume.

$$W = mg = \rho \left(dx\, dy\, dh_G \right) g. \tag{2.1}$$

---

1 There have been many iterations of the U.S. Standard Atmosphere over the years. The original version was published in 1958, and as scientific understanding of the atmosphere advanced, was updated in 1962, 1966, and finally 1976. Some older versions of the standard atmosphere persist today – for example, Anderson (2016) continues to refer to a 1959 definition of the standard atmosphere from the U.S. Air Force. However, the 1976 U.S. Standard Atmosphere and the 1993 ICAO standard atmosphere are widely accepted as the appropriate standards to use today.

Summing all the forces in the vertical direction and setting equal to zero (from Newton's second law applied to a stationary control volume), we obtain

$$p \, dx \, dy - (p + dp) \, dx \, dy - \rho g \, dx \, dy \, dh_G = 0. \tag{2.2}$$

Canceling terms leads to

$$dp = -\rho g \, dh_G, \tag{2.3}$$

which is the hydrostatic equation as a function of geometric altitude. This expression mathematically expresses the physical explanation that we presented earlier for the variation of pressure with altitude. As altitude increases (positive dh_G), the minus sign indicates that the pressure decreases (negative dp). The ρg term is an expression of the weight of the air inside the control volume, which is the reason for the pressure difference.

2.2.2 Gravitational Acceleration and Altitude Definitions

As we proceed with the development of the standard atmosphere, we must consider how gravitational acceleration varies with altitude. From Newton's law of universal gravitation, we know that gravitational acceleration varies inversely with the square of the distance to the center of the earth. Thus, we have

$$g = g_0 \left(\frac{r_{\text{Earth}}}{h_A} \right)^2 = g_0 \left(\frac{r_{\text{Earth}}}{r_{\text{Earth}} + h_G} \right)^2, \tag{2.4}$$

where g is the local gravitational acceleration (varies with altitude), g_0 is the gravitational acceleration at sea level ($9.806\,65\,\text{m/s}^2$ or $32.174\,\text{ft/s}^2$), h_A is the distance from the center of the earth (defined here as the absolute altitude[2]), and r_{Earth} is Earth's mean radius, which is $6356.766\,\text{km}$ (NOAA et al. 1976).

Despite the fact that gravity varies with altitude, it is convenient to derive the standard atmosphere based on the assumption of constant gravitational acceleration. In order to do so, we must define a new altitude, the geopotential altitude, h, which we will use in the hydrostatic equation with the assumption of constant gravity. Referring to Eq. (2.3), we can also write the hydrostatic equation as a function of geopotential altitude and constant gravitational acceleration,

$$dp = -\rho g_0 \, dh. \tag{2.5}$$

Taking the ratio of (2.5) and (2.3), we have

$$1 = \frac{g_0}{g} \frac{dh}{dh_G}, \tag{2.6}$$

since the differential pressure and density terms cancel out for a given change of pressure. The small difference between g_0 and g then leads to a small difference between the geopotential and geometric altitudes. Combining Eqs. (2.4) and (2.6) produces

$$dh = \left(\frac{r_{\text{Earth}}}{r_{\text{Earth}} + h_G} \right)^2 dh_G, \tag{2.7}$$

which can be integrated between sea level and an arbitrary altitude to find

$$h = \left(\frac{r_{\text{Earth}}}{r_{\text{Earth}} + h_G} \right) h_G. \tag{2.8}$$

---

2 There are two different definitions of absolute altitude that we will use in this chapter. The first one, considered here, is for development of the standard atmosphere. The second definition is widely used in aviation as the height above ground level. We will clarify these distinctions at the end of this chapter.

This expression defines the relationship between geopotential altitude, h, and geometric altitude, h_G, which can also be solved for geometric altitude,

$$h_G = \left(\frac{r_{\text{Earth}}}{r_{\text{Earth}} - h} \right) h. \tag{2.9}$$

In our derivation of the standard atmosphere, we will use geopotential altitude, h, and assume constant g_0. Properties of the standard atmosphere such as temperature, pressure, and density, *i.e.*, (T, p, ρ), will be found as a function of geopotential altitude, h, and then mapped back to geometric altitude, h_G, by Eq. (2.9). In this work, we are focused on the lower portions of the atmosphere where most aircraft fly ($h \leq 20$ km or 65, 617 ft). At that upper altitude limit, Eq. (2.9) predicts a maximum difference of 0.31% between the geometric and geopotential altitude. Thus, in many cases related to flight testing, this difference between geopotential and geometric altitudes can be neglected.

2.2.3 Temperature

Temperature at any given point in the Earth's atmosphere will depend not only on the altitude but also on time of year, latitude, and local weather conditions. Since the variation of temperature has spatial, temporal, and stochastic input, the development of the standard atmosphere as a function of only altitude inherently involves many approximations. Thus, we might anticipate that the actual temperature at a given location can deviate significantly from the standard value.

The standard temperature profile has been determined through an average of significant amounts of data from sounding balloons launched multiple times a day over a period of many years, at locations around the globe. The resulting temperature profile is a function of geopotential altitude, with the lapse rate, $a = dT/dh$, representing the linear variation of temperature with altitude for each region (see Table 2.1 and Figure 2.3). In the troposphere ($0 \leq h \leq 11$ km), the standard temperature lapse rate is defined as -6.5 K/km, starting at $T_{\text{SL}} = 288.15$ K. In the lower portion of the stratosphere ($11 < h \leq 20$ km), the temperature is presumed to be constant at 216.65 K. Starting at 20 km, the temperature then increases at a rate of 1 K/km due to ozone heating of the upper stratosphere. Based on the data in Table 2.1, we can write expressions for the temperature profile throughout the standard atmosphere as

$$T = a \left(h - h_{\text{ref}} \right) + T_{\text{ref}}, \tag{2.10}$$

where "ref" refers to the base of the layer (defined by either sea level conditions, or the top of the prior atmospheric layer, working upwards). Output from Eq. (2.10) can be stacked for each

Table 2.1 Definition of temperature lapse rates in various regions of the atmosphere.

Region	h_1 (km)	h_2 (km)	$a = dT/dh$ (K/km)
Troposphere	0	11	−6.5
Stratosphere	11	20	0.0
	20	32	1.0
	32	47	2.8
	47	51	0.0
Mesosphere	51	71	−2.8
	71	84.852	−2.0

h_1 and h_2 are the beginning and ending altitudes of each region, respectively.
Source: Data from NOAA et al. 1976.

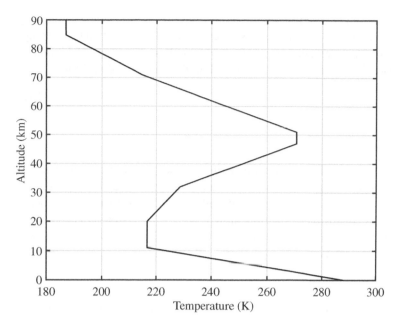

Figure 2.3 Standard temperature profile.

altitude layer, one on top of another, to define the entire standard temperature profile. Since most flight testing, especially for light aircraft and drones, occurs at altitudes below 20 km, we will focus our attention on the troposphere and lower portion of the stratosphere.

2.2.4 Viscosity

We also need to define an expression for dynamic viscosity, μ, which depends on temperature. The most significant impact of viscosity is in the definition of Reynolds number,

$$Re_c = \frac{\rho U_\infty c}{\mu}, \tag{2.11}$$

which is an expression of the ratio of inertial to viscous forces (here, U_∞ is the freestream velocity or airspeed, and c is the mean aerodynamic chord of the wing). Reynolds number has a significant impact on boundary layer development and aerodynamic stall, as we will see in Chapter 12.

The viscosity of air is related to the rate of molecular diffusion, which is a function of temperature (Sutherland 1893). This relationship has been distilled down to Sutherland's Law,

$$\mu = \frac{\beta T^{3/2}}{T + S_{\text{visc}}}, \tag{2.12}$$

where T is the temperature in absolute units, and β and S_{visc} are empirical constants, provided in Table 2.2 for both English and SI units (NOAA et al. 1976). Based on Eq. (2.12), the viscosity

Table 2.2 Constants used in Sutherland's Law.

	SI	English
β	1.458×10^{-6} kg/(s m K$^{1/2}$)	2.2697×10^{-8} slug/(s ft $^\circ$ R$^{1/2}$)
S_{visc}	110.4 K	198.72 $^\circ$ R

Source: Based on NOAA et al. 1976.

of a gas increases with increasing temperature. Thus, the dynamic viscosity decreases gradually through the troposphere, starting with the standard sea level value of $\mu_{SL} = 1.7894 \times 10^{-5}$ kg/(s m) $= 3.7372 \times 10^{-7}$ slug/(s ft) at a temperature of $T_{SL} = 288.15$ K $= 518.67\,°$ R. If kinematic viscosity (ν) is desired instead of dynamic viscosity, it can be found based on its definition,

$$\nu = \frac{\mu}{\rho}. \tag{2.13}$$

2.2.5 Pressure and Density

To derive an expression for the variation of pressure with altitude, we need to integrate the hydrostatic equation. Since density and gravitational acceleration also vary with altitude, we need to cast the hydrostatic equation in terms of only pressure and altitude, with all other variables being constant. We will work with the hydrostatic equation shown in Eq. (2.5), based on geopotential altitude and constant gravity. Density can be expressed as a function of pressure and temperature via the equation of state for a perfect gas,

$$p = \rho RT, \tag{2.14}$$

where $R = 287.05$ J/(kg K) is the gas constant for air. Taking a ratio of Eqs. (2.5) and (2.14), we have

$$\frac{dp}{p} = -\frac{g_0}{RT}\, dh. \tag{2.15}$$

We will work with Eq. (2.15) for two different cases: first, where the temperature is constant with altitude, and then when temperature varies linearly with altitude.

Equation (2.15) can be directly integrated to find pressure as a function of altitude for the isothermal regions of the atmosphere ($11 < h \leq 20$ km and $47 < h \leq 51$ km) since all terms in the equation are constant except pressure and altitude. Performing this integration between the base (h_{ref}) and an arbitrary altitude within that region (h) yields

$$\frac{p}{p_{ref}} = e^{-[g_0/(RT_{ref})](h-h_{ref})}, \tag{2.16}$$

where "ref" indicates the base of that particular region of the atmosphere. When the ideal gas law, Eq. (2.14), is applied to isothermal regions of the atmosphere, we see that density is directly proportional to pressure. Thus, we can also write an expression for density in the isothermal regions as

$$\frac{\rho}{\rho_{ref}} = e^{-[g_0/(RT_{ref})](h-h_{ref})}. \tag{2.17}$$

Equations (2.10), (2.12), (2.16), and (2.17) then form a complete definition of temperature, viscosity, pressure, and density in the isothermal regions of the standard atmosphere.

Portions of the atmosphere with a linear lapse rate, however, require a different approach to integrating Eq. (2.15). In this case, T is no longer constant with respect to altitude, so we must substitute it in the temperature lapse rate. Combining $a = dT/dh$ with Eq. (2.15) yields

$$\frac{dp}{p} = -\frac{g_0}{aR}\frac{dT}{T}. \tag{2.18}$$

Integration of Eq. (2.18) gives the pressure ratio as a function of the temperature ratio, *i.e.,*

$$\frac{p}{p_{ref}} = \left(\frac{T}{T_{ref}}\right)^{-g_0/aR}, \tag{2.19}$$

where p_{ref} and T_{ref} are the pressure and temperature at a reference altitude, respectively. Again applying the ideal gas law, Eq. (2.14), the density ratio is given by

$$\frac{\rho}{\rho_{ref}} = \left(\frac{T}{T_{ref}}\right)^{-(1+g_0/aR)}.$$ (2.20)

Thus, for regions of the atmosphere with linear temperature lapse rates, Eqs. (2.10), (2.12), (2.19), and (2.20) form a complete description of the temperature, viscosity, pressure, and density variation with altitude. The reference condition for the base of each region is simply the values corresponding to the top of the previous (lower) region.

In the flight testing community and elsewhere, we often express the above ratios as specific variables referenced to sea level conditions. Temperature ratio, pressure ratio, and density ratio are defined as

$$\theta = \frac{T}{T_{SL}}, \quad \delta = \frac{p}{p_{SL}}, \quad \text{and } \sigma = \frac{\rho}{\rho_{SL}}.$$ (2.21)

In the standard atmosphere, sea level conditions are defined as $T_{SL} = 288.15\,K$, $p_{SL} = 101.325\,kPa$, and $\rho_{SL} = 1.225\,kg/m^3$, where the subscript "SL" denotes sea level. The ratios defined in Eq. (2.21) still satisfy the ideal gas law, giving

$$\delta = \sigma\theta.$$ (2.22)

It is important to bear in mind that these equations are a function of geopotential altitude, which presumes constant gravitational acceleration. If properties are desired as a function of geometric altitude, then the corresponding geometric altitudes can be found by solving for h_G in Eq. (2.9).

2.2.6 Operationalizing the Standard Atmosphere

Applying the equations developed above, we can take one of several approaches to implementing the standard atmosphere for flight testing work. Most simply, these equations form the basis for tabulated values of the standard atmosphere, which are tabulated by NOAA et al. (1976) or ICAO (1993). In addition, a limited subset of the U.S. Standard Atmosphere (NOAA et al. 1976) is reproduced in Appendix A. Alternatively, pre-written standard atmosphere computer codes may be downloaded and used in a straightforward manner. Popular examples include the MATLAB code by Sartorius (2018) or the Fortran code by Carmichael (2018). If these are not suitable for a particular purpose, then custom code can be written, as described below in a form that simplifies the coding.

In the troposphere where the temperature gradient is $a = dT/dh = -6.5\,K/km$, the temperature distribution in Eq. (2.10) can be expressed as a linear function

$$\theta = 1 - kh,$$ (2.23)

where h is the geopotential altitude and $k = 2.256 \times 10^{-5}\,m^{-1} = 6.876 \times 10^{-6}\,ft^{-1}$ is a decaying rate. According to Eqs. (2.19) and (2.20), the pressure ratio and density ratio in the troposphere ($0 \leq h \leq 11$ km) are given by

$$\delta = \theta^n$$ (2.24)

and

$$\sigma = \theta^{n-1},$$ (2.25)

where $n = -g_0/aR = 5.2559$.

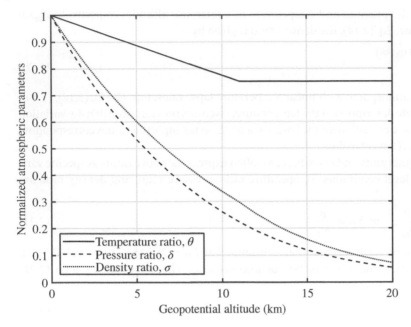

Figure 2.4 The normalized temperature, pressure, and density distributions in the standard atmosphere.

In the lower stratosphere ($11\,\mathrm{km} < h \leq 20\,\mathrm{km}$), the atmospheric temperature is constant at $216.65\,\mathrm{K}$. If we define the critical altitude at the tropopause to be $h_{\mathrm{trop}} = 11\,\mathrm{km}$, then the temperature and pressure ratios at the tropopause are $\theta_{\mathrm{trop}} = 0.7518$ and $\delta_{\mathrm{trop}} = 0.2233$, respectively. Recasting (2.16) in terms of these ratios, we obtain

$$\delta = \delta_{\mathrm{trop}} \exp\left[-nk\left(h - h_{\mathrm{trop}}\right)/\theta_{\mathrm{trop}}\right] \tag{2.26}$$

for the pressure ratio in the lower stratosphere. Finally, the density ratio in the lower stratosphere is simply found by the ideal gas law,

$$\sigma = \delta/\theta_{\mathrm{trop}}. \tag{2.27}$$

Figure 2.4 shows the pressure, density, and temperature distributions normalized by the sea level conditions in the standard atmosphere.

2.2.7 Comparison with Experimental Data

The above equations describe the idealized atmosphere where the parameters are considered as the mean values of the measured quantities. However, as indicated in The U.S. Standard Atmosphere (NOAA et al. 1976), measurement data show considerable variations of the atmospheric parameters depending on time (day and season) and geographic location, which should be considered in flight testing.

Experimental measurements may be compared with the theoretical variation of pressure and temperature derived from the standard atmosphere. Atmospheric data can be collected by a weather balloon (Figure 2.5), which ascends through the atmosphere and measures pressure and temperature throughout the flight. For the case presented here, the balloon ascended to an altitude of 30.161 km (98,953 ft) before bursting and descending via parachute back to Earth. Data throughout the ascent and descent were collected and are presented in Figures 2.6 and 2.7. The temperature data shown in Figure 2.6 show similar trends to the standard temperature profile, but

Figure 2.5 Launch of a high-altitude weather balloon from the oval of The Ohio State University.

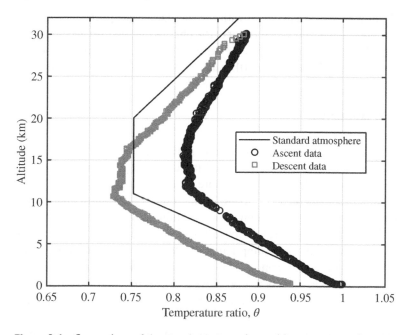

Figure 2.6 Comparison of the standard atmosphere with temperature data measured by a weather balloon.

the agreement is not very good. This is not surprising, since the details of the temperature profile are strongly dependent on local weather, time of year, latitude, etc. However, some of the similarities are noteworthy: the experimental temperature lapse rate is approximately the same as the standard lapse rate, particularly at low altitudes. Also, the location of the tropopause, corresponding to a change to an isothermal temperature profile, is in good agreement. Finally, the slope of the high-altitude lapse rate is also in fairly good agreement with the standard temperature profile.

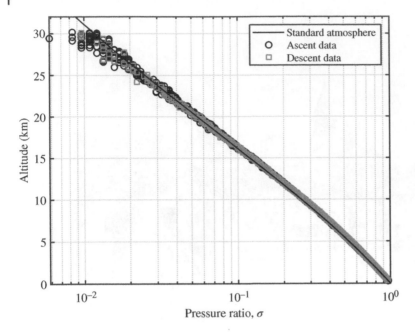

Figure 2.7 Comparison of the standard atmosphere with pressure data measured by a weather balloon.

Pressure data, in Figure 2.7, show excellent agreement with the standard atmosphere. This is also expected since the hydrostatic equation is a good descriptor of the physics of pressure variation with altitude. The good agreement shown here underscores the utility of using pressure measurement for measuring altitude on aircraft (see Chapter 3 for further details on how altimeters operate).

2.3 Altitudes Used in Aviation

We will now conclude this chapter with a discussion of different altitude definitions used in aviation. We have already introduced several definitions of altitude for the preceding discussion on the standard atmosphere. To recap, these include absolute altitude, geometric altitude, and geopotential altitude. Absolute altitude, h_A, is measured from the center of the Earth and is only relevant when determining the value of gravitational acceleration at a particular altitude. Geometric altitude, h_G, is the height of an aircraft above mean sea level. And, geopotential altitude, h, is the height above sea level with the assumption of constant gravitational acceleration. Geopotential altitude is only relevant in the context of deriving the standard atmosphere, so should not be used elsewhere. For the remainder of this book, we will presume that the differences between geometric altitude and geopotential altitude are small and will simply refer to the geometric altitude as h.

However, these altitude definitions are limited to an engineering context. To make things interesting, we also have a set of altitudes that are defined for the aviation community. And, to make things more interesting, some of the aviation altitudes use the same terms but different definitions! The aviation set of altitudes include true altitude, indicated altitude, pressure altitude, density altitude, and absolute altitude. We will discuss each of these as follows.

True altitude is the height above mean sea level. In the aviation community, this altitude is often abbreviated as MSL. When an aircraft altimeter is referenced to the local barometric pressure reading, it indicates true altitude. (Details on altimeter operation are provided in Chapter 3.) Pilots around the world often refer to this setting of the altimeter as QNH. Note that the aviation definition of true altitude is identical to the engineering definition of geometric altitude.

Similarly, indicated altitude is a direct reading from the altimeter, no matter how the altimeter is set. This may or may not be the same as true altitude, depending on the reference pressure used on the altimeter. (The reference pressure essentially shifts the calibration of the altimeter to match local barometric pressure, instead of standard sea level pressure.)

Pressure altitude, in the aviation realm, is defined as the altitude read from the altimeter when it is set to a reference pressure of 29.92 inHg, which is the standard sea level pressure. In many locations around the world, barometric pressure readings are reported in millibars or hPa, where 1013 mbar (=1013 hPa) is equal to 29.92 inHg (both being standard sea level pressure). Pilots refer to this setting of the altimeter – to provide pressure altitude – as QNE. In engineering terms, pressure altitude has essentially the same meaning. An engineer would express pressure altitude as the altitude corresponding to a given pressure in the standard atmosphere. Both the engineering and aviation definitions for pressure altitude are equivalent, since the altimeter referenced to 29.92 inHg (1013 mbar) is calibrated based on the standard atmosphere.

Density altitude is defined, in aviation terms, as the pressure altitude corrected for non-standard temperature. If the temperature on a given day at a particular altitude is hotter than the standard temperature, then the density altitude will be higher. In engineering terms, density altitude is defined as the altitude corresponding to a given density in the standard atmosphere. Aircraft performance depends significantly on local air density, so density altitude is a direct indication of aircraft performance. Higher density altitude (corresponding to lower density) will lead to longer takeoff ground roll, slower rates of climb, higher true airspeed for stall, etc.

Finally, pilots are also concerned with the height of the aircraft above the local terrain, which is termed absolute altitude. In the aviation realm, absolute altitude is often also termed height above ground level, so the acronym AGL is often used. On aviation charts, both true altitude (MSL) and absolute altitude (AGL) are reported for various obstacles. For example, the top of a 500-ft tall radio tower mounted on ground that is 1500 ft above sea level will have a maximum height of 2000 ft MSL or 500 ft AGL. Thus, pilots pay close attention to the absolute altitude (also referred to as QFE) as well as the true altitude (QNH). Note that absolute altitude (AGL) in an aviation context is not the same as absolute altitude (h_A) in an engineering context. The engineering definition of absolute altitude is seldom used in aerospace or aviation, outside of discussions of the standard atmosphere.

A selection of the most significant of these altitudes is illustrated in Figure 2.8. The aircraft depicted in this figure is cruising at a true altitude of 5000 ft (MSL), but because its flight is over

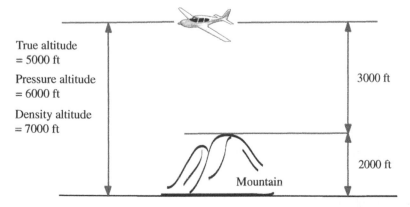

Figure 2.8 Illustration of different altitudes used in aviation.

mountains rising 2000 ft above sea level, the aircraft is at an absolute altitude of only 3000 ft AGL. On this given day, the local barometric pressure reading is lower than standard, causing the pressure altitude to be higher than the true altitude. And, if the temperature on this day is higher than standard, then the density altitude will be even higher than pressure altitude or true altitude. Thus, we could easily have a situation where absolute altitude, true altitude, pressure altitude, and density altitude are all different. In Chapter 3, as we move into instrumentation used for flight testing, we will discuss the operation of the altimeter in greater detail.

Nomenclature

a	temperature lapse rate, dT/dh
c	chord
g	gravitational acceleration
g_0	gravitational acceleration at sea level
h	geopotential altitude
h_A	absolute altitude (height relative to the center of the Earth)
h_G	geometric altitude (height above mean sea level)
k	constant, a/T_{SL}
m	mass of air in the control volume
n	constant, $-g_0/aR$
p	pressure
R	gas constant for air
r_{Earth}	Earth's mean radius
Re_c	Reynolds number based on chord
S_{visc}	Sutherland's constant
T	temperature
U_∞	freestream velocity
W	weight of air in the control volume
x	length of control volume element
y	width of control volume element
β	constant used in Sutherland's Law
δ	pressure ratio, p/p_{SL}
μ	dynamic viscosity
ν	kinematic viscosity
ρ	density
σ	density ratio, ρ/ρ_{SL}
θ	temperature ratio, T/T_{SL}

Subscripts

ref	reference conditions at the base of a given atmospheric layer
SL	sea level
trop	tropopause
1	beginning of an atmospheric layer
2	end of an atmospheric layer

Acronyms and Abbreviations

AGL   height above ground level
ICAO   International Civil Aviation Organization
MSL   height above mean sea level
NOAA National Oceanic and Atmospheric Administration

References

Anderson, J.D. Jr., (2016). *Introduction to Flight*, 8e. New York: McGraw-Hill.

Carmichael, R. (2018). Public domain aeronautical software for the aeronautical engineer. http://www.pdas.com/atmos.html (accessed 28 December 2020).

ICAO (1993). *Manual of the ICAO Standard Atmosphere (Extended to 80 Kilometres (262 500 Feet))*, 3e, ICAO Document 7488. Montréal, QC: International Civil Aviation Organization.

NOAA, NASA, and USAF (1976). *U.S. Standard Atmosphere, 1976*, NOAA-S/T-76-1562, NASA-TM-X-74335. Washington, DC: U.S. Government Printing Office. http://hdl.handle.net/2060/19770009539.

Sartorius, S. (2018). *Standard Atmosphere Functions, v. 2.1.0.0*. MathWorks File Exchange. https://www.mathworks.com/matlabcentral/fileexchange/28135-standard-atmosphere-functions.

Sutherland, W. (1893). LII. The viscosity of gases and molecular force. *The London, Edinburgh, and Dublin Philosophical Magazine and Journal of Science, Series 5* 36 (223): 507–531. https://doi.org/10.1080/14786449308620508.

3

Aircraft and Flight Test Instrumentation

This chapter fundamentally deals with how we will measure the various aircraft performance characteristics. We will cover common instruments used in the aircraft cockpit (both traditional and modern avionics systems), as well as instrumentation found in external data acquisition (DAQ) systems. This fundamental understanding is important for developing an appreciation of the capabilities and limitations of each instrument and sensor system. We will begin with a discussion of various sensors and instrumentation hardware used in flight testing (these instruments can be either the standard cockpit instrumentation or supplemental instrumentation dedicated for flight testing purposes).

A discussion of instrumentation and DAQ must begin by detailing what we wish to measure and how we will measure it. For aircraft flight testing, we essentially need to know the vehicle's state – we need to know the aircraft's position and orientation as a function of time. Practically speaking, this requires measurements of the vehicle's velocity, acceleration, orientation (pitch, roll, and yaw angle relative to a defined set of coordinate axes), altitude, and magnetic heading.

Aircraft instruments are generally calibrated to report quantities in the English unit system, sometimes using nonstandard units. Speed is most often reported in knots (nautical miles per hour), altitude is in feet, vertical speed is in feet per minute, distance is in nautical miles, angles are in degrees, and angular rates are in degrees per second. When analyzing flight test data from these measurements, it is critical to convert to standard consistent units before further analysis. All analysis must be done in a single, consistent unit system with standard units (*e.g.*, English: ft/s, ft, deg, and deg/s, or SI: m/s, m, deg, and deg/s). After completion of the analysis, the results may be converted to any desired units for reporting. This section provides an orientation to the instruments installed in general aviation (GA) aircraft cockpits, including both classical "steam gauge" instruments and more modern glass panel instrument displays. We will first discuss the traditional instruments, followed by glass panel avionics systems.

3.1 Traditional Cockpit Instruments

In traditional flight testing methods at the university level, many of the desired parameters are hand-recorded from cockpit indicators such as the airspeed indicator (ASI), attitude indicator, heading indicator, and altimeter, which are illustrated in Figures 3.1 and 3.2 for a traditional "steam gauge" cockpit. The traditional instruments are arranged in a "six pack" configuration (Figure 3.2): starting at the upper left and moving clockwise, the instruments are the ASI, attitude indicator, altimeter, vertical speed indicator (VSI), heading indicator (also known as the directional gyro), and turn/slip indicator (also known as the turn coordinator). We will group these instruments into two types. The first group, consisting of the attitude indicator, heading

Introduction to Flight Testing, First Edition. James W. Gregory and Tianshu Liu.
© 2021 John Wiley & Sons Ltd. Published 2021 by John Wiley & Sons Ltd.
Companion website: https://www.wiley.com/go/flighttesting

Figure 3.1 Overview of aircraft cockpit instrumentation for traditional "steam gauge" instruments. Source: https://upload.wikimedia.org/wikipedia/commons/thumb/e/e7/Slingsby.t67c.panel.g-bocm.arp.jpg/ 1200px-Slingsby.t67c.panel.g-bocm.arp.jpg.

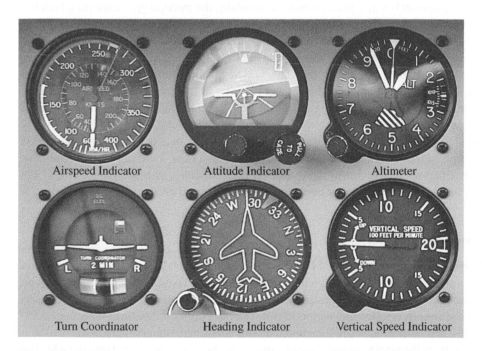

Figure 3.2 Detailed view of the six pack of key instruments in a traditional cockpit. Source: Modified from photo by Mael Balland on Unsplash, https://unsplash.com/photos/V5hAryReZzo.

indicator, and the turn/slip indicator, is based on gyroscopes. The second group, consisting of the ASI, altimeter, and VSI, is based on physical measurements of pressure. We will discuss each group of instruments in more detail as follows. Interested readers can also consult various publications such as Chapter 7 of the *Pilot's Handbook of Aeronautical Knowledge* (2008), Chapter 5 of the *Instrument Flying Handbook* (2012), and the *Advanced Avionics Handbook* (2009) for more details.

3.1.1 Gyroscopic-Based Instruments

Before discussing the first group of instruments, it is helpful to briefly review the ideas behind gyroscopes. The operation of a gyroscope is based on the Newtonian principle of conservation of angular momentum. Due to conservation of angular momentum, the axis of rotation of a spinning object with a given angular momentum ($L = I\omega$, where I is the moment of inertia and ω is the angular speed) will remain fixed unless acted upon by an external applied torque. (Think of how the axis of a child's spinning top remains upright as long as the top is spinning.) Over time, small external forces and moments that are present in any practical implementation of a physical gyro will lead to gyroscopic precession, which results in small changes in the axis of rotation that can grow over time.

The three gyroscopic instruments in a traditional six pack of instruments – the attitude indicator, heading indicator, and turn coordinator – all depend on gyroscopic principles. The rotational speed of the gyros in the attitude and heading indicators are traditionally powered by a vacuum system that flows air over a small turbine connected to the spinning disks in order to maintain rotation. The turn coordinator, on the other hand, is typically powered by the aircraft's electrical system. The attitude indicator provides an artificial horizon that provides an indication of the amount of bank and pitch that the aircraft has at a given moment. Essentially, the internal gyroscope (to which the artificial horizon is mounted) maintains its rigidity in space and the aircraft pitches and rolls about the internal gyro. Arcing along the top of the indicator, the first three white lines on either side of the centerline indicate increments of 10° bank angle (up to 30° bank). The next two lines on either side represent a 45° and 60° bank angles, respectively. The heading indicator is based on a geared gyro and simply indicates the magnetic heading of the aircraft (as long as the gyro has been set to match the magnetic compass). Both the attitude indicator and the heading indicator experience errors due to gyroscopic precession that must be occasionally corrected. For example, periodically in straight and level flight, the pilot may need to reset the heading indicator to match the indication on the magnetic compass. The turn coordinator, also based on a gyro, indicates the instantaneous rate of turn. The two white lines below the level lines each indicate a standard rate of turn of 3°/s to the right or left, which would require two minutes for the aircraft to complete a 360° turn.

3.1.2 Pressure-Based Instruments

The second group of instruments is based on measurement of total pressure and static pressure on the aircraft. Total pressure is measured by a pitot probe, which is often mounted under the wing on GA aircraft (see Figure 3.3(a)). Static pressure is measured by flush-mounted static pressure ports on the side of the aircraft fuselage (see Figure 3.3(b)). The pitot probe and static port are connected via tubing to the ASI, altimeter, and VSI mounted on the instrument panel (Figure 3.4).

Total pressure from the pitot tube and static pressure from the static port are fed into the ASI. The ASI is calibrated based on the assumption of standard sea-level conditions with the isentropic Mach relation to convert a measured "impact pressure" (difference between total and static pressure) into indicated airspeed. For low-speed, low-altitude flight this is equivalent to converting a measured

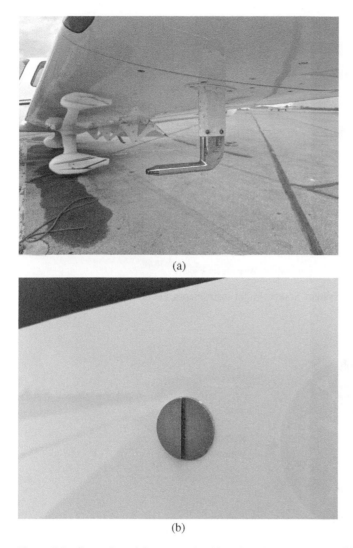

(a)

(b)

Figure 3.3 Examples of the pitot tube (a) and static pressure port (b) on an aircraft.

dynamic pressure to velocity by the Bernoulli equation. Further details on the functioning of the ASI are provided in Chapter 8 on calibration of the ASI.

The VSI, connected to the static pressure port, measures the time rate of change of pressure and converts this to a vertical speed. It is based on a mechanical comparison of the rate of change of static pressure with a known rate of change coming from a calibrated leak. Indications of vertical speed on the VSI tend to fluctuate so it is not as useful for measuring rate of climb or descent in a precise manner. Instead, for measurement of vertical speed in flight test, it is best to establish a steady rate and directly measure altitude from the altimeter and time with a stopwatch.

The altimeter (Figure 3.5) is based on a measurement of static pressure, which is converted into an indicated altitude. Altitude is displayed on the traditional altimeter by three hands, much like an analog clock displays time. The long, thin pointer with a triangle at the end indicates ten thousands of feet; the short, thick hand displays thousands of feet; and the medium length, slender hand displays hundreds of feet. So, the medium, slender hand makes a full revolution of the dial every 1000 ft; the short, thick pointer makes a complete revolution every 10,000 ft; and the long, slender

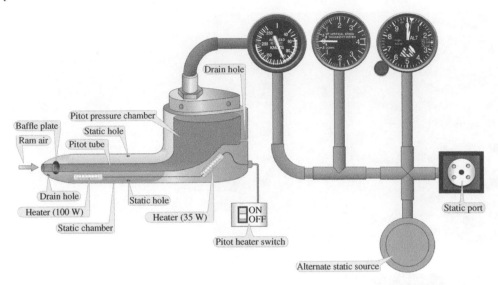

Figure 3.4 Schematic of the pitot-static system. Source: Flight Instruments, Federal Aviation Administration.

Figure 3.5 Diagram of an aircraft altimeter. The reference pressure is set by the knob on the lower left side and indicated in the Kollsman window on the right side of the instrument. Source: Bsayusd, A 3-pointer pressure altimeter. Originally from en.wikipedia.

pointer makes a full revolution every 100,000 ft. For example, the altimeter depicted in Figure 3.5 indicates an altitude of 10,180 ft.

The calibration of the altimeter is based on the pressure lapse rate of the standard atmosphere (assuming standard temperature). As discussed in Chapter 2, the pressure lapse rate is extremely consistent, which enables the altimeter to provide a highly accurate measurement. However, the local barometric pressure can change significantly, which would have a first-order impact on the indicated altitude provided by an altimeter. Thus, an altimeter must account for variations in local barometric pressure reading, which is done by setting a reference pressure. This reference pressure appears in a small window on the right side of the altimeter and is set by a knob on the lower left side of the instrument (see Figure 3.5). It is important to note that the reference pressure is not related in any way to the actual pressure at some arbitrary altitude. Instead, the reference pressure simply shifts the base of the calibration curve, providing an offset above or below standard sea level pressure in order to accommodate the actual local barometric pressure reading and providing a more accurate reading of altitude.

This type of altimeter – the so-called sensitive altimeter – was first invented by Paul Kollsman, leading to the naming of the Kollsman window where the reference pressure is read on the altimeter (*e.g.*, the reference pressure shown in Figure 3.5 is 29.92 inHg). The Kollsman setting essentially shifts the altimeter's calibration curve in order to account for local variations in sea level pressure, which is routinely encountered due to weather variations (*e.g.*, low or high pressure regions moving through a geographical area). Since accurate measurement of altitude is critical in most flight testing work, we will devote some attention to the calibration and correct setting of the altimeter.

The most typical use of the altimeter under routine flight is to set the Kollsman reference pressure to the local barometric pressure, such that the indicated altitude is the aircraft's height above mean sea level (MSL). A pilot can easily obtain up-to-date readings of the local barometric pressure by listening to broadcasts from local weather stations (such as the Automated Weather Observing System, or AWOS, found at many airports). Alternatively, air traffic control (ATC) will often report the local barometric pressure setting, especially when aircraft are transitioning from one controller to another with responsibility for different geographical areas. Reporting of the local barometric pressure is an important function for ATC, since they aim to maintain consistent vertical separation between all aircraft. Thus, for a pilot to ensure an accurate reading of MSL altitude, the altimeter must be set to the correct local barometric pressure reading (viewed in the Kollsman window) by adjusting the Kollsman knob.

However, in flight testing, we often wish to know the pressure altitude, or the pressure at our flight altitude. This can be accomplished by setting the altimeter to a reference pressure of 29.92 inHg (1013 hPa) instead of the local barometric pressure reading. Under this situation, the altimeter will *not* provide an accurate indication of height above sea level; rather, it will indicate *pressure altitude*. A flight test engineer can readily take the measured pressure altitude and convert this to a value of the local freestream static pressure via the standard atmosphere, using the theory described in Chapter 2[1].

3.1.3 Outside Air Temperature

Knowledge of ambient (freestream) temperature of the flight environment is important for calculating the freestream density from the ideal gas law and for determination of true airspeed. The challenge of measuring outside air temperature (OAT) from a moving aircraft is related to the frame of reference of the air (the desired temperature) versus the aircraft reference frame (where temperature is measured). Since the air is moving relative to the aircraft, the temperature measured in the aircraft frame of reference will be higher. If a temperature probe completely stagnates the flow (*i.e.*, with no loss of energy), the isentropic Mach relation,

$$T_\infty = T_0 \left(1 + \frac{\gamma - 1}{2} M_\infty^2\right)^{-1}, \tag{3.1}$$

may be used to find the freestream temperature (T_∞) from measurements of the stagnation temperature (T_0) and the freestream Mach number (M_∞), where $\gamma = 1.4$ is the ratio of specific heats for air. However, most temperature probes do not fully recover the full stagnation temperature, so a recovery factor (k) must be introduced. Furthermore, there may be differences between the local

---

1 Additional details are available in an online supplement, "Effects of Kollsman Setting on Altimeter Reading."

Mach number (M_ℓ) and the freestream Mach number, due to local acceleration or deceleration of the flow around the aircraft body. Thus, we can modify Eq. (3.1) to accommodate these two factors,

$$T_\infty = T_0 \left(1 + \frac{\gamma - 1}{2} k M_\ell^2 \right)^{-1} \left(\frac{1 + \frac{\gamma - 1}{2} M_\ell^2}{1 + \frac{\gamma - 1}{2} M_\infty^2} \right), \tag{3.2}$$

where M_ℓ and k can be found by calibration (Gracey 1980).

3.1.4 Other Instrumentation

Other relevant instruments on a traditional cockpit panel include a clock (for timing various events), the engine tachometer (for measuring engine speed in RPM), manifold pressure for the engine air intake, fuel flow rate, and total fuel burned. In traditional flight testing at the university level, readings from the "steam gauge" instruments must be manually recorded via pen and paper. These manual data recording techniques can be effective, as long as the time scale of any transient characteristic of the flight maneuver is long relative to the speed at which data can be manually recorded. For example, the rate of climb in a sustained, steady climb can be easily recorded by hand with a stopwatch and the altimeter, while the level acceleration of an aircraft would be much more difficult to manually record since velocity is changing quickly over time.

3.2 Glass Cockpit Instruments

Glass panel avionics displays (Figures 3.6–3.8) are becoming increasingly common on small GA aircraft. Glass panels measure and display the same data as traditional instruments, but the

Figure 3.6 Overview of aircraft cockpit instrumentation for a glass panel avionics display in a Cirrus SR20.

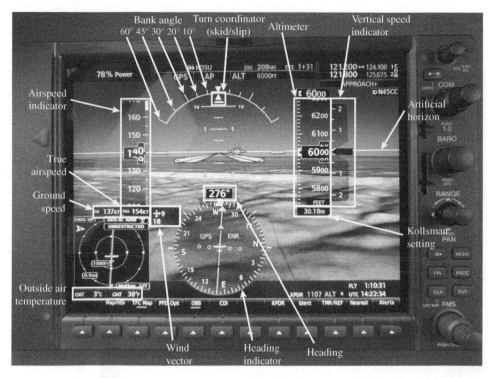

Figure 3.7 Detailed view of the primary flight display (PFD) on a Cirrus SR20, with key indications identified.

organization and display of data is greatly improved for better scanning and interpretation by the pilot. The glass panel display is usually physically segmented into a primary flight display (PFD), where critical data such as airspeed, altitude, heading, and attitude are displayed (left screen in Figures 3.6 and 3.7), and a multifunction display (MFD) where secondary data such as engine performance, navigation, terrain, traffic, etc. are displayed (right screen in Figures 3.6 and 3.8). We will discuss the glass panel display of the same data streams that are found on a traditional six pack of instruments, followed by a few notes on recording data from glass panel avionics systems. The following discussion refers to Figures 3.7 and 3.8.

Measurements such as airspeed and altitude continue to be made in the same manner, where total pressure and static pressure are used to calculate the indicated airspeed. Since the calibration of airspeed is known, and real-time measurements of pressure and OAT are available, the flight computer can calculate the true airspeed in real time. Since global positioning system (GPS) data are also available for measuring ground speed, the flight computer can also calculate the winds aloft, which represent the difference between true airspeed and ground speed. (See Chapter 8 for detailed definitions of these speeds, and details on how they are computed.) Indicated airspeed is prominently displayed on the left side of the glass panel on a linear, sliding scale along with a digital readout in the center of the scale. Altitude is also displayed on a linear sliding scale, but on the right side of the screen. The Kollsman setting (reference pressure) is displayed immediately beneath the altitude scale. On the right side of the altitude ticker is a display of the rate of climb in numerical and graphical form.

The attitude indicator is displayed in much the same way as it is on a traditional attitude indicator; one critical difference is that the artificial horizon completely spans the width of the display for better situational awareness for the pilot. The heading indicator is in the bottom center of

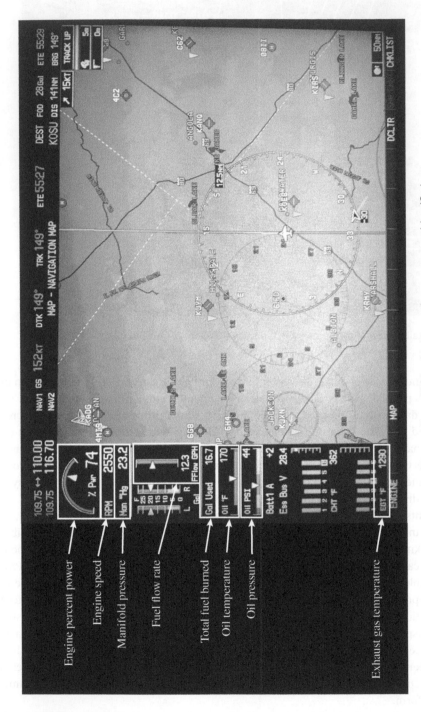

Engine percent power

Engine speed

Manifold pressure

Fuel flow rate

Total fuel burned

Oil temperature

Oil pressure

Exhaust gas temperature

Figure 3.8 Detailed view of the multifunction display (MFD) on a Cirrus SR20, with key engine parameters identified.

the PFD, with heading information displayed in much the same way as on the analog instrument. Both the attitude indicator and the heading indicator are based on microelectromechanical systems (MEMS)-based gyroscopes, magnetometers, and accelerometers. The details of these sensor schemes will be discussed in the following subsections. OAT is typically displayed on the lower left side of the PFD. Engine parameters such as engine speed, percent power, manifold pressure, fuel flow rate, fuel burned, exhaust gas temperature, oil pressure, oil temperature, etc. are displayed on the MFD (see Figure 3.8).

The airspeed and altitude tickers, along with the heading indicator, have accompanying magenta trend bars (on the inside edge of each ticker or along the circumference of the heading indicator) that indicate the projected value that will be true 6 seconds in the future, based on current rates of change. Rate of turn can be inferred from the length of the magenta bar on the heading indicator (e.g., a standard rate turn would have a magenta bar extending 18° from the center).

Data from glass cockpit displays may also be recorded manually (pen and paper), or in some cases, limited data streams are available in digital form from the avionics suite itself. For example, the Avidyne FlightMax Entegra avionics suite records aircraft data including latitude, longitude, pressure altitude, density altitude, exhaust gas temperatures and cylinder head temperatures for all cylinders, oil temperature, oil pressure, engine RPM, OAT, and manifold pressure. While this data stream is predominantly focused on engine parameters (for engine health monitoring), it can be a useful supplement to other data streams used in flight testing. The avionics suite records the data at a rate of one sample per 6 seconds (0.167 Hz) and stores it in internal data storage. The data can be retrieved after the flight via the USB port on the front of the avionics panel.

The fundamental principles for the sensors at the heart of an aircraft's glass panel avionics are the same as those underlying traditional cockpit instruments. Glass cockpit sensors are based on gyroscopic principles and measurement of physical quantities such as pressure and temperature, but there are critical differences. Sensors such as magnetometers, accelerometers, and rate gyroscopes are grouped together into an inertial measurement unit (IMU), known as the attitude and heading reference system (AHRS), that fuses the sensor data streams to provide real-time computations of an aircraft's heading and orientation in space. A second major subsystem is the air data computer (ADC), which computes aircraft speed, altitude, and rate of change of altitude based on measurements of total pressure, static pressure, and OAT. The third major subsystem used in glass panel avionics is the navigation instruments. In modern avionics, this primarily relies upon a global navigation satellite system (GNSS) receiver (i.e., a GPS receiver) but also includes radio receivers for radio-based navigation aids. A significant advantage of glass panel avionics is that derived quantities such as winds aloft, density altitude, true airspeed, etc. can be determined real time in flight via the onboard computer at the heart of the avionics system.

3.3 Flight Test Instrumentation

Recent developments in MEMS have revolutionized sensor design and fabrication, making these sensors available at low cost and in small packages. Complex sensors such as gyroscopes and magnetometers can be fabricated with an extremely small form factor, allowing for them to be installed in very compact devices (see Figure 3.9). Most notably, the rapid development of MEMS has been driven by the proliferation of smartphones, which have many of the same sensors as those found on aircraft. MEMS developments have also enabled integration of these instruments into the cockpit – the modern avionics glass cockpit systems discussed in the previous section rely upon MEMS sensors for determining aircraft state. Alternatively, small external sensor packages such as smartphones, or the custom-built unit, illustrated in Figure 3.9, can be mounted in any convenient

Figure 3.9 Modern flight testing board with built-in GPS, accelerometers, gyroscopes, magnetometers, and pressure transducer. Source: Photo courtesy of Matthew H. McCrink.

location within the aircraft cockpit for simple installation and reasonably accurate DAQ. These comprehensive sensor suites are roughly equivalent to the AHRS that supports glass panel avionics.

The most significant deficiency of these instrumentation packages is the lack of air data measurements such as total pressure, static pressure, and OAT. Measurement of these properties requires direct access to dedicated instrumentation on the aircraft (the pitot-static system), which is not normally possible on GA aircraft commonly accessible to students in an aircraft flight testing course. However, there are techniques ("work-arounds") for inferring these missing flight data. OAT changes very slowly at a given altitude, allowing it to be manually read and recorded at sparse intervals at a particular flight condition. Freestream static pressure can be measured to fairly good approximation by measuring cockpit pressure using a MEMS barometer on the DAQ device (see Gregory and McCrink 2016 for details). Total pressure (and, thus true airspeed) cannot be measured in flight without access to a pitot probe. However, using the techniques described in Chapter 8, true airspeed may be found by measuring ground speed during flight at a single test condition along three separate headings.

Key sensors in a typical external DAQ unit are a satellite navigation receiver and antenna (*e.g.*, the GPS), 3-axis gyroscopes, 3-axis accelerometers, 3-axis magnetometers, and a pressure transducer (barometer). A brief overview of each of these sensors is discussed as follows, but a much more detailed discussion is available in Titterton and Weston (2004).

3.3.1 Global Navigation Satellite System

One example of a global navigation satellite system used for determining position in 3D space (latitude, longitude, and altitude) is the GPS (for the remainder of this book, GPS will be considered synonymous with GNSS, although there are other satellite-based navigation systems in use such as the Russian GLONASS, the European Galileo, or the Chinese BeiDou systems.). GPS signals offer very high positioning accuracy, typically resolving location to within a few meters or less. The basis for the measurement is transmission of a modulated carrier signal with a known pseudorandom code, a time stamp based on a highly stable atomic clock on board the satellite, and the precise location of the satellite in space. The GPS receiver infers the distance to each satellite being received based on the time required for the signal to traverse the distance from the satellite to the receiver (based on the ultrastable clock on each satellite). Since the position of each satellite in space is known with high accuracy and precision, and the speed of light is well known, the receiver can infer the distance to each satellite by phase aligning the pseudorandom code. The receiver then uses multilateration techniques to infer its own location in space based on these distances – the calculated distance from a particular satellite restricts the receiver location to being somewhere on a spherical shell centered on the satellite. The intersection of at least three spherical shells results in position being defined as a single point in space. Note that the clock on the receiver itself generally does not have a high degree of accuracy, so this represents an unknown in the calculation of position. Therefore, there are four unknowns in the computation: latitude, longitude, altitude, and instantaneous time, so the GPS receiver must have at least four satellites in view simultaneously in order to determine an accurate position estimate. GPS receivers can reliably report latitude and longitude with high accuracy, but altitude indications typically have much more error. In particular, smartphone-based GPS receivers lack an external antenna and rely upon algorithms that are optimized for terrestrial applications. Thus, altitude reported by smartphone-based GPS sensors is generally unreliable and should not be used for measurement of altitude in aircraft flight testing (see Gregory and McCrink 2016 for details). General overviews of satellite-based navigation schemes are provided by Misra and Enge (2010) and Kaplan and Hegarty (2017).

Several key error sources limit the accuracy of GPS position estimates. These errors include propagation delays due to dispersion of the signal by the ionosphere (termed ionospheric delay), uncertainty in the time due to drift of the atomic clocks on board the satellites, and uncertainty in the location of the satellites (ephemeris error). Advanced signal processing algorithms have been introduced to correct for these errors, but the most basic correction scheme is to introduce a real-time error correction term from a receiver at a fixed, accurately known position. These correction schemes include differential GPS (DGPS), real-time kinematic (RTK) GPS, and the FAA's Wide Area Augmentation System (WAAS). We will provide a brief overview of each of these as follows.

Differential GPS can provide refined position and velocity histories for flight test applications (Sabatini and Palmerini 2008). A typical DGPS architecture for flight testing is shown in Figure 3.10. The system consists of a reference receiver located at a known location that has been previously surveyed, and one or more DGPS user (mobile) receivers mounted on a test aircraft. The reference receiver antenna, differential correction processing system, and datalink equipment are collectively called the reference station. Both the user receiver and the reference receiver data can be collected and stored for later processing, or sent to the desired location in real time via the datalink. DGPS is

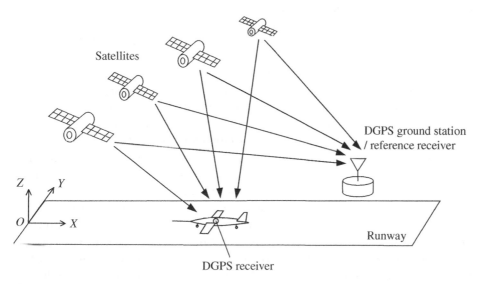

Figure 3.10 Typical differential GPS architecture.

based on the principle that receivers in the same vicinity will simultaneously experience common errors on a particular satellite ranging signal. In general, mobile receivers use measurements from the reference receiver to remove the common errors. The limiting factor for DGPS is that the mobile and fixed receivers need to be in proximity to one another such that they experience the error sources in the same way and to facilitate radio communication from the fixed reference receiver to the mobile receiver on board the aircraft. Thus, DGPS is most applicable to local-area flight operations such as takeoff and landing flight tests.

RTK GPS is a technique similar to DGPS, but offers higher accuracy. Similar to DGPS, RTK provides a correction to the position estimate, which can be transmitted in real time to the mobile GPS receiver or stored for subsequent analysis. The correction factor is determined by measuring the distance to the satellite using a different technique from traditional GPS. Instead of relying solely on the pseudorandom code transmitted by the satellites, RTK GPS uses statistical methods to estimate the number of cycles present in the waveform between the receiver and the satellite and then multiplies the number of cycles by the wavelength (19 cm for the L1 signal) to infer range. There is some resulting error in the estimated distance, due to ambiguity in determining the correct integer number of cycles due to phase differences. The RTK technique can provide remarkable positioning accuracy, improving the position estimate to 1 cm accuracy. However, RTK GPS also requires that the mobile receiver be in the same vicinity as a fixed reference station, which limits its applicability to downrange flight tests.

In the aviation realm, an augmentation system has been recently developed in order to improve upon the baseline accuracy of the satellite-based position measurement system and enable precision instrument approaches without requiring nearby ground-based reference stations. The Wide Area Augmentation System is a satellite-based augmentation system developed by the US Federal Aviation Administration that covers the majority of North America. It is based on a system of ground-based reference stations that calculate the local difference between the GPS-indicated position and the station's actual position (surveyed to very high accuracy). This error data, expressed as a deviation correction, is uplinked in real time to geostationary WAAS satellites at least once every five seconds, which then broadcast the correction to aircraft throughout the national airspace. WAAS GPS corrects for positioning errors predominantly resulting from ionospheric disturbances, which add phase distortion to received GPS signals. The WAAS correction is broadcast over the same frequency bands used for the baseline GPS signal, which reduces system cost and complexity. The resulting accuracy of a WAAS GPS receiver is improved to approximately 2-m in the horizontal and vertical directions, which is an order of magnitude improvement relative to the baseline accuracy of standard GPS.

An important point to recognize when using GPS data for flight test is that the position from a GPS receiver is reported as decimal degrees for latitude and longitude, with a sign convention for positive being North of the equator and East of the prime meridian (*e.g.*, the latitude/longitude coordinates 40.074199 ° , − 83.07968° are in North America). If relative distance is desired (say, in units of ft, m, or nautical miles), then some type of transformation is needed to convert the difference between latitude/longitude coordinate pairs into distance. This transformation is commonly done in the fields of geodesy and navigation, where a model of the Earth's shape must be assumed. Generally speaking, Earth is in the shape of an ellipsoid that bulges near the equator (relative to a perfect sphere) due to the rotation of the planet. One of the most common Earth models is the World Geodetic System 1984 ellipsoid (WGS84), which is periodically refined and revised (National Imagery and Mapping Agency 2000; Pavlis et al. 2012). WGS84 forms the basis of GPS position reporting, with latitude and longitude forming a measure of distance within the WGS84 coordinate system. However, these coordinates must be transformed to local Cartesian coordinates before distance calculations can be made for flight testing. This transformation can be done via the Universal Transverse Mercator conformal projection. This system involves the segmentation of Earth's surface into 60 zones, each measuring 6° of longitude wide. Bands of latitude measuring 8° high are sometimes used to further subdivide the zones. The transformation for each zone from WGS84 to Cartesian coordinates can be performed by methods described by Snyder (1987), and a wide array of MATLAB toolboxes are available for this purpose (*e.g.*, see Wasmeier 2015).

3.3.2 Accelerometers

A three-axis accelerometer provides a direct measurement of acceleration in all three directions, with the acceleration being relative to the local inertial frame of reference. Thus, an aircraft in steady level flight will have an acceleration of 9.81 m/s^2 in the vertical direction (+1g), and zero acceleration in the other directions, since the local inertial frame corresponds with a freely falling object and the aircraft is accelerating upwards relative to the inertial frame. Accelerometers essentially operate as a spring-mass-damper system, where the applied acceleration along a particular axis will cause a displacement of the mass by an amount that depends on the properties of the system. The displacement is then measured by converting it to a voltage, and calibrating this voltage to acceleration. In MEMS devices, the spring-mass-damper system is often a cantilevered beam with a proof mass, where transduction of the displacement to voltage is most often done through capacitive or piezoresisitve schemes. Capacitive transduction involves a gap between the cantilevered beam and a fixed beam, which varies the capacitance of a circuit which can then modulate a measured voltage. The piezoresistive scheme involves a piezoelectric material as the spring in the system, where the voltage drop across the piezoresistor changes with the applied strain due to acceleration.

Accelerometers can be used in flight testing to detect events with sudden changes in acceleration (such as stall), measure the orientation of the aircraft relative to the ground in steady flight, indicate the bank angle in a steady turn by measuring g's in the aircraft's frame of reference, determine the period of dynamic stability phenomena such as the long-period phugoid mode, and other applications. In principle, the measured acceleration can be integrated in order to infer a change in velocity, although the noise in the signal often precludes this in practice. Further, the measured accelerations are integrated into the flight data computer for determining vehicle state in modern avionics systems.

3.3.3 Gyroscopes

Building upon our earlier discussion of gyroscopic principles for traditional flight instruments, we will now consider how they are used for modern avionics systems and instrumentation. Rate gyros used in modern glass panel avionics and DAQ systems are based on MEMS-fabricated gyros. These gyros sense the rate of angular motion, rather than directly measuring the angle itself. A three-axis gyro will have three independent rate gyros mounted along mutually perpendicular planes. One example of a MEMS rate gyro is based on the principle of sensing Coriolis acceleration. In this configuration, a proof mass is mounted on springs and oscillated in a direction perpendicular to the measured axis of rotation. As the proof mass oscillates, its radial distance from the aircraft's center of rotation also changes, leading to time-varying tangential velocity that subjects the mass to varying amounts of Coriolis acceleration while the body rotates. This leads to a time-varying reaction force in a direction perpendicular to the axis of rotation and the direction of oscillation of the proof mass. This reaction force is applied across springs in the lateral direction, which translate the reaction force into a linear motion that is sensed by capacitive elements (interdigitated fingers).

Since rate gyros measure the angular speed – pitch rate, roll rate, or yaw rate – rather than directly measuring the respective angles, the rotation rates must be integrated with respect to time in order to determine the pitch, roll, or yaw angles. Through the integration process, any noise present in the signals accumulates and leads to growing error in time.

3.3.4 Magnetometers

A three-axis magnetometer can be used to sense the aircraft's magnetic heading, much in the same way that a compass provides magnetic heading by always pointing toward magnetic north. On a MEMS-based device, each of the three magnetometers is mounted mutually perpendicular with a reasonably high degree of precision. Each magnetometer senses the local magnetic field via the Hall effect, whereby a voltage difference is induced across an electrical conductor in a direction transverse to both an applied magnetic field and an electric current flowing through the device. For sensing the aircraft's heading, each axis of the magnetometer responds to Earth's magnetic field lines, which are aligned between the magnetic North and South poles. Since the North and South poles are not aligned with the physical poles, there is a difference between magnetic north and true north (referred to as magnetic declination, or variation). The amount of magnetic variation depends on geographic location and time, with a time scale of years (National Centers for Environmental Information 2019).

When using a magnetometer in an aircraft, the sensing of the Earth's magnetic field can be strongly influenced by nearby magnetic fields and ferromagnetic materials, through an effect known as magnetic deviation. Thus, an aircraft's compass must be calibrated to account for these error sources – this is typically done by positioning the aircraft along various points of the compass (typically referenced to a so-called "compass rose" painted on the ramp with 30° heading intervals) and noting the difference between the compass reading and the actual magnetic bearing. The two error sources are referred to as hard and soft iron distortions. Hard iron distortions are due to the presence of nearby magnetic fields, which could be from magnets or electronic circuits, and result in a constant bias offset on the measured magnetic field. Soft iron distortions are due to the presence of nearby ferromagnetic materials (such as the engine block), and result in skewing of the measurement or an offset whose magnitude is heading-dependent. These error sources can be nontrivial when using a three-axis magnetometer on board an aircraft. Thus, a magnetometer that is part of an external DAQ system should be calibrated before flight to remove these effects (the magnetometers built into the aircraft's built-in avionics have already been calibrated).

Fortunately, a compass rose is not required for calibration of a MEMS magnetometer, and the required calibration process is fairly straightforward. Before takeoff, with all electronics operating, the DAQ unit in place, and the magnetometer acquiring data, the aircraft should be swung through two complete 360° circles (a total heading change of 720°). When plotted in time, data from the x- and y- axes of the magnetometer from this maneuver will form an ellipse. If the magnetometer was not subject to any error sources, the acquired data should form a constant radius circle centered on the origin. However, the error sources result in an elliptical pattern, where the offset of the ellipse from the origin (0, 0) depends on the error from hard iron distortions, and the eccentricity of the ellipse depends on the magnitude of error from soft iron distortions. A mapping can be determined to take the acquired data and transform it to a circle. This can be done by the following equation,

$$\begin{bmatrix} x_c \\ y_c \\ z_c \end{bmatrix} = \begin{bmatrix} M_{11} & M_{12} & M_{13} \\ M_{21} & M_{22} & M_{23} \\ M_{31} & M_{32} & M_{33} \end{bmatrix} \left(\begin{bmatrix} x_{nc} \\ y_{nc} \\ z_{nc} \end{bmatrix} - \begin{bmatrix} B_x \\ B_y \\ B_z \end{bmatrix} \right), \tag{3.3}$$

where x_{nc}, y_{nc}, and z_{nc} are the noncalibrated data straight from the magnetometer; the **B** vector contains the bias errors due to hard iron distortions, the **M** matrix corrects for the soft iron distortions; and x_c, y_c, and z_c are the calibrated data. Values of **M** and **B** are determined from a least-squares fit to the data and subsequently applied to all magnetometer data acquired during the flight test.

3.3.5 Barometer

A MEMS pressure transducer can be used as a barometer for an indication of pressure altitude. MEMS barometers are typically piezoresistive devices, where the deformation of a thin silicon diaphragm is measured through piezoresitive principles. The silicon diaphragm is doped at certain locations, thus locally altering the electrical conductivity and imparting resistor-like properties. As the diaphragm deflects in response to an imposed pressure, the resistances of the doped regions change. If these piezoresistors are connected to a Wheatsone bridge, the change in resistance is converted to a change in voltage that can be easily measured. The calibration of these sensors is typically temperature dependent, so MEMS barometers often include an integrated temperature sensor for compensation.

The pressure reading of a MEMS barometer can be converted to altitude via hydrostatics. Necessary input values are the local barometric pressure reading and/or field elevation, along with OAT. All of the necessary theory for this conversion is detailed in Chapter 2 on the definition of the standard atmosphere. The pressure reading (and indicated altitude) of MEMS barometers used in some external DAQ devices can have a nonnegligible bias error (offset) in the reading. This is a minor concern, however, since it is a straightforward matter to determine the offset when the aircraft is on the ground, where a pressure measurement is made and field elevation and/or local barometric pressure are known. For flight testing, the other predominant source of error is a difference between cabin pressure (where the DAQ unit is typically mounted for student flight testing) and freestream static pressure. Gregory and McCrink (2016) evaluated the magnitude of this error for flight in a Diamond DA40 and found that cabin pressure was approximately 140 Pa higher than freestream static pressure at cruise conditions. For this particular aircraft, ram air effects that slightly pressurize the aircraft cabin in flight are the likely cause of the discrepancy. Despite these error sources, altitude indication from cabin pressure tends to be more accurate than GPS-reported altitude.

3.3.6 Fusion of Sensor Data Streams

As noted in the previous subsections, each individual sensor has some limitations to its utility due to various sources of noise or other errors such as temperature drift. This makes direct inference of vehicle state from one set of sensors problematic (*e.g.*, integrating the signals from the rate gyros alone to determine vehicle orientation would introduce substantial errors). Optimal performance, however, is obtained when different data streams can be fused together in a manner such that the collected set provides a much more accurate and reliable estimate of vehicle state. One very common method for fusing together data streams is Kalman filtering. The Kalman filter is an algorithm that takes multiple time records of sensor data and produces estimates of vehicle state with improved accuracy compared to the estimate if only a single data stream were available. In essence, it gives a solution or prediction of vehicle state with an accuracy that is much improved over the estimate provided by any subset of sensors. The basis of the Kalman filter is described as follows.

First, we need a precise definition of the vehicle's state. It is the set of data that completely describes the vehicle's position and orientation, along with the rates of change of position and orientation. Thus, the position estimate and its derivatives would be the spatial location of the vehicle, its velocity, and its acceleration in some frame of reference. The orientation estimate is the vehicle's pitch, yaw, and roll angles, along with rates and accelerations of those same angles. These estimates of position and orientation have some uncertainty associated with them, which the Kalman filter assumes to be random and Gaussian distributed.

At each time step, a prediction of the vehicle's state at the next time step is formulated based on knowledge of the current state, an estimate of the uncertainty of the knowledge of that state, and a model based on the vehicle dynamics. When that next time step arrives, the prediction of the vehicle state is compared with measured data of the vehicle's state (along with the associated uncertainty in both the prediction and the measurements of the state). A refined estimate of the vehicle's state is generated by a weighted sum of the predicted state and measured state. Since the Kalman filter relies on both measurements of the vehicle state and model predictions of the vehicle state, the accuracy of the resulting state estimate is dramatically improved.

This is just a very brief overview of data fusion using Kalman filtering, for the purposes of providing perspective. These filters are routinely used in cockpit avionics systems, unmanned aerial vehicle (UAV) flight control systems, and in DAQ systems to improve the reliability and accuracy of vehicle state estimation. A much more detailed discussion of Kalman filtering is available in the literature (*e.g.*, Rogers 2007; Zarchan and Musoff 2015).

3.4 Summary

We have covered a broad range of content related to instrumentation in this chapter. The central theme of this chapter is a description of how various aspects of aircraft performance can be measured in flight, and the key operating principles of standard aircraft instrumentation and avionics, as well as dedicated flight test instrumentation.

Standard aircraft instruments that are commonly used in flight testing include the airspeed indicator, altimeter, vertical speed indicator, engine tachometer and manifold pressure, and the heading indicator. MEMS-based sensors include GPS, magnetometers, accelerometers, gyroscopes, and barometer. The general utility of these sensors across the range of flight tests presented in this text is mapped out in Table 3.1, assuming that raw data from each sensor are used individually (rather than the robust estimation of vehicle state via Kalman filtering, which may likely be beyond the scope of an undergraduate course on flight testing). Table 3.1 clearly shows that GPS, accelerometers, and the barometer are the most broadly useful sensors. Each particular sensor has strengths and weaknesses, as discussed earlier in this chapter, making them more or less useful for acquiring

Table 3.1 DAQ sensor utility matrix.

Flight test	GPS	Magnetometer	Accelerometer	Gyros	Barometer
Power required	✈		✈		
Airspeed calibration	✈				
Climb	✈				✈
Glide	✈				✈
Takeoff and landing	✈		✈	✈	✈
Stall	✈		✈	✈	✈
Turning flight	✈	✈	✈	✈	✈
Longitudinal stability	✈		✈	✈	✈
Lateral/directional stability	✈		✈	✈	✈

flight test data. This concise summary of the relevancy of each sensor provides an overview of the relative importance of each sensor and a quick reference for test planning for a given flight test. Specific details for various performance flight tests are covered in detail in Chapters 7–16.

GPS provides ground speed, heading, and location at a maximum sample rate of about 1 Hz. Altimetry provided by GPS can have substantial errors: the vertical dimension is the least accurate for GPS positioning due to the geometry of the position estimation problem, the fact that antenna reception may be poor from within an aircraft (particularly for a smartphone-based GPS receiver), and because the position estimation algorithms are optimized for terrestrial applications. Also, measurements of distance with GPS must involve transformation of latitude/longitude coordinates into an appropriate local Cartesian plane before further analysis can be done. Location, ground speed, and heading information are useful for nearly every flight test described in this text.

The magnetometer, in practice, is of marginal utility for aircraft performance flight testing. This is primarily because heading information is also available from the GPS receiver, the fact that the magnetometer signal is relatively noisy, and calibration is required. The magnetometer indication must be calibrated for hard and soft iron sources via a quick pre-flight maneuver consisting of two complete turns on the ground. Postprocessing of this calibration and application to the magnetometer data for derivation of heading are more involved. However, the magnetometer provides heading information at a more rapid rate than GPS, and it is highly advantageous to incorporate magnetometer data into a state estimation algorithm such as the Kalman filter.

Three-axis gyroscopes are also of marginal utility as a stand-alone sensor, since MEMS gyros measure angular rates instead of the absolute angles that are more relevant for determination of aircraft attitude. Further, significant noise can be introduced when integrating these signals in order to determine aircraft attitude.

Accelerometers provide a good indicator of any transient event found in flight testing, making it straightforward to find that event in a long data record. Features in the flight such as the stall event, takeoff point, measurement of load factor in a turn, and identification of the characteristic frequency in dynamic longitudinal stability are relatively straightforward to find and measure in accelerometer data. The accelerometers are also useful for measuring the frequency of engine vibrations, which is an indicator of engine speed (rpm), as long as the sampling rate is high enough to avoid aliasing (discussed in the next chapter). Accelerometer signals also facilitate identification of dynamic stability characteristics of the aircraft, such as the phugoid or Dutch roll modes. One key limitation of a 3-axis accelerometer is that the DAQ device axes may not be coaligned with the aircraft body axes, but this can be calibrated by comparing accelerometer signals during steady (1 g) flight.

A barometer can provide a measurement of cabin pressure, which serves as a reasonable proxy for freestream static pressure in most situations. Both altitude and vertical speed can be reliably estimated from barometry data. Pressure altitude can be inferred using the local barometric pressure reading and refined with a measurement of outside static temperature.

Finally, we will conclude with a note of caution about comparing data from different sensors to infer flight test results. Some sensors can have inherent temporal delays, which could lead to timing or phase mismatch between sensor streams. For example, the computation of GPS position estimates requires processing time on board the receiver, so it is important to make sure the correct time base is used for comparison of GPS data with other sensor data streams. Similarly, different sensors (even those installed on the same DAQ device) can have different sample rates. For example, GPS data are often sampled at a rate on the order of 4 Hz, while accelerometers may be sampled at 100 Hz. Again, it is important to use the correct time base for comparing data streams. The next chapter will discuss in detail how data streams are digitized, along with analysis techniques such as filtering and spectral analysis.

Nomenclature

B	magnetometer calibration, bias error vector
g	gravitational acceleration
I	moment of inertia
k	recovery factor
L	angular momentum vector
M	magnetometer calibration matrix
M_ℓ	local Mach number
M_∞	freestream Mach number
t	time
T_∞	freestream static temperature
T_0	stagnation temperature
y_c	calibrated magnetometer data, y-component
z_c	calibrated magnetometer data, z-component
x_{nc}	uncalibrated magnetometer data, x-component
y_{nc}	uncalibrated magnetometer data, y-component
z_{nc}	uncalibrated magnetometer data, z-component
γ	ratio of specific heats
ω	angular velocity vector

Subscripts

c	calibrated
nc	uncalibrated

Acronyms and Abbreviations

ADC	air data computer
AHRS	attitude and heading reference system
ATC	air traffic control
AWOS	automated weather observing system
GLONASS	Globalnaya Navigazionnaya Sputnikovaya Sistema (Russian GNSS)
GNSS	global navigation satellite system
GPS	Global Positioning System
IMU	inertial measurement unit
MEMS	microelectromechanical systems
MFD	multifunction flight display
MSL	mean sea level
OAT	outside air temperature
PFD	primary flight display
RTK	real-time kinematic
UAV	unmanned aerial vehicle
WAAS	Wide Area Augmentation System
WGS84	World Geodetic System 1984 ellipsoid model

References

Gracey, W. (1980). Measurement of Aircraft Speed and Altitude. NASA-RP-1046, https://ntrs.nasa.gov/citations/19800015804.

Gregory, J.W. and McCrink, M.H. (2016). Accuracy of smartphone-based barometry for altitude determination in aircraft flight testing. AIAA 2016-0270, Proceedings of the 54th AIAA Aerospace Sciences Meeting, San Diego, CA.

Kaplan, E.D. and Hegarty, C. (2017). *Understanding GPS/GNSS: Principles and Applications*, 3e. Boston, MA: Artech House.

Misra, P. and Enge, P. (2010). *Global Positioning System: Signals, Measurements, and Performance*, 2e. Lincoln, MA: Ganga–Jamuna Press.

National Centers for Environmental Information, and British Geological Survey (2019). World magnetic model. In: *National Ocean and Atmospheric Administration*. http://ngdc.noaa.gov/geomag/WMM/.

National Imagery and Mapping Agency. (2000). Department of Defense World Geodetic System 1984, Its Definition and Relationships With Local Geodetic Systems. NIMA Technical Report TR8350.2, 3e, Amendment 1, http://earth-info.nga.mil/GandG/publications/tr8350.2/wgs84fin.pdf.

Pavlis, N.K., Holmes, S.A., Kenyon, S.C., and Factor, J.K. (2012). The development and evaluation of the Earth Gravitational Model 2008 (EGM2008). *Journal of Geophysical Research, vol.* 117 (B04406): 1–38. https://doi.org/10.1029/2011JB008916.

Rogers, R.M. (2007). *Applied Mathematics in Integrated Navigation Systems*, 3e. Reston, VA: American Institute of Aeronautics and Astronautics.

Sabatini, R. and Palmerini, G.B. (2008). Differential global positioning system (DGPS) for flight testing. In: *RTO AGARDograph 160*, Flight Test Instrumentation Series, vol. 21.

Snyder, J. P. (1987). Map Projections – A Working Manual. U.S. Geological Survey Professional Paper 1532, http://pubs.er.usgs.gov/publication/pp1395.

Titterton, D.H. and Weston, J.L. (2004). *Strapdown Inertial Navigation Technology*, 2e. Stevenage, Hertfordshire, UK: The Institution of Electrical Engineers https://doi.org/10.1049/PBRA017E.

Wasmeier, P. (2015). *Geodetic Transformations Toolbox*. Matlab Central, http://www.mathworks.com/matlabcentral/fileexchange/9696-geodetic-transformations-toolbox.

Zarchan, P. and Musoff, H. (2015). *Fundamentals of Kalman Filtering: A Practical Approach*, 4e. Reston, VA: American Institute of Aeronautics and Astronautics.

4

Data Acquisition and Analysis

This chapter fundamentally deals with how we'll digitally represent the various aircraft performance characteristics for further analysis. A critical element of this chapter is a detailed discussion of how digital data acquisition (DAQ) systems work. Our discussion of DAQ techniques is motivated by industry and military flight test programs, which rely on complex data acquisition systems. Instrumentation on board the aircraft can involve hundreds or thousands of sensors of various kinds, and an equivalent number of channels to digitize and record this data. Data can be sampled at high rates in order to effectively capture transient phenomena, leading to vast quantities of data which require high bandwidth and storage. Furthermore, these data are often streamed in real time to ground stations via radio telemetry link, such that flight test engineers can monitor the data as the test is being conducted. Live telemetry of data and real-time analysis adds to the safety and efficiency of the flight test program, enabling the flight test team to avoid hazardous test conditions or to adapt to test events as they develop.

While the typical student will not be able to work with such high-end systems in the university environment, the basic principles of DAQ are still relevant in low-cost, small-scale data acquisition systems. With the continued evolution and miniaturization of digital electronics, these systems are now readily accessible even to students. Simple data acquisition systems in contemporary flight testing use include smartphones (Gregory and Jensen 2012; Gregory and McCrink 2016), LabVIEW- or MATLAB-based digital DAQ systems (Muratore 2012), Arduino microprocessors (Koeberle et al. 2019), commercially available systems, and even the standard avionics onboard the aircraft (see Chapter 3). Thus, students can readily get exposure to the basic principles of DAQ systems, methods, and data analysis employed in flight testing.

This chapter provides an overview of the foundations related to DAQ and processing, such that the capabilities and limitations of DAQ methods can be appreciated. Our discussion of DAQ will begin with defining how signals may be represented as a function of time or frequency. This will directly lead to a discussion of filtering, whereby unwanted frequency content can be attenuated in a signal. Following this, we'll focus on the essential characteristics of DAQ, with an overview of the methods used to digitally represent an analog signal. We'll then conclude with an example of how DAQ techniques are applied to flight testing.

4.1 Temporal and Spectral Analysis

We'll start our discussion of DAQ by considering how signals can be represented in either the time domain or the frequency domain. Both representations of a signal are of use in the flight testing environment. The time domain representation is the most intuitive, where the signal is plotted as a function of time. The frequency domain representation is less intuitive, but no less powerful.

Introduction to Flight Testing, First Edition. James W. Gregory and Tianshu Liu.
© 2021 John Wiley & Sons Ltd. Published 2021 by John Wiley & Sons Ltd.
Companion website: https://www.wiley.com/go/flighttesting

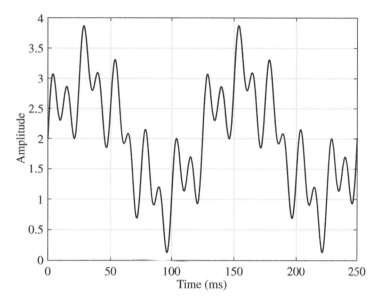

Figure 4.1 Sample signal in the time domain.

This form of presenting a signal shows the relative significance of different frequency components found in the signal, facilitating visual separation of different frequency peaks in the spectrum. We'll examine both approaches of data representation as follows, and establish the link between the two.

A time-domain representation of a signal is our intuitive view of signal waveforms, which is a plot of voltage as a function of time. For example, Figure 4.1 shows a plot of the function

$$f(t) = c_0 + c_1 \sin(\omega_1 t) + c_2 \sin(\omega_2 t) + c_3 \sin(\omega_3 t). \qquad (4.1)$$

in the time domain, where $c_0 = 2$, $c_1 = 1$, $c_2 = 0.5$, $c_3 = 0.5$, $\omega_1 = 50.3$ rad/s ($f_1 = \omega_1/2\pi = 8$ Hz), $\omega_2 = 5\omega_1$, and $\omega_3 = 10\omega_1$. A representation of the same signal in the frequency domain, however, will plot the amplitude of each frequency component of the signal versus the corresponding frequency. Inspection of Eq. (4.1) reveals that the signal has three non-zero frequency components: a dominant, low-frequency component with amplitude of c_1 at a frequency of ω_1; and two smaller, higher-frequency components with amplitudes c_2 and c_3 at frequencies ω_2 and ω_3. We can take these three amplitudes (c_1, c_2, and c_3), along with the time-averaged amplitude of the signal (c_0, which has a frequency of $\omega_0 = 0$) and plot them versus the corresponding frequencies, resulting in the frequency domain plot shown in Figure 4.2.

Plotting the signal represented in Figure 4.1 in the frequency domain is straightforward in this case since we know the amplitude and frequency of each component from Eq. (4.1). But what if we don't know the frequency content of a signal? How would we generate a frequency domain representation of some arbitrary signal that we have acquired? The answer lies in a computational technique known as the fast Fourier transform (FFT), which is a straightforward and fast algorithm for computing spectral content of digitized signals. This section will provide only a very abbreviated overview of spectral analysis, which is the process of representing a time-based signal in the frequency domain. The interested reader is encouraged to consult other sources such as Wheeler and Ganji (2003) or Bendat and Piersol (2010) for further details.

The concept behind Fourier analysis is that an arbitrary, periodic signal may be represented by a summation of a constant with an arbitrary number of sine and cosine functions (Powers 1999).

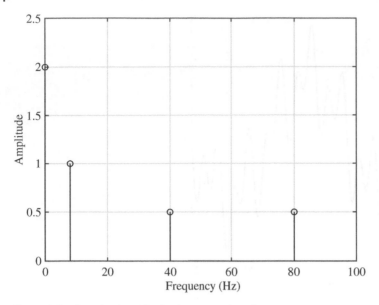

Figure 4.2 Sample signal in the frequency domain.

The mathematical representation of a Fourier series is given as

$$f(t) = a_0 + \sum_{n=0}^{\infty} a_n \cos(n\omega t) + \sum_{n=0}^{\infty} b_n \sin(n\omega t). \tag{4.2}$$

Note that our example function defined earlier (Eq. (4.1)) is of the same form as Eq. (4.2), which made it straightforward for us to pick out the coefficients of the Fourier series by inspection and plot the signal in the frequency domain (Figure 4.2). For an arbitrary periodic waveform, the coefficients in Eq. (4.2) may be determined from

$$a_0 = \frac{1}{T} \int_0^T f(t) dt, \tag{4.3}$$

which is the time-average of the signal over one period of the waveform, $T = 1/f = 2\pi/\omega$. The other coefficients are given by

$$a_n = \frac{2}{T} \int_0^T f(t) \cos(n\omega t) dt$$

$$b_n = \frac{2}{T} \int_0^T f(t) \sin(n\omega t) dt. \tag{4.4}$$

Note that Eq. (4.2) is an infinite series, implying that an infinite number of coefficients may be required to fully represent an arbitrary periodic waveform. In practice, the number of coefficients used to represent a signal is truncated to some reasonable number of computationally determined coefficients. If Eqs. (4.2)–(4.4) are evaluated for the sample function given by Eq. (4.1), with $T = 2\pi/\omega_1 = 0.125$ seconds, we could directly compute the integrals and find that the coefficients of the Fourier series are $a_0 = c_0$, $b_1 = c_1$, $b_5 = c_2$, and $b_{10} = c_3$, with all other coefficients being zero.

Figure 4.3 shows another example of the application of a Fourier series representation to a triangle waveform,

$$f(t) = -\left(\frac{2}{\pi}\right) \sin^{-1}[\sin(\omega t)], \tag{4.5}$$

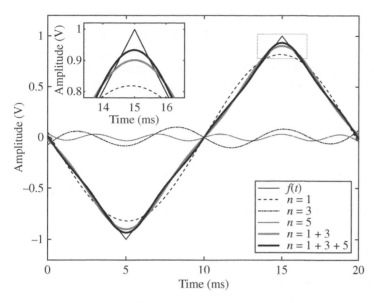

Figure 4.3 Fourier series approximation of a triangle waveform.

with frequency $\omega = 314.2$ rad/s ($f = 50$ Hz). Note that the function represented in Eq. (4.5) is odd, meaning that $f(t) = -f(-t)$. The implication of this for the Fourier series is that the a_n coefficients are zero and the function can be represented entirely by the sine terms in Eq. (4.2), making it a Fourier sine series. The coefficients of this sine series are $b_1 = -0.8106$, $b_3 = 0.09006$, $b_5 = -0.03242$, $b_7 = 0.01654$, ... Inclusion of each successive term in the sine series improves the fidelity of the Fourier approximation to the original waveform (Figure 4.3), with higher frequency coefficients improving the fit at the peaks of the triangle waveform. This is because the amplitude of the higher order terms decays rapidly with frequency, as shown in Figure 4.4.

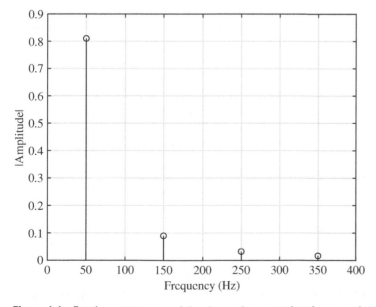

Figure 4.4 Fourier components of the sine series approximation to a triangle waveform.

While the definition of the Fourier series is useful for illustrating the representation of signals in the frequency domain, the development thus far is not yet useful for frequency representation of an arbitrary waveform that we might encounter in flight testing. Actual analysis of signals via Fourier techniques is done through the Fourier transform, which relaxes the constraint on periodicity. The steps in frequency ($\omega_n = n\omega$) in Eq. (4.2) are reduced until ω becomes a continuous function of frequency and we have

$$F(\omega) = \int_{-\infty}^{\infty} f(t)e^{-j\omega t}dt, \tag{4.6}$$

which is the Fourier transform of a function $f(t)$. Here the exponential function is another way of representing the sine and cosine terms, and $j = \sqrt{-1}$ such that $F(\omega)$ is a complex-valued function (see Wheeler and Ganji 2003 or Bendat and Piersol 2010 for further details). Similarly, if one has a defined Fourier transform, $F(\omega)$, the original signal $f(t)$ can be determined from the inverse Fourier transform,

$$f(t) = \frac{1}{2\pi} \int_{-\infty}^{\infty} F(\omega)e^{j\omega t}d\omega. \tag{4.7}$$

In applying the Fourier transform to a digital representation of a signal (sampled at discrete time intervals) the discrete Fourier transform (DFT) is used. The DFT is represented by

$$F(k\Delta f) = \sum_{n=0}^{N-1} f(n\Delta t)e^{-j2\pi k\Delta f(n\Delta t)} \quad k = 0, 1, 2, \ldots, N-1, \tag{4.8}$$

where N is the total number of samples in the data record, Δt is the time interval between samples, Δf is the increment in frequency (equal to the inverse of the period), and there are k frequency components in the transform. (Note that only values up to $k = N/2$ are unique). In the same way that we defined an inverse Fourier transform, we can also define an inverse DFT,

$$f(n\Delta t) = \frac{1}{N} \sum_{k=0}^{N-1} F(k\Delta f)e^{j2\pi k\Delta f(n\Delta t)} \quad n = 0, 1, 2, \ldots, N-1, \tag{4.9}$$

for recovering the original signal.

The DFT and inverse DFT can be determined numerically, but this process tends to be computationally expensive since the number of computations is on the order of N^2 real-valued multiply-add operations. Due to this computational expense, the FFT technique has been developed, which requires on the order of $N\log_2 N$ computations. (For a record length of $N = 2^{16}$, the FFT requires 2^{12} fewer computations compared to the DFT). Bendat and Piersol (2010) may be consulted for complete details on the derivation and implementation of the FFT algorithm. Returning to our analysis of a triangle waveform (Eq. (4.5)), spectral analysis based on the FFT algorithm results in the power spectrum shown in Figure 4.5. The dominant frequency and higher harmonics are faithfully captured by the FFT algorithm, as is evident when we compare Figure 4.5 with Figure 4.4.

Note that the power spectral density is often represented on a logarithmic scale, which helps the higher-frequency components of the signal appear more prominently on the plot (this is in contrast to the linear scaling employed in Figures 4.2 and 4.4). Power spectra often have broader frequency peaks along with lower-amplitude ripple between the peaks, which are characteristic of the DFT and FFT algorithms. This phenomenon, referred to as spectral leakage, results from a non-integer number of waveforms being present within the data record analyzed by the FFT, with the end effects being the primary culprit. Leakage may be reduced through windowing, where the window is a tapering function (high in the center region, with low values at the ends) multiplied with a subset of the data. Since the window size is a subset of the full data record, the full FFT is then the

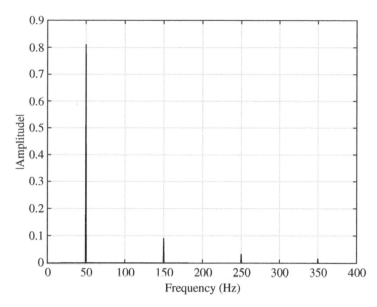

Figure 4.5 Power spectrum based on the FFT of the triangle waveform.

average of the computed FFTs of the windowed portions of the signal. An amount of overlap may be specified, which determines how much the window is shifted along the length of the record. Welch's modified periodogram method (Welch 1967) is one common approach to implementing a windowing function with overlap (see the `pwelch` function in MATLAB). Welch's method reduces the noise in the power spectrum, at the expense of reduced frequency resolution.

4.2 Filtering

We'll now discuss filtering techniques, and how they can be used to improve the interpretation of the signal by removing unwanted frequency content. When considering a signal represented in the frequency domain, filtering suppresses the amplitude of signal content over a select range of frequencies. This is done by defining and applying a transfer function, which is a frequency-dependent weighting value that is multiplied with the signal in the frequency domain.

Common types of filters include low pass, high pass, band pass, and band stop schemes, which are illustrated in Figure 4.6. A low pass filter will attenuate signal content at frequencies above a specified cutoff frequency, and preserve signal content below that cutoff frequency. Low pass filtering is useful for removing high-frequency noise in a signal, which may obscure the desired low frequency data. High pass filtering does just the opposite – it attenuates signal content at frequencies below the cutoff, and preserves signal content at higher frequencies. High pass filtering is particularly useful for removing the steady-state voltage (a DC mean value) from a signal. The band pass filter is essentially a combination of the two, where the high-pass cutoff frequency is at a lower frequency than the cutoff for the low-pass filter. The region between the two cutoff frequencies is the passband. Finally, a band stop filter is designed to selectively remove a range of frequencies – it is the logical inverse of a band pass filter, where the low-pass cutoff frequency is below the high-pass cutoff frequency.

Figure 4.6 illustrates the effect of each of these filters on a signal (Eq. (4.1)) in both the time domain (left column) and in the frequency domain (center column). The right-most column

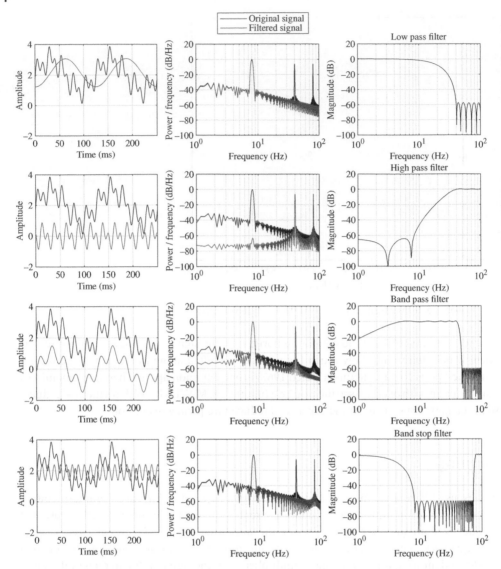

Figure 4.6 Examples of various filtering schemes applied to signals in the time domain (left column) and frequency domain (center column). The right column shows the transfer function associated with each defined filter: low pass, high pass, band pass, and band stop.

represents the transfer function for each filter. A filter can be applied to a given signal by transforming that signal into the frequency domain (via FFT), linearly multiplying the signal's power spectrum with the filter's frequency response function in the frequency domain, and then transforming the signal back into the time domain (via the inverse FFT). In Figure 4.6, the baseline signal is the same waveform that we considered in an earlier example (see Eq. (4.1)). The effects of the low pass filter are to suppress the higher frequency components at 40 and 80 Hz, leaving only the DC level and the sinusoidal component at 8 Hz. In contrast, the high pass filter removes the DC level and the lowest frequency component (8 Hz), leaving both high frequency components unattenuated. The band pass and band stop filters have equivalent performance, where the filter suppresses signal content in portions of the spectrum where the filter's transfer function has a high level of attenuation.

Beyond the definition of the various filter types, there are other filter characteristics that are important to consider. Common filter classes include Butterworth, Bessel, Chebyshev, and elliptical – each of which has varying characteristics. One important parameter is the filter order, which governs the stopband attenuation rate – a description of how much increase in attenuation can be achieved over a given frequency interval. Attenuation rates are typically specified as dB/octave or dB/decade, where an octave is a factor of two change in frequency and a decade is an order of magnitude change in frequency. Filter order is related to attenuation rate for the Butterworth filter by $6m$ dB/octave, where m is the filter order. A third important parameter is the amount of ripple allowed in the passband or the stopband. Figure 4.6 illustrates ripple throughout the stopband, where the transfer function is not flat across a range of frequencies. There is typically a tradeoff between ripple and attenuation rate, where high attenuation rate is achieved at the expense of increased ripple, and the filter class has a significant impact on the attenuation rate.

One final, critical characteristic of filters is the phase lag induced by the filter. In the same way that a filter exhibits attenuation as a function of frequency, there will be induced phase delay that is also frequency-dependent. In most data processing applications this is an undesirable feature of the filter, but can be worked around through careful filter design or creative application of the filter (*e.g.*, the `filtfilt` function in MATLAB's Signal Processing Toolbox, which feeds a signal forward and then backward through the filter in order to cancel the phase effect). Phase delay in processed flight test data can be important when comparing a filtered signal with an unfiltered signal. A detailed discussion of filter design is beyond the scope of this text, but appropriate resources may be consulted for further details (Wheeler and Ganji 2003; Bendat and Piersol 2010).

Filtering can be applied before digitization or afterwards. Typically pre-DAQ analog filtering is used to prevent aliasing, which will be discussed in Section 4.4. Analog filtering involves the use of dedicated circuitry, and once the signal has been digitized there is no longer any flexibility to change the filter cutoff frequency or characteristics. Digital filtering, on the other hand, can be changed at will during post-processing, allowing an interactive and adaptive approach to data analysis.

4.3 Digital Sampling: Bit Depth Resolution and Sample Rate

Let's now consider the details of how a signal is actually captured in digital form. The fundamental principle of DAQ is creating a digital representation of an analog signal. An analog signal is defined as one where the signal level (*e.g.*, voltage) is a continuous function of time. Digital signals, however, are always a discretized representation of that continuous function, with the fidelity of that representation depending on how many discretization levels are used across amplitude and time. Resolution of the signal amplitude depends on the bit-depth resolution of the data acquisition device, and the defined input range. The input range essentially determines the minimum and maximum voltages that can be recorded for a given measurement, where any input values exceeding those limits will be clipped. Typical data acquisition ranges can be unipolar, where all of the input voltages are of the same sign (*e.g.*, 0–5 V, or 0–10 V), or bipolar, where positive or negative voltages can be measured (*e.g.*, −5 to 5 V, or −10 to 10 V). Bit depth resolution is a measure of how many discretization levels are used to subdivide the input range. This resolution is typically expressed as a power of 2, due to the architecture of the data acquisition hardware. For example, a 12-bit data acquisition device will have 2^{12} (4096) discretization levels spanning the input range.

A combination of input range and the bit depth resolution defines the minimum change in voltage that can be resolved in a digital waveform. The maximum error between a given analog voltage and its digital representation is the quantization error,

$$Q = \pm 0.5R/2^B, \tag{4.10}$$

where R is the input range, and B is the number of bits of the DAQ device. Digitization of a desired signal should use an input range that is as close to the limits of the anticipated signal as possible (with little risk of the signal exceeding that input range) and a bit depth resolution as high as possible. The downsides of increased bit depth resolution are increased cost of the data acquisition hardware, and larger file sizes required to store the digitized signals. In practice, the bit depth resolution should be sufficiently high such that the discretization error is small relative to the smallest voltage change in the desired signal.

Figures 4.7 and 4.8 illustrate the effects of bit depth resolution on the digital representation of an analog signal. For this example, a simple sine wave with frequency of 1000 Hz (628 rad/s) is defined as

$$y = 2.2 + \sin(\omega t). \tag{4.11}$$

A digital representation of that signal with an input range of 0–5 V and a 4-bit converter (2^4, or 16 steps) is shown in Figure 4.7, compared to the original analog function. The step-stair appearance of the signal is due to quantization error, where the digital representation of the continuous waveform is rounded off to the nearest quantization level. If the bit depth resolution is increased to 2^{12}, as shown in the zoomed-in waveform in Figure 4.8, the signal representation is much more faithful to the original analog signal. The increase in bit depth resolution from 2^4 to 2^{12} provides a factor of 256 more levels to represent the analog waveform than the 4-bit case shown in Figure 4.7.

Sample rate, defined as the number of digital samples acquired per second, is the other predominant factor that dictates the fidelity of the digital representation of the analog waveform. Sampling rate is determined by the time required to perform the analog-to-digital conversion process, limiting how many samples can be digitized in a given amount of time. If the sample rate is low – *i.e.*, there is a long period of time between each sample of the analog signal – the acquisition process could miss important changes in the analog signal in the intervening time. Not only is important information missed by an insufficient sample rate, but the resulting digital representation can be misleading.

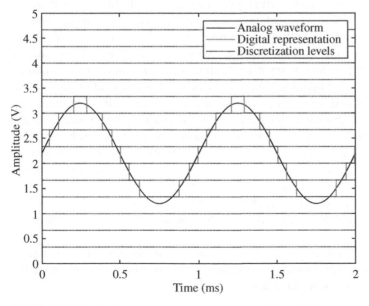

Figure 4.7 Digital representation of an analog waveform with input range of 0–5 V, a 4-bit converter, and very high sample rate.

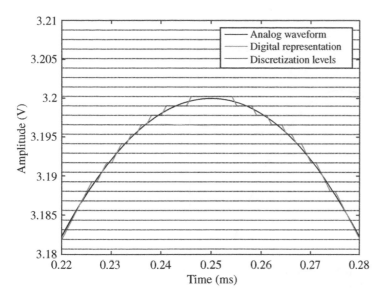

Figure 4.8 Digital representation of the analog waveform with input range of 0–5 V, a 12-bit converter, and very high sample rate.

The sample rate must be sufficiently high relative to the highest frequency present in the signal. Note that signals can be a superposition of many constituent frequencies. As a general rule of thumb, a sample rate of at least 10 to 20 times the highest frequency present in the signal will provide a reasonable representation of the analog signal in the time domain. For example, Figure 4.9 shows a generally insufficient representation of the analog signal, where the sample rate of 4400 Hz is only 4.4 times higher than the 1000 Hz frequency being measured. In contrast, Figure 4.10 shows a fairly good representation of the signal, with a sample rate of 22 000 Hz

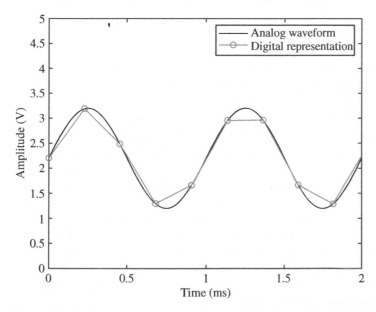

Figure 4.9 Digital representation of the analog waveform with a sample rate of 4400 Hz (4.4 samples per cycle of the 1000 Hz analog waveform), with very high bit depth resolution.

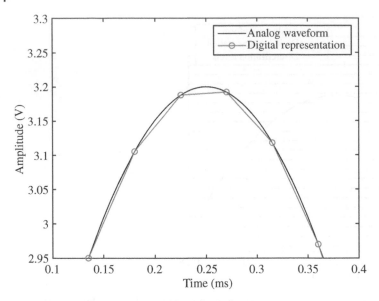

Figure 4.10 Digital representation of the analog waveform with a sample rate of 22 000 Hz (22 samples per cycle of the 1000 Hz analog waveform), with very high bit depth resolution.

(22 times the frequency of the sampled waveform), with only minor residual error in capturing the magnitude of the waveform peaks (approximately 80 mV error).

4.4 Aliasing

A theoretical minimum sampling rate needed to capture all frequency content is given by the Nyquist criterion (Nyquist 1928). This limit states that the sampling rate (f_s) must be greater than twice the maximum expected frequency component (f_{max}),

$$f_s > 2f_{max}.$$ (4.12)

Notice that the criterion dictates that the sample rate must be *greater than*, not *greater than or equal to*, the maximum frequency. If a sample rate equal to twice the maximum frequency were employed, the digital representation of that frequency (if the signal were a single-frequency waveform) could be a straight line, a situation illustrated by Figure 4.11. Based on this understanding of the Nyquist criterion, it is clear that the scenario illustrated in Figure 4.9, while not capturing the magnitude of the waveform peaks very well, is sufficiently high to identify the frequency of the signal being measured.

The consequences of not meeting the Nyquist criterion can be significant. If the Nyquist limit is not met, then frequencies in the signal beyond $f_s/2$ cannot be resolved. This situation alone would be somewhat benign if the high frequencies were simply omitted. However, these high frequencies beyond the cutoff are actually represented as lower frequency content in the digital misrepresentation of the waveform. This can be insidious if the flight test engineer is unaware of the presence of this high-frequency content and misinterprets these false low-frequency representations as being real. This phenomenon of false representation of high frequency content (beyond the Nyquist cutoff) as low frequency content is referred to as *aliasing*.

The effects of aliasing can be represented in the time and frequency domain, as shown in Figure 4.12. The top row of the figure illustrates the actual waveform to be represented in digital

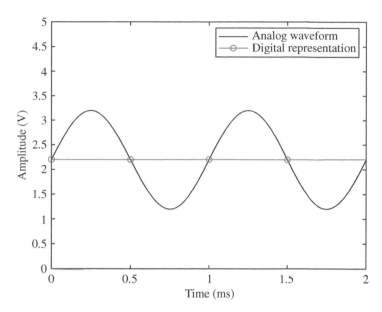

Figure 4.11 Digital representation of the analog waveform with a sample rate of 2000 Hz (2 samples per cycle of the 1000 Hz analog waveform), which does *not* satisfy the Nyquist criterion.

form (the analog signal). The frequency of this signal is 20 Hz (left column), 80 Hz (center column), or 120 Hz (right column). Below this, the second row shows the digital representation of each signal when sampled at 100 Hz. Note that only the left-most column satisfies the Nyquist criterion. Even though the indicated sample rate (100 Hz) is higher than the signal's fundamental frequency of the center column (80 Hz), the Nyquist criterion in this case ($f_s > 2f_{max} = 160$ Hz) is not met. The signal frequency of right-most column (120 Hz) is well beyond the Nyquist limit of 50 Hz for a sample rate of 100 Hz. The third row shows the power spectra associated for each measurement, and the fourth row is a plot of the actual frequency of the analog signal versus the indicated frequency at a sample rate of 100 Hz. This illustrates a very interesting phenomenon due to aliasing: in all three cases the indicated frequency is 20 Hz. The zig–zag line in each of the bottom row figures represents the indicated frequency as a function of the actual frequency if the sample rate is held fixed at 100 Hz. The first portion of the line (from 0 to 50 Hz, the Nyquist limit) shows a one-to-one relationship between actual and indicated frequency, as we would expect. Beyond the Nyquist limit, however, the indicated frequency proportionally *decreases* as the actual frequency continues to go up. This happens until the actual frequency reaches twice the Nyquist limit, where the indicated frequency again increases with the actual frequency. In all instances, however, the indicated frequency always falls between 0 and 50 Hz. It is impossible for the power spectrum to indicate a frequency higher than the Nyquist limit. The interesting nature of this frequency folding diagram illustrates how all three actual frequencies (20, 80, and 120 Hz) can have an indicated frequency of 20 Hz when sampled at 100 Hz. This illustrates the significant challenge associated with aliasing.

The only way to truly prevent aliasing is to somehow remove the unwanted high-frequency content before the signal is digitized, through analog filtering. Anti-alias filtering of the signal involves attenuation of the amplitudes of any frequency peaks beyond the Nyquist limit such that they fall below the noise floor and no longer pose a concern. This filtering must be done to the analog signals themselves *before* digitization – digital filtering is ineffective at addressing the aliasing problem. The filter cutoff and attenuation rate must be selected in order to guarantee

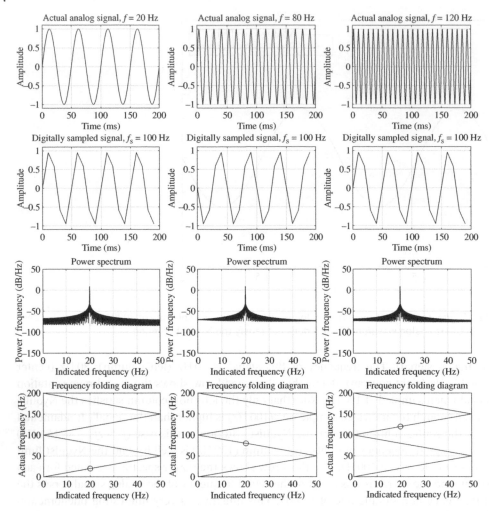

Figure 4.12 Illustration of aliasing due to insufficiently high sample rate. Each column corresponds to an analog signal of varying frequency. The top row depicts the original analog signal to be sampled (20, 80, and 120 Hz), the second row shows the digital representation of those signals with a sample rate of 100 Hz, the third row shows the corresponding power spectra, and the fourth row shows the relationship between the actual and indicated frequencies on frequency folding diagrams.

that any higher frequency content will be sufficiently attenuated. This anti-alias filtering criterion is met when the signal is attenuated to a level below the minimum detectable level of the DAQ hardware, which is based on the selected range and bit depth resolution of the DAQ hardware.

In flight testing, one commonly encountered challenge associated with aliasing is related to the frequency of engine vibrations relative to the sampling rate. In many situations the flight test engineer may wish to keep the size of digital data files at a manageable level. For example, if the sample rate is maintained at 100 Hz for a one-hour test flight while sampling a 3-axis accelerometer alone, the file size is on the order of 10 MB. On the other hand, engine vibrations can be at relatively high frequencies. For a horizontally opposed 6-cylinder engine such as the Continental IO-360, there are three combustion events for each revolution of the engine. When the engine is operating at a speed of 2400 RPM (40 Hz), we would expect a dominant frequency in the accelerometer signals of 120 Hz. However, this is well beyond the Nyquist limit, and would result in an aliased frequency appearing at 20 Hz (folded twice). The best way to handle this situation is to use an analog low-pass Nyquist filter before digitizing the accelerometer signals, but some basic data acquisition devices may not have built-in anti-aliasing filters.

4.5 Flight Testing Example

Let's now take a moment to consider how spectral analysis can be applied to flight test data, in order to determine the dominant frequencies in an arbitrary waveform more readily than by inspection of time-domain plots. For example, consider a case where we wish to determine the period of the phugoid mode from the accelerometer time signal (see Chapter 14 for more details on measurement of the phugoid mode in flight testing). Figure 14.10 (in Chapter 14) displays such a signal from the z-axis accelerometer from a data acquisition device mounted on the dashboard of a Cirrus SR20 in flight. Due to the presence of high frequency noise sources (many frequency components simultaneously present in the signal), it is difficult to directly and precisely infer the dominant period of the waveform that would indicate the period of the phugoid mode. However, a frequency domain representation of the same signal (Figure 14.11) shows the dominant frequency much more clearly. This power spectrum was calculated using the `periodogram` function in MATLAB, which provides an estimate of the power spectral density via the FFT of the signal. In this case, the highest-amplitude peak is at 0.032 Hz, which corresponds to a period of 31.2 seconds.

The time domain representation of the raw accelerometer data shown in Figure 14.10 has a lot of high-frequency noise relative to the fundamental frequency of interest (0.032 Hz long-period mode). This high-frequency noise is likely due to vibrations coming from the engine and other sources. Even if the level and spectral content of that noise were not known in advance, a low pass digital filter can be applied in post-processing to aid in the interpretation of the signal. This signal has been filtered by a low pass, 8th-order infinite impulse response (IIR) filter with a cutoff of 0.1 Hz and passband ripple of 0.01, with results shown by the thick line in Figure 14.10. Design of digital filters can be performed in the MATLAB Signal Processing Toolbox (*e.g.*, `designfilt`).

4.6 Summary

We'll now summarize the key points of DAQ and outline how they pertain to flight testing in the typical university flight test curriculum. The sample rate for a data acquisition device must be set high enough such that the Nyquist criterion is satisfied for the highest-frequency component of the signal to be measured. It's worth bearing in mind that some events in aircraft flight testing can actually result in relatively high frequencies (100's of Hz). Thus, the DAQ sample rate should be set as high as needed to meet the Nyquist criterion, but not so high that the user is overloaded with too much data that leads to large file sizes and unwieldy data analysis.

A common mistake made by students is to plot the entire digital data record for an hour-long flight. This can lead a student to throw up their hands in dismay due to the overwhelming size of the file and from feeling lost in the data. It can be challenging to locate specific flight events within a large data file without some guideposts or reference points. To help with this, it can be very helpful to record the clock time at key events during the flight test, to assist in finding those key events in the data set (which will be correlated with elapsed time). Thus, it is important to record the time when data acquisition begins (to the nearest second), and the time at which each event occurs. Also, it can be very helpful to know what to look for in the data set – to anticipate the characteristic features that will be evident in a particular data stream for a given flight maneuver. These can be found by interactively scaling the extent of the time record used for plotting (essentially 'zooming in' on the region of interest to identify these critical features). Finally, once a particular region of interest within a large data set is identified, it can be helpful to use digital filtering with an appropriate filter cutoff to remove unwanted high-frequency content that distract from identification of key performance features.

Nomenclature

a_n	Fourier series coefficients
b_n	Fourier series coefficients
B	bit depth resolution of a DAQ device
c_n	amplitude coefficients for example signal definition
f	physical frequency
$F(\omega)$	Fourier transform of a periodic function $f(t)$
f_s	sample rate
j	imaginary number $(= \sqrt{-1})$
k	index
m	filter order
n	index
N	total number of samples in a data record
Q	quantization error
R	DAQ input range
t	time
T	period of a waveform
ω	angular frequency $(=2\pi f)$

Subscripts

max	maximum

Acronyms and Abbreviations

AHRS	attitude and heading reference system
DAQ	data acquisition
DFT	discrete Fourier transform
FFT	fast Fourier transform
GPS	Global Positioning System
IIR	infinite impulse response filter
IMU	inertial measurement unit
MEMS	microelectromechanical systems
RPM	revolutions per minute
UAV	unmanned aerial vehicle

References

Bendat, J.S. and Piersol, A.G. (2010). *Random Data: Analysis and Measurement Procedures*, 4e. Hoboken, NJ: John Wiley & Sons.

Gregory, J. W. and Jensen, C. D. (2012). Smartphone-based data acquisition for an undergraduate course on aircraft flight testing. AIAA 2012-0883, Proceedings of the 50th AIAA Aerospace Sciences Meeting, Nashville, TN.

Gregory, J. W. and McCrink, M. H. (2016). Accuracy of smartphone-based barometry for altitude determination in aircraft flight testing. AIAA 2016-0270, Proceedings of the 54th AIAA Aerospace Sciences Meeting, San Diego, CA.

Koeberle, S. J., Rumpf, M., Scheufele, B., and Hornung, M. (2019). Design of a low-cost RPAS data acquisition system for education. AIAA 2019-3658, Proceedings of the AIAA Aviation 2019 Forum, Dallas, TX.

Muratore, J. F. (2012). A family of low cost personal computer based data acquisition systems for flight test. AIAA 2012-3166, Proceedings of the 28th AIAA Aerodynamic Measurement Technology, Ground Testing, and Flight Testing Conference, New Orleans, LA.

Nyquist, H. (1928). Certain topics in telegraph transmission theory. *Transactions of the American Institute of Electrical Engineers* 47 (2): 617–644. https://doi.org/10.1109/T-AIEE.1928.5055024.

Powers, D.L. (1999). *Boundary Value Problems*, 4e. San Diego, CA: Harcourt Academic Press.

Welch, P.D. (1967). The use of Fast Fourier Transform for the estimation of power spectra: a method based on time averaging over short, modified periodograms. *IEEE Transactions on Audio and Electroacoustics* 15 (2): 70–73. https://doi.org/10.1109/TAU.1967.1161901.

Wheeler, A.J. and Ganji, A.R. (2003). *Introduction to Engineering Experimentation*, 2e. Upper Saddle River, NJ: Prentice Hall.

5

Uncertainty Analysis

Any measurement is subjected to unknown or uncontrollable error sources that lead to a deviation between the measured value and the true value. In some cases, these differences may be negligibly small, but in other situations the error in the measurement can be significant enough to obscure the true physical phenomena or lead to incorrect conclusions. The domain of uncertainty analysis uses statistical tools to quantify the level of uncertainty in the measurement. Uncertainty analysis places bounds on the range of values that the true value may fall between, referenced to a given confidence level – in other words, it quantifies how confident we are in the measured result. In order to be truly useful, any measurement must include uncertainty bounds. For example, a reported measurement of airspeed could be expressed as $V = 120 \pm 1$ knots within a 95% confidence interval, which indicates that we are 95% confident that the true value of indicated airspeed falls between 119 and 121 knots.

An example of the utility of uncertainty analysis is provided in Figure 5.1, where a decision must be made on whether or not the experimental data support the presence of a laminar drag bucket. At first glance, Figure 5.1(a) appears to indicate the polar with a drag bucket more accurately describes the trend of the data. However, it is not possible to evaluate the validity of one drag polar over the other without quantifying the uncertainty. Figure 5.1(b,c) express the uncertainty through a set of error bars on each data point. If the uncertainty is low, as illustrated in Figure 5.1(b), then the data clearly show the presence of a laminar drag bucket. However, if the uncertainty is higher, as shown in Figure 5.1(c), then it may not be possible to assess whether the drag bucket is indeed present since both drag polars fall within the uncertainty range expressed by the error bars. To support this type of analysis and decision-making, we'll present methods in this chapter for estimating the uncertainty in measured experimental data.

In flight testing work, assessment of the uncertainty in flight vehicle performance measurements can be difficult and, in some cases, expensive. Errors may arise from deficiencies in the instrument calibration, random perturbations of the measured variable, suboptimal piloting technique, differences in flight vehicle component performance, and atmospheric perturbations (*e.g.*, turbulent gusts). While uncertainty values are usually not reported in documentation such as the Pilot's Operating Handbook (POH) for an aircraft, an assessment of the uncertainty will inform the determination of the final reported values. Uncertainty analysis also guides decision-making on whether a manufacturer's delivered aircraft system meets the performance requirements agreed upon with the customer or contracting government agency.

The field of uncertainty analysis is a comprehensive one, but we will focus only on the high points in this chapter. For reference, detailed discussions of uncertainty analysis are provided by Coleman and Steele (1999) and Taylor (1997), whereas Woolf (2012) provides a detailed assessment of statistical techniques to flight testing. Here, we'll start with definitions of the two main types of errors: bias error and precision error. We'll then provide a brief introduction to

Introduction to Flight Testing, First Edition. James W. Gregory and Tianshu Liu.
© 2021 John Wiley & Sons Ltd. Published 2021 by John Wiley & Sons Ltd.
Companion website: https://www.wiley.com/go/flighttesting

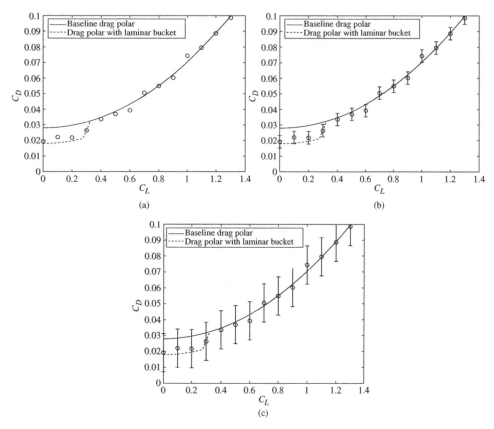

Figure 5.1 Illustration of the importance of uncertainty analysis in guiding interpretation of data and decision-making on model validity. (a) No uncertainty estimate, which is ambiguous, (b) Low uncertainty, where the data validates the laminar drag bucket model, and (c) High uncertainty, which is insufficient for determining which model is valid.

the statistics necessary for uncertainty analysis, followed by an overview of how uncertainty in each measurand propagates through the data reduction equation to yield the uncertainty of the final measured result. The uncertainty in a final result can be estimated by direct analysis (using partial derivatives of the data reduction equation), or through Monte Carlo analysis of more complicated systems of data reduction equations. We'll provide an assessment of typical uncertainty values for some of the instruments commonly used in flight testing and conclude with some examples that illustrate both the direct analytical and the Monte Carlo approaches to uncertainty analysis.

5.1 Error Theory

5.1.1 Types of Errors

There are two types of measurement errors: bias error (also known as systematic error) and precision error (also known as random error). These error types are shown in Figures 5.2 and 5.3. Bias error (systematic error) is characterized by the deviation of a measured value of a physical quantity from its true value. The bias error is usually provided as a fixed value associated with the instrument calibration. Contributions to systematic error may be from (1) intrinsic limitations

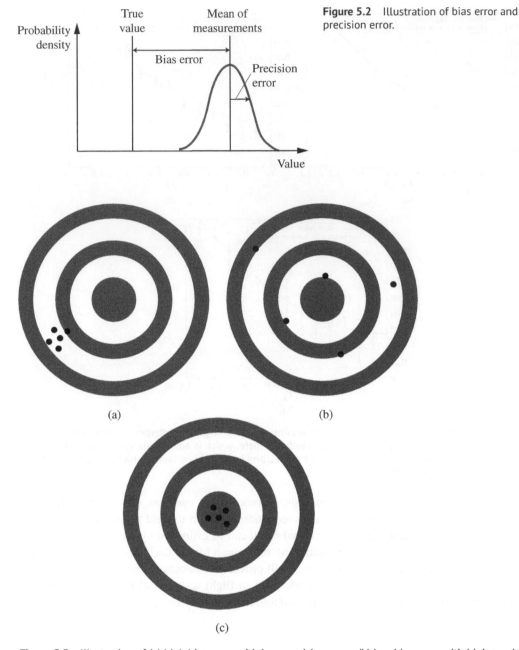

Figure 5.2 Illustration of bias error and precision error.

Figure 5.3 Illustration of (a) high bias error with low precision error; (b) low bias error with high precision error; and (c) low bias error and low precision error.

and flaws of the working principles and designs of measuring instrument or experimental technique, (2) instrument calibration uncertainties, (3) improper handling or operation of an instrument, or (4) neglecting other unaccountable factors. Bias error can be reduced by improving the quality of the calibration in an effort to zero out the fixed error. This approach is equivalent to a sharpshooter adjusting the sights or scope on a rifle to remove the fixed bias error evident in Figure 5.3(a).

In some measurements, two types of the features of the systematic error are given with instruments having a linear response:

(1) Zero-shifting (offset) error which occurs if the instrument has nonzero reading when the quantity to be measured should be zero,
(2) Scale factor (slope) error which occurs when the instrument consistently reads the change in the measured quantity that is more or less than the actual change proportionally.

These errors are illustrated in Figure 5.4 as the linear distribution of the systematic error which represents the simplest case in measurements. For relatively simple measurements like length measurements, a systematic error can be determined by careful calibrations against the standard defined by The National Institute of Standards and Technology (NIST). Here, the standard is a relatively accurate value measured by a more deliberately designed and built instrument to reduce the systematic error. In general, distributions of the systematic errors for instruments are nonlinear and complicated, making it very difficult to determine the error since the true values are simply not known. Statistical methods cannot be applied to analyze and determine the systematic error since it is deterministic (Liu and Finley 2001).

Systematic (bias) errors are most often estimated by manufacturer-specified performance of the instrument or from characteristics of the instrument itself. The quoted accuracy of an instrument (*e.g.*, ±1.0 °C for a thermistor) can be used as a reasonable estimate of the systematic error. Or, if a manufacturer-specified estimate is not available, half the spacing between the least significant digit in the measurement can be used, whether the reading is from an analog dial or a digital display.

Precision error, on the other hand, is associated with random, unpredictable variations in the measurement process or in the value of the measurand. Examples of precision (random) error sources include inherent fluctuations of the instrument or uncontrollable and dynamic environmental effects. Precision error is due to unpredictable and uncontrollable sources that

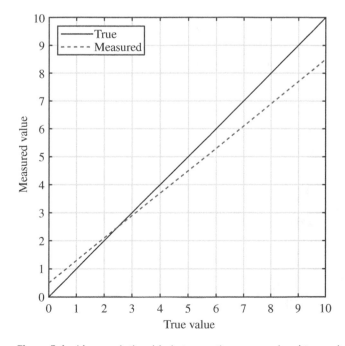

Figure 5.4 Linear relationship between the measured and true values, which illustrates the zero-shifting error and scale factor error.

are statistically independent. Thus, a distribution of the random error is generally described by a Gaussian distribution according to the central limit theorem (Cramer 1999), as illustrated in Figure 5.2. Random error can be reduced by refining the measurement technique to better control the randomly fluctuating parameters or by repeating the experiment multiple times and determining the true value through statistical methods (see Bevington and Robinson 1992; Taylor 1997; or Coleman and Steele 1999).

5.1.2 Statistics of Random Error

Random errors are statistical fluctuations in measured data, which are associated with the precision limitations of a measurement system and uncontrollable factors in measurements. Therefore, the properties of random errors are usually described by statistical methods. From a statistical perspective, a random error is contributed by a large number of statistically independent sources in a linear fashion (*e.g.*, the sum of a large number of independent random variables). According to the central limit theorem (Cramer 1999), the probability density function (pdf) of x should obey the Gaussian distribution:

$$p(x) = \frac{1}{\sqrt{2\pi}\sigma} \exp\left(-\frac{(x - \langle x \rangle)^2}{2\sigma^2}\right), \tag{5.1}$$

where $\langle x \rangle$ and σ are the mean and standard deviation of x, respectively.

Figure 5.5 shows the Gaussian distribution that fits the histogram of 200 measurement values. The standard deviation σ determines the broadness of the Gaussian distribution. In probability theory, based on the pdf, the mean and the standard deviation are defined by:

$$\langle x \rangle = \int_{-\infty}^{\infty} x\, p(x)\, dx, \tag{5.2}$$

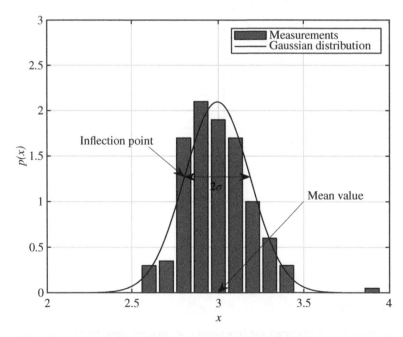

Figure 5.5 The Gaussian distribution and histogram of a series of measurements.

and

$$\sigma^2 = \int_{-\infty}^{\infty} (x - \langle x \rangle)^2 \, p(x) \, dx. \tag{5.3}$$

In data processing, Eqs. (5.2) and (5.3) are approximated by the ensemble average and the standard deviation of M samples given by:

$$\langle x \rangle = \frac{1}{M} \sum_{i=1}^{M} x_i, \tag{5.4}$$

and

$$\sigma = \sqrt{\frac{1}{M-1} \sum_{i=1}^{M} (x_i - \langle x \rangle)^2}. \tag{5.5}$$

The probability that a measurement value will fall within the range $D = [\langle x \rangle - t\sigma, \langle x \rangle + t\sigma]$ is given by:

$$\Pr(D) = \int_{\langle x \rangle - t\sigma}^{\langle x \rangle + t\sigma} p(x) dx = \frac{1}{\sqrt{2\pi}} \int_{-t}^{t} \exp\left(-z^2/2\right) dz, \tag{5.6}$$

where $z = (x - \langle x \rangle)/\sigma$ and the number t denotes the t standard deviation (for example, 2σ). The function in Eq. (5.6) is called the error function, which has been tabulated in numerous sources (*e.g.*, Appendix A of Taylor 1997; or Appendix A of Coleman and Steele 1999). For example, it is expected that 68% of all measurements of x will fall within the range of $-\sigma \le x - \langle x \rangle \le \sigma$, and therefore it corresponds to the confidence level of 68%. Similarly, we expect 95% of a measurement to fall within the range of $-2\sigma \le x - \langle x \rangle \le 2\sigma$. The standard deviation estimation based on the Gaussian distribution provides a reference for the bounds of a random error.

5.1.3 Sensitivity Analysis and Uncertainty Propagation

The contribution to the total error from elemental error sources can be generally evaluated based on an error propagation equation in an uncertainty analysis (Bevington and Robinson 1992; Taylor 1997; Colman and Steele 1999). Here, some basic concepts and notations in the uncertainty analysis are summarized. We consider a measured quantity X that is a function of the input variables $\{\xi_i, i = 1 \cdots N\}$, *i.e.*,

$$X = f\left(\xi_1, \xi_2, \xi_3 \cdots \xi_N\right). \tag{5.7}$$

The uncertainties of X and ξ_i are given by, respectively,

$$\delta X = X - \langle X \rangle \text{ and } \delta\xi_i = \xi_i - \langle\xi_i\rangle, \tag{5.8}$$

where $\langle X \rangle$ denotes the statistical ensemble average (or the mean, see Eq. (5.4)). The elemental error could be also a systematic error, and in this case $\delta X = X - X_{\text{best}}$ and $\delta\xi_i = \xi_i - \xi_{i,\text{best}}$, where the subscript 'best' denotes the best estimated value.

Based on the Taylor series expansion, the uncertainty of X is given by a linear approximation:

$$\delta X \approx \sum_{i=1}^{N} \frac{\partial f}{\partial \xi_i} \delta\xi_i. \tag{5.9}$$

The uncertainties are usually characterized by the variances of X and ξ_i defined as:

$$\sigma_X^2 = \langle \delta X^2 \rangle \text{ and } \sigma_i^2 = \langle \delta \xi_i^2 \rangle. \tag{5.10}$$

In addition, the covariance is defined by:

$$\sigma_{i,j} = \langle \delta \xi_i \delta \xi_j \rangle, \tag{5.11}$$

which describes the correlation between $\delta \xi_i$ and $\delta \xi_j$. In other words, Eq. (5.11) describes the effect of a single error source on multiple input variables. We often use the standard deviations σ_X and σ_i in an uncertainty analysis.

Taking the square of Eq. (5.9) and then the ensemble average, we have the relative uncertainty:

$$\frac{\sigma_X^2}{X^2} = \sum_{i,j=1}^{N} S_i S_j \rho_{ij} \frac{\sigma_i \sigma_j}{\xi_i \xi_j}, \tag{5.12}$$

where the correlation coefficients between ξ_i and ξ_j are defined as:

$$\rho_{ij} = \frac{\sigma_{i,j}}{\sigma_i \sigma_j}, \tag{5.13}$$

and the sensitivity coefficients S_i are defined as:

$$S_i = \frac{\xi_i}{X} \frac{\partial f}{\partial \xi_i}. \tag{5.14}$$

The relative standard deviation is given by:

$$\frac{\sigma_X}{|X|} = \sqrt{\sum_{i,j=1}^{N} S_i S_j \rho_{ij} \frac{\sigma_i \sigma_j}{\xi_i \xi_j}}. \tag{5.15}$$

Eq. (5.12) or Eq. (5.15) describes the error propagation in which the contribution of the elemental errors leads to the total uncertainty of X. It is noted that ρ_{ij} could be either positive or negative for $i \neq j$.

The complexity of the equations is reduced when the variables are statistically independent, and the correlation coefficients become $\rho_{ij} = \delta_{ij}$, where δ_{ij} is the Kronecker delta ($\delta_{ij} = 0$ if $i \neq j$ and $\delta_{ij} = 1$ if $i = j$). In this case, Eq. (5.12) reduces to

$$\frac{\sigma_X^2}{X^2} = \sum_{i=1}^{N} S_i^2 \left(\frac{\sigma_i}{\xi_i} \right)^2, \tag{5.16}$$

and the relative standard deviation of X is given by:

$$\frac{\sigma_X}{|X|} = \sqrt{\sum_{i=1}^{N} S_i^2 \left(\frac{\sigma_i}{\xi_i} \right)^2}. \tag{5.17}$$

In an uncertainty analysis, the first step is to establish the functional relation between the measured quantity and the input variables by modeling a measurement system. The mathematical model could be an algebraic, differential, integral, or differential-integral equation. Then the sensitivity coefficients S_i can be evaluated and the uncertainty $\sigma_X/|X|$ can be estimated by substituting the relative elemental errors $\sigma_i/|\xi_i|$ into Eq. (5.17). For a complicated measurement technique that has multiple instruments and extensive data processing, the functional relation between X and ξ_i is usually obtained using numerical methods, and in this case simulations and systematic tests are required to determine the sensitivity coefficients (*e.g.*, Monte Carlo simulations are discussed in section 5.1.6).

5.1.4   Overall Uncertainty Estimate

The final reporting of overall uncertainty in a measurement needs to combine the estimates of both the bias (systematic) error and the precision error. The combined uncertainty may be expressed as the root-sum square of the bias and precision error:

$$U = \sqrt{B^2 + P^2}, \tag{5.18}$$

where B is the bias error and $P = 2\sigma_X / \sqrt{M}$ is the precision error. Inherent in this estimation are assumptions that the error is normally distributed (*i.e.*, Gaussian) that at least 10 independent samples have been acquired for generating the statistics ($v = M - 1 \geq 9$ degrees of freedom) and that the uncertainty estimate corresponds with a 95% confidence interval (see Coleman and Steele 1999).

5.1.5   Chauvenet's Criterion for Outliers

Observing a set of data obtained in repeated measurements, we may find a few data points that significantly deviate from the mean value. A question is whether an anomalous data point should be rejected as a nonphysical outlier. First, the reasons for the anomaly should be sought from a physical standpoint. Is it caused by mishandling instruments in measurements? Is it caused by certain unexpected physical mechanisms (*e.g.*, a turbulent gust encountered during a climb in flight testing)? If the cause is an evident undesirable and anomalous influence, then the data point can be considered for omission from calculation of the statistics. If no particular physical reason can be identified for the anomaly, however, some statistical criteria could be used for rejecting the suspected result.

Chauvenet's criterion provides a simple method to determine whether an anomalous data point can be deleted based on the Gaussian statistics (Taylor 1997). We consider a set of M measurements $(x_1, x_2, x_3 \cdots x_M)$. The mean $\langle x \rangle$ and the standard deviation σ are first estimated from Eqs. (5.4) and (5.5), respectively. The suspicious point is denoted as x_{sus}, and the relative difference between x_{sus} and $\langle x \rangle$ is calculated by:

$$t_{sus} = \frac{|x_{sus} - \langle x \rangle|}{\sigma}. \tag{5.19}$$

According to Eq. (5.6), the probability that x_{sus} will fall outside of the range $D = [\langle x \rangle - t_{sus}\sigma, \langle x \rangle + t_{sus}\sigma]$ is given by:

$$\Pr(\text{outside of } D) = 1 - \Pr(D). \tag{5.20}$$

Then, the expected number of data points as deviant as x_{sus} is estimated by:

$$m = M \times \Pr(\text{outside of } D). \tag{5.21}$$

According to Chauvenet's criterion, if $m < 0.5$, x_{sus} could be rejected.

For example, we can consider a case where 200 measurements are made, with a mean value of $\langle x \rangle = 2.995$ and a standard deviation of $\sigma = 0.1903$. If a suspected point is $x_{sus} = 3.9$, we know $t_{sus} = |x_{sus} - \langle x \rangle| / \sigma = 4.756$, $\Pr(D) = 0.9999916$, and $\Pr(\text{outside of } D) = 8.394 \times 10^{-6}$ based on the table of the error function. Thus, since $m = M \times \Pr(\text{outside of } D) = 1.679 \times 10^{-3} < 0.5$, $x_{sus} = 3.9$ could be rejected according to Chauvenet's criterion. After rejecting the suspect data point, new statistics (mean and standard deviation) should be computed for the final data set.

Finally, we should note some cautions about how Chauvenet's criterion is implemented on a particular data set. Before blindly applying the criterion, careful care and attention should be devoted to inferring any possible reasons for the suspect data point being an outlier. If these reasons are true aberrations (*e.g.*, a pilot notes an atmospheric gust occurring at the same time the data point is being recorded), then there is greater confidence in ruling out the outlier. However, if there is a physical reason that is expected to and should have an impact on the measurement, then it would be inappropriate to rule out the suspect data point from the statistics. Also, it would be inappropriate to repeatedly apply Chauvenet's criterion in a recursive manner as the statistics of the data set improve.

5.1.6 Monte Carlo Simulation

In the conventional uncertainty analysis described above, several underlying assumptions are made. The nonlinear functional relation between the uncertainties of the measured quantity and the input variables is linearized based on a Taylor series expansion when the uncertainties are small enough. Therefore, the error propagation can be described by a simple root-sum square expression with the given sensitivity and correlation coefficients. Furthermore, since the uncertainties of the statistically independent input variables propagate and contribute to the total uncertainty in a linear fashion, the statistics of the uncertainties is Gaussian in light of the central limit theorem.

However, there may be situations where the probability distributions of the input variables are not Gaussian, the propagation of uncertainties may be nonlinear, or there may be cases where the data reduction equation is complicated enough that the partial differentiation becomes laborious. In these situations, it can be advantageous to implement a numerical scheme for estimating the uncertainty in a measured result – this can be done through Monte Carlo simulation, where the probability distributions (not just the mean and standard deviation) are propagated through the data reduction equation.

Monte Carlo simulations take the values of each of the input variables and allow them to vary according to the estimated uncertainty of each variable. For each instance (a set of input variables with random perturbations), the values are fed through the data analysis process (the data reduction equations) and a result is computed. Statistics on the ensemble of resulting values are estimated, leading to uncertainty estimates for the final values.

To operationalize a Monte Carlo approach for uncertainty estimation, a complete data analysis code (or set of codes) is needed, which accept a series of input variables and provide one or more output variables that result from the analysis. The data reduction code should be exactly what is used for analyzing actual flight test data (all computations). A Monte Carlo simulation can be created by wrapping a new code around the data reduction code, where the key steps of the outer code are outlined as follows:

(1) Define mean values and statistics (pdfs) for each input variable.
(2) Iterate on a for loop that computes for each of M instances:
    (a) Using a random number generator, along with the mean and input statistics defined in step 1, compute instantaneous values for each of the input variables.
    (b) Feed the input variables into the data reduction code and perform computations.
    (c) Record the output variable(s).
(3) Following completion of M iterations, compute statistics (*e.g.*, mean and standard deviation) on the ensemble of value(s) for each output variable.

In the first step of the Monte Carlo simulation, the input statistics are determined by the physical characteristics of the instrumentation or properties being measured. These are usually described by well-known distributions such as a Gaussian (normal), log-normal, Poisson, beta, and/or gamma distribution, depending on the underlying physical principles of measurements. Generation of random instantaneous values in step 2a can be done with a random number generator such as `rand` or `randn` in MATLAB. The number of iterations, M, computed by the Monte Carlo code should be sufficiently large that statistical convergence is achieved. Several applications of Monte Carlo simulation in flight testing were described by Woolf (2012), and further details on Monte Carlo simulation approaches are provided by Coleman and Steele (1999).

5.2 Basic Error Sources in Flight Testing

We'll now pivot our discussion to how uncertainty analysis can be applied to flight test measurements. This section begins with estimated values of the uncertainty of a selection of the most critical instruments often used in flight testing. We'll then conclude with two examples of how uncertainty analysis is applied to flight test measurements, using the analytical approach with partial derivatives of the data reduction equation as well as the Monte Carlo simulation approach.

5.2.1 Uncertainty of Flight Test Instrumentation

Uncertainty in flight test data could be due to four primary sources: variability of the flight environment, variation in individual aircraft performance, differences in piloting skills or technique, and errors inherent to the instrumentation. We'll briefly consider the implications of the first three sources here and focus most of our attention on instrumentation error.

First is natural variability in the properties of the flight environment. This does not include deviation of actual pressure or temperature from the standard atmosphere values, since these properties can be directly measured. Instead, atmospheric variation may include the effects of turbulent gusts, wind shear, or convective currents due to flight over different thermal masses. Second, there may be variation in aircraft performance between different engines or airframes for a specific aircraft type. These variations could result from different rigging of the flaps or control surfaces, slight differences in manufacturing tolerances, engine installation differences, lubrication level, or operational factors such as the number of dead insects present on the leading edge of the wing. Third is variability in piloting technique – pilots of varying skill level will be able to conduct the flight test with more or less precision. Even an individual pilot will exhibit varying piloting skills due to factors such as fatigue. All three of these factors – the flight environment, aircraft variation, and pilot skill – are difficult or even impossible to control. In order to generate representative flight test results, in an ideal sense, a large number of repeated flight tests must be conducted. Within this collection of independent flight test data points, each of the parameters (and other, unknown parameters) must be allowed to vary independently. Thus, in an ideal sense, a test matrix would select (at random) from a pool of qualified test pilots and test aircraft, with the flights conducted under a wide array of atmospheric conditions. This approach to flight testing is rarely justified for cost considerations. Thus, the practical approach to handling these variations is to control as many parameters as reasonably possible (*e.g.*, conduct test flights only under calm atmospheric conditions) and to estimate the variability due to piloting technique through controlled small sample testing.

The fourth major source of uncertainty in flight testing is error inherent to the instrumentation. The magnitude of this error source can be quantified through the methods discussed in this chapter, which will be illustrated in the next two examples. Here, we'll provide an estimate of the uncertainty of three key instruments in the cockpit – the airspeed indicator (ASI), outside air temperature (OAT) gauge, and the altimeter (used for recording pressure). For a complete understanding of the uncertainty associated with these measurements, it may be helpful to review the relevant material presented in the chapters on instrumentation (Chapter 3) and airspeed calibration (Chapter 8). In our discussion here, we'll presume that position error – which affects both the altimeter and the ASI – has already been accommodated through flight test calibration (see Chapter 8 for details). Gracey (1980, Chapter XI) is an excellent resource that documents the errors inherent to the pitot-static system and should be consulted for more details.

The ASI directly measures the impact pressure, which is the same as dynamic pressure if we ignore compressibility effects. After correcting for position error, the main source of error is instrument error, which is due to small errors in applying the calibration to the ASI dial, as well as uncertainty in reading the indicated airspeed. The markings on a typical analog ASI provide velocity resolution down to 5 knot intervals, although some ASIs have resolution to the nearest 1 knot. The position of the needle on the ASI can perhaps be inferred to the nearest ±2 knots if the ASI resolution is 5 knots. Gracey (1980) provides data for US Air Force calibration standards that range from ±2 to ±5 knots across a range of airspeeds (see Table 5.1).

The principles of freestream temperature measurement are discussed in Chapter 3, where the uncertainty is a function of the calibration of the recovery factor as well as the local Mach number. Any error in the calibration of these factors will contribute to the systematic error of the measurement. Also, the reporting of temperature in the cockpit is usually rounded off to the nearest degree Celsius, so at a minimum there is ±0.5 °C uncertainty in temperature measurement. The combination of resolution error and calibration error may lead to a combined uncertainty of approximately ±1.0 °C.

The third major flight measurement is pressure altitude, which directly provides a measurement of pressure by conversion through the standard atmosphere (see Chapters 2 and 3). The most significant source of error in the measurement of pressure altitude is position error, which is the central theme of the calibration process described in Chapter 8. Remaining error sources in altitude measurement include scale error, hysteresis, after effect, friction, case leaks, and barometric scale error. Fortunately, these instrument errors are bounded by regulatory limits. The Federal Aviation Administration (FAA) requires calibration of the pitot-static system at least once every 24 calendar months (see 14 CFR 91.411 and 14 CFR 43 Appendix E). In order for an aircraft to remain airworthy for flight under instrument flight rules, the reported altitude must fall within certain tolerances stipulated by the calibration procedure documented in 14 CFR 43 Appendix E. The following error sources are primarily applicable to mechanical altimeters. (Digital pressure transducers integrated into an air data computer (supporting glass panel avionics) will have much lower instrument error, which will be specified by the manufacturer.) We'll discuss each of the altimeter instrument errors in turn by describing the calibration procedure stipulated by the FAA – this is important context for determining which of the instrument errors is relevant for a given flight maneuver.

Scale error relates to the accuracy of the markings on the altimeter. Since the relationship between pressure and altitude is nonlinear (exponential), the error between the indicated and

Table 5.1 Typical uncertainty values for an airspeed indicator.

Calibrated airspeed (knots)	Tolerance (knots)
50	±4.0
80	±2.0
150	±2.5
250	±3.0
300	±4.0
550	±5.0
650	±5.0

Source: Gracey (1980).

actual altitude will increase with altitude for a fixed altimeter scale. Thus, the allowable scale error also increases with altitude. Scale error is calibrated by applying increasingly lower pressure (simulating higher altitude) and allowing the instrumentation to stabilize at each test point (nominally, 2 minutes according to Gracey, 1980). The difference between the indicated altitude and the altitude corresponding to the applied pressure is documented across all tested altitudes. Table 5.2 provides the allowable limits on scale error specified by 14 CFR 43 Appendix E.

Hysteresis error describes the error in the reported altitude when the altimeter is subjected to a decrease in altitude, compared to an increase in altitude. After the scale error calibration is performed (a series of increasing altitudes), the simulated altitude is rapidly decreased to an altitude set point at 50% of maximum altitude and down again to 40% of maximum altitude. At each of these set points, the indicated altitude is recorded and compared with the indicated altitude for the corresponding test points on the scale error calibration test. The difference in corresponding indicated altitudes must be no more than ±75 ft.

Following the hysteresis test, the after effect calibration is performed. This involves a measurement of the indicated altitude at the initial condition (vented to atmosphere) prior to the scale error calibration test. This value of indicated altitude must be within ±30 ft of the initial reading.

Friction errors result from friction in the gears used to transduce pressure changes into movement of the needles around the dial indicator. The significance of friction error is measured by

Table 5.2 Maximum scale error permitted by 14 CFR 43 Appendix E.

Altitude (ft)	Equivalent pressure (in. Hg)	Tolerance ± (ft)
0	29.921	20
500	29.385	20
1,000	28.856	20
1,500	28.335	25
2,000	27.821	30
3,000	26.817	30
4,000	25.842	35
6,000	23.978	40
8,000	22.225	60
10,000	20.577	80
12,000	19.029	90
14,000	17.577	100
16,000	16.216	110
18,000	14.942	120
20,000	13.750	130
22,000	12.636	140
25,000	11.104	155
30,000	8.885	180
35,000	7.041	205
40,000	5.538	230
45,000	4.355	255
50,000	3.425	280

Table 5.3 Maximum friction error permitted by 14 CFR 43 Appendix E.

Altitude (ft)	Tolerance ± (ft)
1,000	70
2,000	70
3,000	70
5,000	70
10,000	80
15,000	90
20,000	100
25,000	120
30,000	140
35,000	160
40,000	180
50,000	250

recording an indicated altitude at each set point before and after the altimeter instrument has been vibrated. The maximum allowable friction error (between these two readings) is shown in Table 5.3 as a function of altitude.

A properly functioning altimeter should form a perfect seal between the surrounding air and the case. However, the movable diaphragm (aneroid wafer) inside the altimeter, as well as the altimeter case itself, may be subject to leaks. A case leak calibration is performed at a pressure corresponding to an altitude of 18,000 ft, and the indicated altitude is monitored over a period of 1 minute. The indicated altitude must not deviate more than ±100 ft during this 1-minute constant pressure hold.

Finally, barometric scale error captures the error associated with setting the reference pressure (the Kollsman knob discussed in Chapter 3). Normal operation of the Kollsman knob adjusts the indicated altitude higher when the reference pressure is increased (see discussion in Chapter 3). Table 5.4 documents the specific values of altimeter change that should be achieved for each setting of the reference pressure. Calibration for barometric scale error compares the change in indicated

Table 5.4 Maximum barometric scale error permitted by 14 CFR 43 Appendix E.

Reference pressure (inches Hg)	Altitude difference (ft)
28.10	−1,727
28.50	−1,340
29.00	−863
29.50	−392
29.92	0
30.50	+531
30.90	+893
30.99	+974

altitude at each of the reference pressure set points, with the expected values provided in Table 5.4. The indicated altitude changes must be within 25 ft of the expected values at each set point.

Of the error sources presented here, scale error, after effect, friction error, and barometric scale errors are static errors, while hysteresis and case leak errors are dynamic. Error sources can be combined in a root-sum square manner to determine an overall error in indicated altitude. For steady-altitude flight testing where the freestream static pressure is desired, the dynamic error terms (hysteresis and case leak) may be safely neglected. For example, if a nominal test altitude of 3000 ft is selected, the combined error becomes $(30^2 + 30^2 + 70^2 + 25^2)^{0.5} = \pm 86$ ft. If dynamic altitude changes are being measured (e.g., climb or glide flight test), then the hysteresis and case leak error sources should also be included in the uncertainty estimate. Once an uncertainty in altitude is found, the corresponding uncertainty in measured pressure can be found from the standard atmosphere.

5.2.2 Example: Uncertainty in Density (Traditional Approach)

Measurements of air pressure, temperature, and density in freestream at a given flight altitude are of fundamental importance, since they are the physical quantities contained in all major equations of aircraft flight dynamics. In this example, we wish to find the uncertainty in local freestream density based on measurements of pressure altitude and OAT. The measured values of pressure altitude and OAT are $h = 3000$ ft and $T = 10\,°C$, respectively. Based on the preceding discussion regarding uncertainty of the altimeter and the OAT gauge, we'll take the altimeter bias error to be $U_h = 2\sigma_h = \pm 86$ ft and the OAT bias error to be $U_T = 2\sigma_T = \pm 1\,°C$.

First, we'll need to develop our data reduction equation. Density will be found by the equation of state for a perfect gas,

$$\rho = p(RT)^{-1}, \tag{5.22}$$

where $R = 1716$ ft lb/(slug $°R$) is the specific gas constant for air, p is pressure, and T is temperature. Since we are measuring pressure altitude, rather than pressure, we need to also incorporate an equation for the standard atmosphere. Referring to Chapter 2, we combine the standard temperature lapse rate with the pressure variation with temperature to obtain

$$p = p_{SL}\left(\frac{a}{T_{SL}}h + 1\right)^{-g_0/aR}, \tag{5.23}$$

where $p_{SL} = 2116.2$ lb/ft^2, $T_{SL} = 518.69\,°R$, $g_0 = 32.17$ ft/s^2, and the standard temperature lapse rate at this altitude, $a = dT/dh = -6.5\times10^{-3}$ K/m $= -3.566\times10^{-3}\,°R$/ft, is assumed constant. Combining Eqs. (5.22) and (5.23) we have

$$\rho = \frac{p_{SL}}{RT}\left(\frac{a}{T_{SL}}h + 1\right)^{-g_0/aR}, \tag{5.24}$$

where the only input variables are pressure altitude and OAT. With the measured values of h and T, we find that the freestream density is 2.1685×10^{-3} slug/ft^3.

Now, we'll develop an expression for the overall uncertainty by applying Eq. (5.17):

$$\frac{U_\rho}{\rho} = \sqrt{S_h^2\left(\frac{U_h}{h}\right)^2 + S_T^2\left(\frac{U_T}{T}\right)^2}. \tag{5.25}$$

The relative uncertainty in pressure altitude and temperature can be calculated from what we already know, $U_h/h = 0.0287$ and $U_T/T = 0.0035$.

Applying Eq. (5.14), we find that the sensitivity coefficients at this flight condition are

$$S_h = \frac{h}{\rho}\frac{\partial\rho}{\partial h} = \frac{h}{\rho}\frac{\partial}{\partial h}\left(\frac{p_{SL}}{RT}\left(\frac{a}{T_{SL}}h + 1\right)^{-g_0/aR}\right) = h\left(\frac{-g_0}{RT_{SL}}\right)\left(\frac{a}{T_{SL}}h + 1\right)^{-1} = -0.1107,$$

$$(5.26)$$

and

$$S_T = \frac{T}{\rho}\frac{\partial\rho}{\partial T} = \frac{T}{\rho}\frac{\partial}{\partial T}\left(\frac{p_{SL}}{RT}\left(\frac{a}{T_{SL}}h + 1\right)^{-g_0/aR}\right) = -\frac{T}{\rho}\frac{p_{SL}}{RT^2}\left(\frac{a}{T_{SL}}h + 1\right)^{-g_0/aR} = -1. \quad (5.27)$$

Combining the sensitivity coefficients and relative uncertainties into Eq. (5.25), we have

$$\frac{U_\rho}{\rho} = \sqrt{(-0.1107)^2(0.0287)^2 + (-1)^2(0.0035)^2} = 0.0047, \quad (5.28)$$

or a relative uncertainty of about half a percent. This gives an absolute uncertainty in density of $U_\rho = 1.0297 \times 10^{-5}$ slug/ft^3. Note that even though the relative uncertainty in pressure altitude (0.0287) is higher than that for temperature (0.0035), the sensitivity coefficient for pressure altitude (−0.1107) is lower than that for temperature (−1). Thus, in this case, the uncertainties in pressure altitude and temperature contribute about equally to the overall uncertainty in density.

5.2.3 Example: Uncertainty in True Airspeed (Monte Carlo Approach)

In this example, we'll explore the use of Monte Carlo methods for estimating the uncertainty in a calculation of true airspeed, based on measurements of indicated airspeed, pressure altitude, and OAT. Chapter 8 provides a detailed discussion of the different types of airspeed, which may be helpful to review. Briefly, indicated airspeed is found by measuring impact pressure, which is the difference between the total pressure and the static pressure. The impact pressure is converted to indicated airspeed through the isentropic Mach relation and the assumption of standard sea-level conditions for pressure and temperature. True airspeed, however, must be based on actual measurements of freestream static pressure and temperature, along with the measured impact pressure.

We'll start by assuming that position error has already been corrected for, and any instrument errors in the ASI propagate through the uncertainty analysis. Thus, in this case, calibrated airspeed can be assumed to be equal to indicated airspeed. Based on the isentropic Mach relations, the impact pressure is

$$q_c = p_{SL}\left(1 + \frac{\gamma - 1}{2}\frac{V_{cal}^2}{\gamma RT_{SL}}\right) - p_{SL}, \quad (5.29)$$

where V_{cal} is the calibrated airspeed, γ is the ratio of specific heats, R is the specific gas constant for air, and T_{SL} and p_{SL} are the standard sea-level values of temperature and pressure, respectively. Once impact pressure has been calculated from the measured calibrated airspeed, the true airspeed is found from

$$V = \sqrt{\frac{2\gamma RT}{\gamma - 1}\left[\left(\frac{q_c}{p} + 1\right)^{(\gamma-1)/\gamma} - 1\right]}, \quad (5.30)$$

where T and p are the actual values of freestream static temperature and pressure. (See Chapter 8 for complete details on airspeed measurement and conversion between airspeed types.)

For this example, the measured value of indicated airspeed is 120 knots at a pressure altitude of 3000 ft where the measured OAT is a relatively warm 16 °C. The indicated airspeed is read from the glass panel avionics display, with an estimated uncertainty of ±1 knot. As noted previously, the estimated uncertainty in pressure altitude is ±86 ft, and the uncertainty in OAT is ±1.0 °C.

To perform the Monte Carlo analysis, the basic data reduction code is created. This code accepts input values of indicated airspeed (in knots), pressure altitude (in ft), and OAT (in °C). The code then converts the input values to standard consistent units and converts pressure altitude to pressure through a standard atmosphere subroutine. The analysis then proceeds with calculating impact pressure from Eq. (5.29) and true airspeed from Eq. (5.30). The resulting value of true airspeed is converted back to units of knots and is the output of the code.

An outer loop is then wrapped around the data reduction code, where the nominal input values and statistics are provided in the outer code. The outer loop will iterate over a large number of samples, allowing the instantaneous input values to vary according to the statistics. Each output value of the data reduction code is stored, and the statistics of the true airspeed values are calculated at the end in order to determine the uncertainty.

In this example, normal pdfs of indicated airspeed and pressure altitude are assumed, and a uniform distribution of OAT is assumed. The standard deviations of these distributions are half of the provided uncertainty values, since $U = 2\sigma$ for a large sample size with a 95% confidence interval. After the code was run for 10,000 cases, the statistics of the input values are illustrated in Figure 5.6, and the resulting distribution of true airspeed values is shown in Figure 5.7. The resulting value of true airspeed is 126.89 ± 1.08 KTAS, where $U_{TAS} = 2\sigma_{TAS}$ and the relative uncertainty is 0.86%.

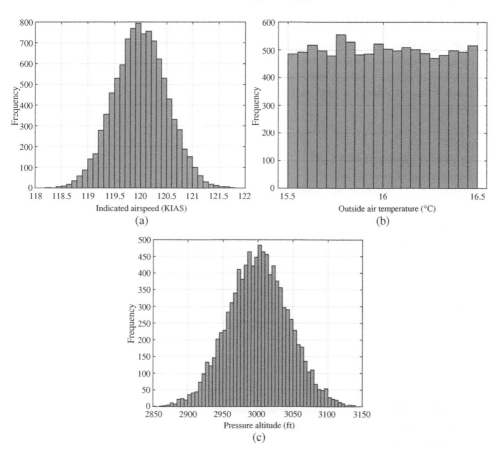

Figure 5.6 Histograms of the input variables: (a) indicated airspeed, (b) outside air temperature, and (c) pressure altitude.

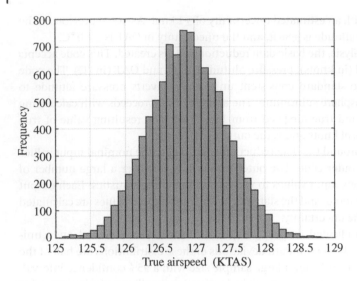

Figure 5.7 Histogram of the calculated true airspeed.

Nomenclature

a	standard temperature lapse rate
B	systematic error estimate (bias error)
D	range of values that a sample may fall between
f	function to define X with ξ_i input variables
g_0	gravitational acceleration (constant)
h	pressure altitude
i	index
j	index
N	number of input variables in X
m	expected number of points
M	number of independent measurements made
p	probability density function
p	pressure
P	precision error estimate
q_c	impact pressure
R	gas constant
S	sensitivity coefficient
t	student's t distribution
T	temperature
U	overall uncertainty
V	true airspeed
V_{cal}	calibrated airspeed
x	arbitrary variable
X	arbitrary quantity being analyzed for uncertainty
z	variable to represent $(x - \langle x \rangle)/\sigma$ for convenience
δ_{ij}	Kronecker delta
γ	ratio of specific heats
ν	degrees of freedom

ρ	density
$\rho_{i,j}$	correlation coefficient between variables ξ_i and ξ_j
σ	standard deviation
σ_i^2	variance
$\sigma_{i,j}$	covariance
ξ	input variable

Subscripts

best	best estimated value
instrum	instrument error
p	precision error
position	position error
SL	sea level
sus	suspicious data point

Acronyms and Abbreviations

ASI	airspeed indicator
NIST	National Institute of Standards and Technology
OAT	outside air temperature
pdf	probability density function
POH	Pilot's Operating Handbook
Pr	probability

References

Bevington, P.R. and Robinson, D.K. (1992). *Data Reduction and Error Analysis for the Physical Sciences*, 2e. New York: McGraw-Hill.

Coleman, H.W. and Steele, W.G. (1999). *Experimentation and Uncertainty Analysis for Engineers*, 2e. New York: John Wiley & Sons.

Cramer, H. (1999). *Mathematical Methods of Statistics*. Princeton, NJ: Princeton University Press.

Gracey, W., 1980, *Measurement of Aircraft Speed and Altitude*, NASA-RP-1046, https://ntrs.nasa.gov/citations/19800015804.

Liu, T. and Finley, T.D. (2001). Estimating bias error distributions. *Review of Scientific Instruments* 72 (9): 3561–3571. https://doi.org/10.1063/1.1394188.

Taylor, J.R. (1997). *An Introduction to Error Analysis: the Study of Uncertainties in Physical Measurements*, 2e. Sausalito, CA: University Science Books.

Woolf, R.K. (2012). Applications of Statistically Defensible Test and Evaluation Methods to Aircraft Performance Flight Testing. AIAA Paper 2012-2723. New Orleans, LA: 28th Aerodynamic Measurement Technology, Ground Testing, and Flight Testing Conference https://doi.org/10.2514/6.2012-2723.

6

Flight Test Planning

Careful planning and preparation are essential to both safety of flight and ensuring that the desired flight test objectives are achieved. This chapter details the considerations involved with planning flight test programs as well as individual flight tests. The first half of this chapter provides an overview of how flight tests are planned and conducted, including preparation of detailed test cards, as well as careful risk assessment and risk management. Following this, we will also discuss some of the typical pre-flight planning steps that pilots perform on a regular basis for individual flights, as a point of introduction to students and non-pilot readers. These pre-flight activities include selection of a specific flight location (relative to local airspaces), checking of notices to airmen and temporary flight restrictions, evaluation of aircraft performance relative to the available runway length, consultation of current and forecast weather, determination of aircraft weight and balance information, and finally a pre-flight inspection of the aircraft to ensure that it is in an airworthy condition. (Note that the discussion of weather products and airspace definitions presented in this chapter is largely focused on policies and procedures found in the United States; similar regulations and procedures are found in Europe, Asia, and elsewhere.) The discussion on flight test planning provided here is just an overview; other sources such as Harney (1995), Mondt (2014), Ward et al. (2006), Kimberlin (2003), or Corda (2017) can be consulted for other details on flight test planning.

6.1 Flight Test Process

The flight testing community typically follows a standard process by which flight tests are planned and executed. This process has been converged upon over years of practice and learning from various mistakes. A well-defined process for planning and conducting the flights is an important safeguard for risk; it optimizes efficiency and minimizes budget requirements. This flight test process is almost always a team effort, where each member of the team is integrally involved. Flight test teams are often composed of pilots, flight test engineers, instrumentation engineers, data analysts, ground support crew, program managers, and others. A cohesive, well-functioning team is critical for safe and successful flight testing.

Planning of flight tests must begin with a careful and clear definition of the goal or objective of the flight test program, which then flows down to clear and unambiguous definition of the objectives for each flight. These flight test objectives should be framed as some type of question that must be answered, which is often grounded in prior data from design, wind tunnel testing, or flight testing. Predictions of the aircraft's performance based on these data sources must be verified through flight testing. Thus, the relevant performance calculations will yield predictions of the actual flight performance which must be verified. Pre-flight predictions are also important for anticipating the capabilities of an aircraft and defining the range of test conditions that should be

Introduction to Flight Testing, First Edition. James W. Gregory and Tianshu Liu.
© 2021 John Wiley & Sons Ltd. Published 2021 by John Wiley & Sons Ltd.
Companion website: https://www.wiley.com/go/flighttesting

flown. For example, if an aircraft has a predicted climb performance of 1000 ft/min at sea level, then the flight test matrix should be planned to cover that range of anticipated climb rates.

Based on this definition of a flight test's objectives, the flight test team then prepares a specific plan for an individual flight or set of flights. Every single element of the flight test plan must be aligned with the overall test objective – no extraneous aspect can be allowed, due to concerns about safety and cost. It may be helpful to include optional test points at the end of the test plan, in case the primary test points are completed more quickly than anticipated, and there is remaining available time in the flight period. At a minimum, the flight test plan should include the following aspects:

(1) A listing of all the conditions to be flown, which is informed by the standard flight test techniques that are detailed in the remainder of this book. These desired test conditions must be limited by safety considerations (to be discussed in the next section on risk management) and by appropriate regulatory limits for the authority responsible for the aircraft and airspace.
(2) A detailed list of all measurements to be recorded, and a list of all required instrumentation needed to make those measurements. The measurements and instrumentation are determined by the required uncertainty in the measurement, in order to successfully address the test objectives.
(3) The order of the test points must also be determined, such that the safety and efficiency of the test program can be optimized. For example, if there is some risk associated with expansion of the flight envelope (*e.g.*, with flutter flight testing), then the test points must be ordered in a way that the test risk increases slowly over time, such that any incipient problems can be identified, and the test aborted in a safe manner if needed. The development of the flight test plan must be informed by a careful risk assessment, which will be discussed in the next section. Aside from ordering for risk, the test points should be structured in a way that minimizes wasted time and cost. This can be achieved by identifying synergy between different test elements. For example, it may be possible to combine flight tests that require the same flight conditions, such as steady, level flight for evaluating trim conditions, and calibration of the airspeed indicator. Or, in another example, climb flight tests could be combined with glide flight tests, since a series of climbs can be interlaced with a series of glides in order to minimize wasted flight time.

Following the development of a clear test plan, the flight test team will translate the information to a series of test cards. Test cards are used by the pilot, test engineer, and ground crew as a set of step-by-step instructions for conducting the flight test, and for recording relevant hand-written data. Test cards are critical for keeping the entire team synchronized and referencing the same information. The test cards are almost like a contract that everyone agrees upon before flight. A disciplined, conscientious, and careful flight test team will never depart from the procedure detailed in the test cards unless an emergency situation emerges and it is clearly no longer safe to follow the test card procedure. Test cards must clearly detail the flight conditions to be flown for each test point including: altitude, airspeed, flap setting, landing gear configuration, power setting, airspeed limits, etc. The test cards must also clearly identify the data to be acquired and detail how it will be acquired or measured. A sample test card from the USAF Test Pilot School is shown in Figure 6.1. The format of the card is designed such that it can be easily read in flight, it fits on a pilot's kneeboard for quick reference, and the data can be recorded in an easy manner. This sample test card is used for calibration of the airspeed indicator (see Chapter 8). The left side of the card is for the pilot's use, with guidance on establishing the flight maneuvers, maintaining the correct conditions, and matters related to safety of flight. The right side of the card is for the flight test engineer's use, where instrument settings and data acquisition details are provided and space is located for hand-recording data.

AIRMANSHIP ENGINEERING TEST DEMO FLIGHT

AIR DATA CALIBRATION

CARD No: RUN 7		
DATA BAND	ALT: 6000 FT MSL ± 200 FT A/S: 100 KIAS ± 2 KT FLAPS: UP POWER: A/R	ALTIMETER: LOCAL STATIC SOURCE: **NORMAL**
INITIAL CONDS	TEMP INDICATOR: DEG F HEADING: 360 recommended	

1. RESET HEADING INDICATOR

2. FLY TRIANGLE PATTERN
   a) CONSIDER A/P HEADING HOLD

   ☐ **SET INITIAL HDG** A/R

   ☐ **± 120 DEG** FROM INITIAL HDG

   ☐ **± 240 DEG** FROM INITIAL HDG

LIMITATIONS
REMAIN ABOVE 1000 FEET AGL
FLAP SPEEDS < 110 KIAS FOR 10 DEGREE FLAPS
            < 85 KIAS FOR FLAPS > 10 DEGREES

PILOT CALLS
ON CONDITION; AIRSPEED WHEN STABLE

AIRMANSHIP ENGINEERING TEST DEMO FLIGHT

DATA REQUIREMENTS
Flight Instruments
Engine Instruments
Aircraft GPS Indications

GO/NO-GO
Day VMC
Max. Peak Wind – 24 kts (PVT) 28 kt (COM)

OBJECTIVE
Determine airspeed calibration for C-172S using GPS wind triangle method.

APPROACH AND PROCEDURES
1. Set temperature indicator to Deg F.
2. Realign heading indicator with mag. Compass (analog gages only)
3. Record altimeter setting _____
4. While maintaining altitude, fly a triangle pattern with the heading of each leg having approximately 120 degrees difference from the other legs. Airspeeds at the center of the three legs should be as close to the target airspeed as possible.
5. At the center of each leg (when stable), record data.

Leg	DE Total Indicated Temp (deg F)	DE/PILOT Indicated Airspeed (KT)	TC GPS Ground Speed (GS) (KT)	TC GPS Track Angle (TK) (deg)	GARMIN G1000 ONLY DE/PILOT PFD Wind Speed (KT)	DE/PILOT PFD Wind Direction (deg)	DE/PILOT PFD True Airspeed (KT)
1							
2							
3							

Figure 6.1 Sample flight test card for an airspeed calibration test. Source: USAF Test Pilot School.

Once the detailed test plan and test cards have been prepared, the entire flight test team meets to conduct a thorough safety and technical review. In this environment, the test team is invited to think critically about the safety and objectives of the entire test. This level of detached, critical thinking is important for the team to assess any emergent problems that may have been overlooked until this point. It is important to have wise, experienced, senior colleagues as a part of this review, since these individuals have the experience and judgment needed to assess the reasonableness and risk of the test plan.

Finally, the test team transitions into a pre-flight check of the aircraft and immediate preparations for flight. The pilot will perform a careful walk-around inspection of the aircraft, to ensure that it is in an airworthy condition and that all systems are functioning as designed. More details on the pre-flight inspection are provided later in this chapter.

Following the flight test, a post-flight briefing is conducted in order to review any anomalies that emerged, such as weather conditions that may have had an impact on a test point, or an aircraft response to a test condition that did not proceed as planned. The post-flight briefing then immediately flows into a careful post-flight analysis of the data, which may take substantial time depending on the volume and complexity of the recorded data. Standard procedures for analyzing and reducing flight test data are a key topic for the remainder of this book. Once the data analysis is complete, the findings from the flight test are compared with the initial objectives of the test. For example, this may involve a comparison of measured climb performance with the predicted performance. At this point, the team can assess whether the flight test was a success. If there is a difference between flight test data and the predictions, the team of engineers should determine whether additional flight testing is needed to improve data quality, or if the models and predictive tools should be updated and improved to match reality. This post-analysis assessment of the data often takes the form of a formal written report that will permanently document the findings and be circulated to various technical and management members of the flight test team.

The process described above is a cycle, which becomes part of a larger flight test program which leads to the overall program goals. Throughout this process, careful attention is given to how to safely and efficiently conduct the flight test in order to optimize data output under the given budgetary constraints. In summary, there are two key principles of flight test planning, which must be pursued in order to ensure safety:

(1) *Forethought*: The flight test team must anticipate, to the maximum extent possible, any contingency or possible event that could affect the safety of flight. This requires a number of people working together to brainstorm and think through ideas.
(2) *Clear objectives*: The team needs to clearly define what questions the program must answer, as precisely as possible. A clearly defined objective is required in order to align all flight test plans and to determine whether the outcome is successful. Clear objectives allow the test conditions to be appropriately defined in order to answer the test questions in a safe manner that minimizes risk.

6.2 Risk Management

Commercial and government flight test programs of new aircraft or aircraft systems must involve a detailed assessment of risk and development of risk mitigation plans. The details of risk assessment are beyond the scope of this text, but a general overview is provided here. Despite our cursory review here, risk assessment and test safety are of paramount importance and the flight testing professional must adopt a careful, rigorous, disciplined, and methodical approach to maintain acceptable levels of risk. Otherwise, the possible consequences include loss of life, property, and aircraft sales. The details of risk management vary from one organization to another, but the general approach follows these general themes. An interested reader can consult Appleford (1995), Ward et al. (2006), Mondt (2014), or Corda (2017) for discussions of risk management of flight testing programs. Other, more detailed, discussions of risk management are available from FAA (2012a, 2012b), Department of Defense (2012), International Civil Aviation Organization (2018), and NASA (2010, 2011).

At the outset of our discussion, it is helpful to define a few important terms. In common vernacular, these terms are often used interchangeably, but it is helpful for us to use more precise definitions here. A *hazard* is anything (event, circumstance, object, etc.) that could lead to an undesirable outcome. In aircraft operations, an example of a hazard might be a blown tire during landing, which could result in loss of control. *Risk*, on the other hand, is the impact or outcome of a hazard if it is not mitigated, controlled, or eliminated. An assessment of risk is formed based on estimates of the *severity* of a given hazard's outcome, and the *probability* or *likelihood* that the given hazard will occur. Risk assessment is based on a process of identifying hazards and estimating the risk associated with each.

First, a testing team must identify all possible hazards that could result in a given test or testing program. One successful approach to hazard identification is an open brainstorming session with people present from various backgrounds and experience levels. Both the experiences of a seasoned engineer who has seen many unexpected hazards emerge in testing programs, and the wide-eyed view of an inexperienced team member who has not been biased by group think should be present. Everyone who participates in hazard assessment must feel comfortable to speak freely. Test leads should cultivate an open discussion where no idea is shot down as silly or irrelevant. The team must work as a cohesive unit to identify as many hazards as possible, thinking creatively about various failure modes. A part of this initial hazard assessment should include clear identification of the causes of each hazard, or the multiple issues working in conjunction to cause a particular hazard.

A helpful resource for identifying hazards for a particular flight test is the reference materials and workshops organized by the Flight Test Safety Committee (formed by members of the Society of Experimental Test Pilots, the Society of Flight Test Engineers, and the American Institute of Aeronautics and Astronautics). These resources provide an easily accessible treasure trove of experience on flight test maneuvers, hazards that have been encountered, and effective mitigation strategies for reducing the risks of those hazards. Of particular note is the Flight Test Safety Database, which provides a searchable interface for mining the collective wisdom of flight test professionals (see Maher 2019; Flight Test Safety Committee 2020). These resources represent the collective knowledge of the flight test community and should be consulted in the planning of any flight test. Another strategy for learning from the experience of others is to read the stories documented by Stoliker et al. (1996), Merlin et al. (2011), Mondt (2014), and others, where hazards for widely varying flight test programs have been documented.

Once a list of hazards has been identified, the likelihood and severity of each hazard must be estimated. Likelihood can range from certain or nearly certain, down to unlikely or remote. The severity of consequences of each hazard can range from catastrophic or deadly, down to minor or benign. One example of a useful likelihood and consequence taxonomy is given in Table 6.1 (similar to one provided by Ward et al. (2006)). Each likelihood and consequence category is assigned an identifier to assist in categorization of each identified hazard. The consequence and likelihood categories are then arranged in a risk assessment matrix, with high likelihood and high severity placed in the upper left corner, and improbable, benign hazards in the lower right corner. An example of such a risk assessment matrix is shown in Figure 6.2, which is from the Department of Defense Standard Practice on System Safety (2012). Identification or estimation of likelihood and

Table 6.1 Taxonomy for assigning likelihood and severity levels in risk assessment.

Probability	Code	Severity	Code
Frequent	A	Catastrophic: resulting death, serious injury, destruction of the aircraft or other test equipment	1
Probable	B	Critical: severe injury (involving hospitalization or lost work), aircraft or equipment damage requiring major repair	2
Occasional	C	Marginal: minory injury (outpatient, no significant lost work), some repair effort that could span multiple shifts	3
Remote	D	Negligible: no injury, minor damage that is easy to repair (within hours), minimal loss of testing time	4
Improbable	E		

Risk assessment matrix				
Severity / Probability	Catastrophic (1)	Critical (2)	Marginal (3)	Negligible (4)
Frequent (A)	High	High	Serious	Medium
Probable (B)	High	High	Serious	Medium
Occasional (C)	High	Serious	Medium	Low
Remote (D)	Serious	Medium	Medium	Low
Improbable (E)	Medium	Medium	Medium	Low
Eliminated (F)	Eliminated			

Figure 6.2 Risk assessment matrix. Source: US Department of Defense.

consequence of each hazard is truly an estimation and must rely upon the experience of more seasoned members of the flight test team. Many organizations also maintain historical documentation that can be consulted for data on various hazards or emergent hazards.

Following risk assessment, a set of mitigation strategies should be developed to reduce the various risks involved. These risk mitigations should primarily focus on the most severe, most likely hazards, but all hazards should be included. Mitigating strategies can include hardware modifications, software modifications, procedural modifications, or all three. Examples of hardware modifications for risk mitigation are the installation of an airframe parachute for an aircraft undergoing spin testing, or installation of additional sensors to provide real-time data to the flight test team for identification of emergent hazards. An example of a software modification may be the addition of robust, automatic switchover logic between communication links if the primary control link on an unmanned aircraft fails. Procedural modifications for risk mitigation could involve tailoring the sequence and spacing of test points in the test plan in order to approach high-risk test points gradually. Flight testing professionals must be careful, however, to recognize that unexpected behaviors can rapidly emerge due to nonlinear behavior that is characteristic of the system. A classic example of this is expansion of the flight test envelope for identification of flutter boundaries. Clearly, the testing team does not wish to actually fly at the condition where significant flutter occurs, so the expected boundary must be approached cautiously.

After a series of mitigating strategies have been developed, the team performs a residual risk assessment to identify the probability and severity of latent hazards, resulting in a modified risk assessment matrix that includes the effects of the mitigation approach. The team must be careful to consider any unanticipated consequences of risk mitigation for one hazard impacting the severity or likelihood of another hazard. The flight testing organization's management must identify the boundary for acceptable and unacceptable risk, which will represent a boundary across the table in Figure 6.2. This risk acceptance boundary typically extends from the lower left to the upper right of the matrix, with the specific location of the line determined by an organization's risk tolerance. If unacceptable risks remain on the risk assessment matrix, then additional mitigation strategies must be developed (often at higher cost), the test point eliminated, or the entire test program cancelled.

6.3 Case Study: Accept No Unnecessary Risk

A critical tenet of flight testing is to assiduously adhere to the plan developed prior to flight. Everyone involved in the flight test team must avoid any changes "on the fly," since this inherently involves greater risk. Planning of the flight requires great care and careful consideration of risk factors. If an in-flight adaptation to the flight plan is made, there is insufficient time to deliberate among the team and consider the various implications to the safety of flight. Thus, there is a mantra within the flight test community to "plan the flight, and fly the plan." Furthermore, it is critical to maintain focus on program objectives and not get lured into non-essential testing and the extra risks associated with that extraneous work.

There have been many scenarios when those responsible for a flight test program can be tempted to deviate from the agreed-upon test plan. Consider the demise of the XB-70, for example. The XB-70 Valkyrie program in the 1950's was crafted to create a long-range supersonic tactical bomber that could cruise at Mach 3. The program was extremely expensive and was ultimately deemed unnecessary due to the advent of intercontinental ballistic missiles. Following cancellation of the program, the two XB-70 aircraft that were built were reverted to a flight test research program for supersonic flight.

In mid-1966, GE approached officials with North American Aviation (manufacturer of the XB-70) and the US Air Force to request permission for a photo shoot with the XB-70. GE wanted to capture publicity photos of a series of aircraft, all powered by GE engines, flying in formation. Through a series of decisions, officials within NASA and the Air Force ultimately approved the photo shoot with an Air Force T-38A, a Navy F-4B, and Air Force YF-5A, and a NASA F-104N. Difficult questions about the ultimate purpose of the flight, or whether the formation flight was necessary to accomplish test objectives, were left unaddressed. So, on June 8 1966, the XB-70 was taken up by North American pilot Al White, with USAF Col. Carl Cross as co-pilot.

The flight amounted to little more than a publicity stunt, where the chase plane photographer took some amazing photos of the fleet of GE-powered aircraft in formation (see Figure 6.3). As the photographer snapped away, the F-104N piloted by Joe Walker strayed too close to the XB-70 and

Figure 6.3 XB-70 (large aircraft, center), flying with four other GE-powered aircraft in formation on June 8 1966. Source: US Department of Defense.

got caught in the powerful tip vortex of the heavy bomber. As later surmised by the USAF accident investigation report, the tip vortex interacted with the tail of the F-104N in an adverse manner, reducing its longitudinal stability. Walker's F-104 flipped over and collided directly with the XB-70, taking out the tail of the larger aircraft and instantly destroying the F-104N. Upon seeing the tragic collision, Col. Joe Cotton who was test director and riding in a T-38 chase plane, immediately called out "Midair! Midair! Midair!" The XB-70 did not immediately spin out of control, but maintained straight and level flight for another 16 seconds before slowly turning to the right and entering a spiral. Cotton then called out over the radio, "The B-70 is turning over on its side and starting to spin. I see no chute. Bail out! Bail out!" In the precious few seconds after the collision, White was able to safely eject from the XB-70, but unfortunately Cross was unable to initiate the eject sequence before g-forces built up and prevented the ejection mechanism from working correctly. The XB-70 ultimately entered an uncontrollable spin and crashed, claiming the life of Cross in addition to Walker.

This midair tragedy starkly illustrates the importance of careful flight planning, thoughtful review of the test objectives, and strict adherence to the agreed-upon flight plan. Any deviations from the flight plan that have not been thought through in advance must be avoided at all costs. This is the only acceptable way to mitigate the risks associated with flight test programs.

6.4 Individual Flight Planning

The fundamental starting point for planning a particular flight test is to have a clear and unambiguous objective defined for the test. For example, the objective could be to determine the stall speed of an aircraft (see Chapter 12) or to calibrate the airspeed indicator (see Chapter 8). Then, flowing directly from the test objective, a specific flight profile can be defined. The specifics of the required flight maneuvers can be determined from detailed knowledge of the theory for a particular flight test (see the individual chapter in this text for a specific flight test in order to develop this understanding). The required conditions for each maneuver (time, airspeed, and approximate throttle setting) must be defined, which enables estimation of fuel burn for each maneuver. The total fuel required should be summed and compared with the fuel capacity of the aircraft, allowing for sufficient reserve. (Note that most flight tests discussed in this book can be completed with 1–2 hours of flight time, which is well within the limits of typical general aviation aircraft.) Flight profiles should be planned to save on fuel consumption and test time by taking advantage of synergies wherever possible. For example, flight testing for climb performance and glide distance can be effectively combined, taking advantage of the fact that the aircraft must climb back up after conducting a glide test. Thus, data are acquired on both the climbs and glides, making better use of the flight time.

6.4.1 Flight Area and Airspace

The test area for a particular flight test must be clearly defined at the outset, which can be done by consulting appropriate aviation charts available from the relevant civil aviation authority (*e.g.*, see Figure 6.4). In the United States, airspace classifications defined by the Federal Aviation Administration (FAA) are A, B, C, D, E, and G. Flight in class A airspace (extending between 18,000 ft and approximately 60,000 ft MSL) is limited to aircraft on an instrument flight plan. Classes B, C, and D are controlled airspaces associated with airports, with varying restrictions for each (ranging from highly-controlled class B around large commercial airports, to relatively lightly controlled class D around smaller airports with control towers). Class E is considered controlled airspace (with radar coverage) and is the most pervasive type of airspace – no specific approval is needed from air traffic

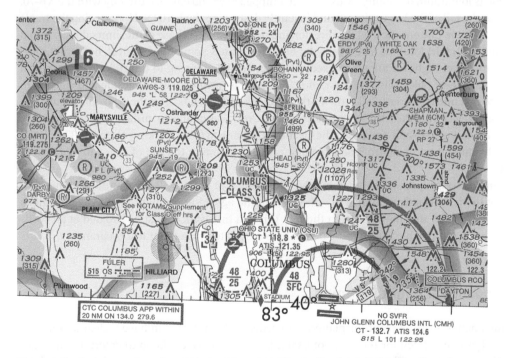

Figure 6.4 Sample portion of an FAA chart for navigation under visual flight rules (for reference only; not to be used for navigation). Source: Federal Aviation Administration.

control to fly in class E. Otherwise, all remaining airspace is class G, which is uncontrolled airspace with the fewest restrictions and no radar coverage is available.

The flight test area should generally be outside controlled airspace associated with airports (class B, C, or D), such that flight can be conducted at will without concerns of complying with air traffic control instructions. The immediate vicinity of uncontrolled airports should also be avoided in order to deconflict with traffic operating in and out of airfields. (Note that there are many uncontrolled airports throughout the United States that lie under class E or G airspace.) Furthermore, the flight area must be away from areas where general aviation traffic tends to congregate – this is for safety purposes in order to minimize the risk of mid-air collision, since the flight maneuvers often involve relatively sudden changes in altitude, airspeed, and heading, making the flight path of the test aircraft unpredictable for surrounding aircraft. Also, it is best that the flight profile be conducted over a relatively unpopulated area in order to minimize noise and disruption to people on the ground.

Aviation charts such as the segment of a sectional chart shown in Figure 6.4 may be consulted for determining the best flight area to use. The class C airspace around the John Glenn Columbus International airport (CMH) is denoted by two concentric magenta rings – the inner ring with a radius of 5 nautical miles extends from the surface up to an altitude of 4800 ft above mean sea level (MSL), while the other ring (10 nm radius) extends from a base of 2500 ft MSL up to 4800 ft MSL. (Note: these altitude limits are noted in magenta text within each ring, with the upper altitude listed in hundreds of feet and placed over the floor, also indicated in hundreds of feet – *e.g.*, "48" over a horizontal line, over "SFC.") The class D airspace around the Ohio State University airport (OSU) is denoted by the dashed blue line with a 4 nautical mile radius and extends from the surface up to 3400 ft MSL (indicated by "34" in a blue box). (Note that not all class D airspace has the same lateral dimensions; the radius is typically 4–5 nm, as required by local conditions.) Uncontrolled airports such as Marysville (MRT) or Delaware (DLZ) are indicated by small magenta circles with an unfilled rectangle to indicate the length and orientation of the runway. The large, lightly shaded magenta circles around each smaller airport indicate class E airspace that extends down to an altitude of 700 ft above ground level (AGL); otherwise, class E extends down to only 1200 ft AGL. Yellow-filled regions on the chart indicate populated areas, which should generally be avoided for flight testing. The light blue lines crisscrossing the map are established airways for navigation of cross-country flights and also should be avoided. More details on aviation charts, and different types of airspaces, are available from the FAA (2016).

6.4.2   Weather and NOTAMs

Immediately prior to flight, the pilot must perform a pre-flight briefing to collect all information necessary for ensuring a safe flight. This includes weather reports and forecasts for the flight area, details on runway length at all intended airports, fuel requirements, notices to airmen (NOTAMs), and temporary flight restrictions (TFRs). Each of these critical items is discussed as follows.

Weather is one of the most critical factors governing the go/no-go decision for a given flight. For safety of flight and for quality flight test data, it is important to avoid hazards posed by severe weather such as thunderstorms, icing, and strong gusting winds. Generally speaking, any convective activity should be at least 20 mi away, aircraft without icing protection systems must stay out of visible moisture or rain when the temperature at altitude is below zero, the component of surface winds across the runway must be below the crosswind limit of the aircraft (typically around 20 knots), and regions of low-level strong wind shear should be avoided. Visibility also matters – to ensure safe separation with other aircraft in flight, pilots must be able to see far enough to detect another aircraft that maybe on a converging course. Visual meteorological conditions (VMC) require lateral visibility of at least 3 statute miles within Classes B, C, D, and E airspace (below 10,000 ft). Beyond the visibility limits, pilots flying under visual flight rules (VFR) must also maintain a safe distance away from the clouds, to minimize the possibility of collision with an aircraft flying out of the clouds on an instrument flight. This safe separation is 1000 ft above, 500 ft below, and 2000 ft lateral distance from clouds within Classes C, D, and E airspace (below 10,000 ft MSL).

Pilots and flight test engineers can find weather information from a variety of sources; one of the most helpful tools is the National Weather Service's Aviation Weather Center (NOAA 2019). Current weather conditions are found by checking the meteorological aerodrome reports (METARs), where conditions are provided in a cryptic format such as:

```
KOSU 062053Z 22010KT 10SM CLR 30/19 A3002
```

Decoding these cryptic abbreviations, this weather report is for the Ohio State University airport (KOSU) at 20 : 53 UTC on the sixth day of the month (062053Z). The winds are coming from a magnetic heading of 220° at a speed of 10 knots (22010KT), the visibility is 10 statute miles (10SM), the skies are clear (CLR), the temperature is 30 °C and the dewpoint is 19 °C (30/19), and the local barometric pressure reading is 30.02 inHg (A3002). Thankfully, this information can also be provided in an interpreted or translated format, making the text a bit more comprehensible.

Similarly, terminal area forecasts (TAFs) provide an outlook of the weather forecast, generally over a period of 24 hours (or, in some cases, 36 hours). These forecast reports are structured using a similar syntax to the METAR, but provide a forecast of conditions over defined periods of time. The TAF is a generally reliable short-range forecast that can inform flight planning. When consulting the METARs and TAFs, the most significant hazards to flight are clouds or visibility, thunderstorms (also known as convective activity), icing, and turbulence.

Broad-area forecasts of significant meteorological conditions for aviation are published as AIRMETS (Airmen's Meteorological Information) and SIGMETS (Significant Meteorological Information), with a suffix letter that indicates the type of hazard. The Sierra ("S") suffix is related to poor visibility, the Tango ("T") suffix is related to turbulence, and the Zulu ("Z") suffix is related to icing. For example, an AIRMET Sierra is a report of a defined region of poor visibility and mountain obscuration, with ceilings <1000 ft and visibility <3 mi over a large area. AIRMET Tango indicates a region of light to moderate turbulence, or sustained surface winds >30 knots. AIRMET Zulu indicates a region with anticipated moisture and temperature conditions that have the potential for ice accretion.

Other critical information needed for flight includes the takeoff and landing runway length of the aircraft (documented in the pilot's operating handbook, or POH) relative to the available runway length, as well as an estimate of the required fuel for the flight (again available from fuel burn numbers available in the POH) relative to fuel available on board the aircraft. The flight test team must also be aware of any published hazards to flight, which may include radio towers or other ground obstructions that are unlit, closed runways or taxiways, radio communications or navigation aids that are out of service, etc. This information is published as NOTAMS, or notices to airmen. Finally, the flight test team must be aware of TFRs, which are essentially no-fly zones associated with major sporting events or VIP visits (*e.g.*, the President of the United States).

6.4.3 Weight and Balance

Pre-flight planning must also include a careful analysis of the weight and balance characteristics of the aircraft. Through the design process, there are clearly defined limits on the maximum takeoff weight of an aircraft, as well as forward and aft limits on the aircraft's center of gravity (CG). It is a simple matter to calculate the takeoff weight of an aircraft – this involves starting with the

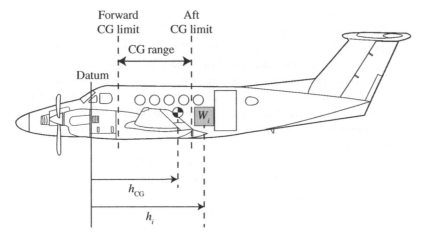

Figure 6.5 Aircraft center of gravity limits.

basic empty weight of the aircraft (defined as including the baseline weight of the airframe, full engine oil, and unusable fuel) and adding the weights of the pilots and passengers, baggage, payload instrumentation, and fuel for the flight to find the total takeoff weight of the aircraft. This weight must be less than the designed maximum takeoff weight of the aircraft.

The balance of the aircraft (*i.e.*, determination of aircraft center of gravity) requires a little more analysis, but is nothing more than summing the moments about some reference location. For the purposes of this discussion, a notional aircraft is shown in Figure 6.5, where the CG location is a distance h_{CG} from a datum plane, and forward and aft limits of the net CG location are shown. As the aircraft is loaded with fuel and passengers, the CG location will naturally shift, but must be maintained at a location between the forward and aft limits.

The selection of the reference datum plane is somewhat arbitrary; the aircraft manufacturer determines this location and provides baseline moment data for the aircraft about this location. The reference location is often defined as an easily identifiable point that will not change with any subsequent aircraft modifications, such as the engine firewall (as shown in Figure 6.5) or a point in space in front of the propeller spinner.

Starting with the baseline moment for the aircraft about this reference point, the pilot will add moments for each additional weight added to the aircraft. Referring to Figure 6.5, the increment in moment is found by multiplying an additional weight (W_i) by the moment arm (h_i) from the reference datum, such that $M_i = h_i W_i$. The pilot's operating handbook publishes moment arm data for various key locations in the aircraft such as the pilot and co-pilot seats, rear passenger seats, baggage compartment, and fuel storage location. Armed with the weight and location of pilots, passengers, equipment, and fuel, the additional moments for each component can be calculated and summed with the baseline moment. The total moment is simply the sum of all contributing moments, including the net moment of the empty aircraft, fuel, pilot, co-pilot, passengers, and baggage: $\sum_i M_i = \sum_i h_i W_i$.

Finally, the total moment about the center of gravity is divided by the total takeoff weight to determine a net moment arm,

$$h_{CG} = \frac{\sum_i M_i}{\sum_i W_i}. \tag{6.1}$$

This moment arm is the distance between the aircraft center of gravity and the reference location. CG limits published in the POH will include minimum and maximum distances for this CG location. The pilot must ensure that the CG location remains within these limits throughout the entire flight, including the effects of fuel burn and any mid-flight configuration changes.

These weight and balance characteristics may be best illustrated by an actual example. Consider the weight and balance of a light general aviation aircraft similar to a Cirrus SR20. (The following values are not to be used for actual flight planning! These numbers are fictional and are included here for illustrative purposes only. For flight planning, consult the pilot's operating handbook for the specific aircraft to be flown, which will include details on the CG moment arm distances for various loading locations.) This notional aircraft has properties as shown in Table 6.2, with an allowable CG envelope shown in Figure 6.6.

The aircraft has a basic empty weight of 2100 lb with a moment arm of 139.5 in. from the reference datum plane (arbitrarily set at a location 100 in. forward of the engine firewall in this case).

Table 6.2 Sample weight and balance calculations for a typical general aviation aircraft.

Description	Weight (lb)	Location (in.)	Moment (in.-lb)
Basic empty weight	2,100	139.5	292,950
Pilot	175	143.0	25,025
Front seat passenger	180	143.0	25,740
Rear passenger	160	185.3	29,648
Fuel (50 gal at 6 lb/gal)	300	153.5	46,050
Takeoff weight	2,915	143.9	419,413
Fuel burn (40 gal)	−240	153.5	−36,840
Landing weight	2,675	143.0	382,573

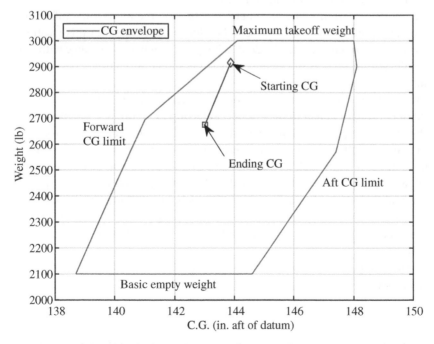

Figure 6.6 Example of a CG envelope, showing the forward and aft limits of the envelope, as well as the CG movement during flight due to fuel burn.

This produces a basic empty weight moment of 292,950 in.-lb. Now, let us consider the pilot, who weighs 175 lb and sits at a moment arm of 143.0 in. for a moment of 25,025 in.-lb. For passengers, we have one front-seat passenger (180 lb at a location 143 in. from the datum) and a rear seat passenger (160 lb at a location 185.3 in. aft of the datum), adding moments of 25,470 and 29,648 in.-lb, respectively. Now, if the aircraft is loaded with a full load of fuel (50 gal at 6 lb/gal), we add 300 lb at a moment arm of 153.5 in. for an additional moment of 46,050 in.-lb.

With full fuel, pilot, and passengers on board, we get the takeoff weight and moment of the aircraft by simply summing the values in each category – this condition has a weight of 2915 lb and a net moment of 419,413 in.-lb. Then, dividing the net moment by the total weight (Eq. (6.1)), we find the takeoff CG location of 143.9 in. Meanwhile, the limits of this aircraft are a maximum takeoff weight of 3000 lb, with minimum and maximum CG locations shown in Figure 6.6. Since the takeoff weight is less than the maximum takeoff weight, and the takeoff CG is between the limits shown in Figure 6.6, the loaded aircraft as described here is within the operating limits. For each flight, a pilot must compute these weight and balance characteristics and carry this documentation on board.

The above discussion has not accounted for fuel burn, but this will clearly have an impact on both the weight and balance of the aircraft in flight. The amount of fuel burn depends on the throttle setting throughout flight and the fuel consumption characteristics of the engine. Typical general aviation aircraft have a fuel burn rate on the order of 10 to 20 gallons per hour. Since the fuel CG is positioned at a moment arm of 153.5 in. (well aft of the aft CG limit of the aircraft), fuel burn during flight will progressively move the CG location forward: for every 10 gal of fuel burned in this configuration, the CG will move forward by approximately two-tenths of an inch. Thus, it is important to also check the CG location for the condition when all of the fuel has been burned, to verify that the aircraft CG will not go out of limits during the flight. Figure 6.6 shows a visual description of the CG limits, along with the starting and ending CG locations for this notional example. If 40 gal are burned in flight (240 lb removed from a location 153.5 in. from the datum plane), then 36,840 in.-lb are removed from the net moment. This shifts the CG location from a takeoff condition of 143.9 in. to a landing condition of 143.0 in..

6.4.4 Airplane Pre-Flight

Immediately prior to flight, careful pilots will take time to do a thorough pre-flight inspection of the aircraft (sometimes referred to as a "walk-around"). During this walk-around, the pilot is gathering information about the condition of the aircraft, looking for any anomalies that may affect the safety of flight or inform how the flight is conducted. Some aspects are absolutely critical, while others may be more benign or advisory in nature. In the initial stage of a walk-around, it is most helpful to get a view of the big picture, assessing the general condition of the aircraft. Are there any obvious structural deficiencies in the aircraft, such as a flat tire, bent landing gear, or creases in the fuselage skin? What is the level of oil in the engine, and how long has it been since an oil change? How much fuel is in the tanks, and is it the right kind? (The gas commonly used in general aviation aircraft in the United States is 100-octane leaded fuel, known as 100LL or Avgas. This gas formulation is intentionally tinted blue to make it easy to differentiate from other gas types.)

Following a quick overview of the aircraft, the pilot should conduct an orderly and detailed inspection of the aircraft, generally following the same order each time. This allows the pilot to form a mental checklist during the walk-around inspection, which is then backed up by the checklist published in the POH. Common items that a pilot will inspect are provided in the following representative list:

- condition of the freestream static pressure port(s) on the side of the aircraft (are they free of obstructions or clogs?)
- control surfaces move freely and in the correct direction relative to stick position
- attachment hardware for the control surfaces is secure
- radio antennas are in place and unbroken
- all screws and rivets on the fuselage, engine nacelle, and propeller spinner are secure
- there is no excessive oil leakage on the fuselage or fuel spillage on the ground
- fuel vents on the wings are unclogged (vents are critical for allowing air into the tanks as the fuel level is drawn down)
- the stall warning port is unclogged (the stall warning horn is an audio device connected to the leading edge of the wing, which senses imminent stall by the local pressure on the wing and automatically sounds a warning to the pilot)
- pitot probe is unclogged, with no visible damage
- proper operation of pitot heat
- all lights are operational (strobe, navigation, taxi, and landing lights)
- all necessary paperwork is on board (airworthiness certificate and aircraft registration)
- engine compartment is free of debris and air inlets provide unobstructed flow for cooling
- engine cowling is secure
- no significant nicks or dings on the propeller (feel it!), which is especially important for composite material propellers
- all fasteners on the propeller spinner cone are secure (we do not want a bolt to wiggle loose and enter the engine!)
- the alternator belt is tight
- all static wicks are present and undamaged (static wicks are typically found on newer, composite material aircraft)
- tires are properly inflated and in good condition (no significant bald spots)
- wheel fairings are securely attached
- flaps are securely attached and have no obstructions along retraction path
- cables for control surfaces freely move, have no excessive slack, and run freely along pulleys or other connecting mechanisms
- sufficient fuel of the correct type is on board (verified by visual inspection)
- no water or particulates are in the fuel (verified by draining fuel from each sump location)
- fuel filler caps are secure
- sufficient oil is present in the engine, as verified on the dip stick
- cowl plugs and pitot tube covers are removed (items typically with a red "remove before flight" ribbon attached)
- tie-down ropes and wheel chocks removed and stowed

There are many other aspects of a pre-flight inspection that a pilot may wish to consider – this is just a brief list of some of the most significant or most common items that a pilot will address. The overall objective of a preflight inspection is for a pilot to obtain as much information as possible about the condition of the aircraft, in order to make an informed go/no-go decision, or to be aware of any operating limitations that may affect flight.

6.5 Conclusion

This chapter has provided an overview of the most significant elements of planning flight tests in industry and government programs. Risk management is a critical aspect of these programs, where the hazards associated with flight testing must be mitigated in order to maintain a reasonably safe program. We also provide here an introduction to certain elements of flight planning for a general aviation aircraft – including obtaining weather briefings, weight and balance calculations, and pre-flight inspections – to help orient the non-pilot student to the world of aviation. Now, we will turn our attention to a range of performance, stability, and control flight test topics. Each of the following chapters will address one type of flight test, providing an overview of the theory, piloting techniques, and data analysis methods for each test.

Nomenclature

h_{CG} distance from the CG to the datum
h_i moment arm between an incremental weight and the datum
M_i moment produced by an incremental weight
W_i incremental weight included in weight and balance calculation

Acronyms and Abbreviations

AGL   above ground level
AIRMET   airmen's meteorological information
CG   center of gravity
FAA   Federal Aviation Administration
METAR   meteorological aerodrome report
MSL   mean sea level
NOTAM   notice to airmen
POH   pilot's operating handbook
SFC   surface
SIGMET   significant meteorological information
TAF   terminal area forecast
TFR   temporary flight restriction
VFR   visual flight rules
VMC   visual meteorological conditions

References

Appleford, J.K. (1995). Safety aspects. In: *Introduction to Flight Test Engineering*, AGARDograph 300 Flight Test Techniques Series, vol. 14 (ed. F.N. Stoliker), 10-1–10-9. Neuilly-Sur-Seine Cedex, France: Research and Technology Organization, North Atlantic Treaty Organization https://apps.dtic.mil/sti/pdfs/ADA444990.pdf.

Corda, S. (2017). *Introduction to Aerospace Engineering with a Flight Test Perspective*. Chichester, West Sussex, UK: John Wiley & Sons Ltd.

Department of Defense. (2012). *DoD Standard Practice: System Safety*. MIL-STD-882E (11 May 2012).

Federal Aviation Administration. (2012a). *Aircraft Certification Service Flight Test Risk Management Program.* FAA Order 4040.26B (31 January 2012).

Federal Aviation Administration. (2012b). *FAA Safety Risk Management Policy.* FAA Order 8040.4A (30 April 2012).

Federal Aviation Administration (2016). Chapter 15: Airspace. In: *Pilots Handbook of Aeronautical Knowledge.* FAA-H-8083-25B (ed. J.S. Duncan). Flight Standards Service, Federal Aviation Administration, Department of Transportation.

Flight Test Safety Committee. (2020). Flight Test Safety http://www.flighttestsafety.org/ (accessed 8 February 2020).

Harney, R.J. (1995). Preparation of the flight test plan. In: *Introduction to Flight Test Engineering,* AGARDograph 300 Flight Test Techniques Series, vol. 14, 8-1–8-10.

International Civil Aviation Organization (2018). *ICAO Safety Management Manual,* 4e, Document 9859-AN/474. Montreal, Quebec, Canada: International Civil Aviation Organization.

Kimberlin, R.D. (2003). *Flight Testing of Fixed-Wing Aircraft,* 4–6. Reston, VA: American Institute of Aeronautics and Astronautics, Chapter 1.

Maher, R. (2019). *Flight Test Safety Database.* National Aeronautics and Space Administration, http://ftsdb.grc.nasa.gov/ (accessed 8 February 2020).

Merlin, P. W., Bendrick, G. A., and Holland, D. A., 2011, *Breaking the Mishap Chain: Human Factors Lessons Learned from Aerospace Accidents and Incidents in Research, Flight Test, and Development,* NASA Aeronautics Book Series, NASA SP-2011-594.

Mondt, M.J. II (2014). *The Tao of Flight Test: Principles to Live By.* Boone, IA: J. I. Lord.

NASA. (2010). NASA Risk-Informed Decision Making Handbook, NASA/SP-2010-576. http://hdl.handle.net/2060/20100021361.

NASA. (2011). NASA Risk Management Handbook, NASA/SP-2011-3422. http://hdl.handle.net/2060/20120000033.

NOAA. (2019). Aviation Weather Center. http://aviationweather.gov.

Stoliker, F., Hoey, B., and Armstrong, J. (1996). *Flight Testing at Edwards: Flight Test Engineers' Stories 1946–1975.* Lancaster, CA: Flight Test Historical Foundation.

Ward, D., Strganac, T., and Niewoehner, R. (2006). *Introduction to Flight Test Engineering,* 3e, vol. 1, 7–12. Dubuque, IA: Kendall/Hunt, Chapter 1.

7

Drag Polar Measurement in Level Flight

In this chapter, we will discuss the level flight performance and flight testing methods to determine the drag polar of an aircraft, which is comprised of the parasitic drag coefficient and the Oswald efficiency factor. Determination of the drag polar is a critical starting point in aircraft performance flight testing, since so many other performance parameters depend on the drag characteristics. Climb performance, glide performance, takeoff and landing distance, turning flight, aircraft range, endurance, maximum speed, and many other parameters depend directly on the drag polar.

Our flight testing methods focus on measuring the engine power and airspeed over a range of level flight conditions, since the power required to sustain level flight is equal to the product of aircraft drag and airspeed. First, based on the relation between the power required for level flight and velocity, we introduce the generalized power (PIW) and the standard airspeed (VIW) to normalize the effects of the air density change and the weight change in flight. Then, the PIW–VIW method is developed to analyze power and speed data obtained at different altitudes and weights. Estimation of the engine shaft power (brake horse power) is described based on theory and engine performance charts. Next, theory and estimation methods for propeller performance are discussed. Our chapter concludes with flight testing procedures, data acquisition methods, and a flight test example.

7.1 Theory

7.1.1 Drag Polar and Power Required for Level Flight

The level flight performance of aircraft is mainly determined by the balance between the thrust and drag while the weight is balanced by the lift. As shown in Figure 7.1, the force balance in level flight is described by

$$L = W - T \sin \alpha_T, \tag{7.1}$$

and

$$D = T \cos \alpha_T, \tag{7.2}$$

where L is the lift, D is the drag, W is the weight, T is the thrust, and α_T is the thrust angle relative to the flight direction. The drag is decomposed into the parasite drag and the induced drag, i.e., $D = D_0 + D_i$, where D_0 is the parasite drag that is mainly contributed by the skin-friction drag and the pressure drag associated with flow separation, and D_i is the induced drag associated with the downwash velocity induced by the wake vortices. According to lifting-line theory, the induced drag

Introduction to Flight Testing, First Edition. James W. Gregory and Tianshu Liu.
© 2021 John Wiley & Sons Ltd. Published 2021 by John Wiley & Sons Ltd.
Companion website: https://www.wiley.com/go/flighttesting

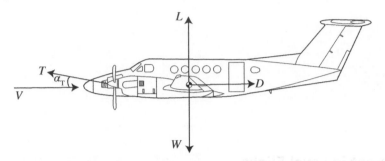

Figure 7.1 Force diagram for level flight.

coefficient (C_{D_i}) is proportional to the square of the lift coefficient (C_L), giving us the drag polar in coefficient form as

$$C_D = C_{D_0} + C_{D_i} = C_{D_0} + \frac{C_L^2}{\pi\,\mathrm{AR}\,e}, \tag{7.3}$$

where e is the Oswald efficiency factor and AR is the aspect ratio of the wing. The parasite drag and induced drag are also referred to as the zero-lift drag and the drag due to the lift, respectively. When we assume that Eq. (7.3) adequately represents the drag characteristics, then the parasite drag coefficient (C_{D_0}) and Oswald efficiency factor (e) define the drag polar for an aircraft.

Aerodynamic drag can be very difficult to predict with any accuracy, since non-linear viscous phenomena and complex flow interactions dominate aircraft drag. Viscous drag, separation, and three-dimensional flows are all first-order contributors to drag, but substantial computational processing power and extensive experimental validations are needed in order to make accurate predictions. However, in the process of aerodynamic design, there are some lower-order estimation techniques, primarily based on historical data, which may be used to predict the drag polar. These approaches are briefly summarized here, with further details available from McCormick (1995). Other references for aircraft drag polar estimation include Roskam (1971, 2008), Smetana et al. (1975), Gur et al. (2010), and Sloof (1986).

Estimation of the parasitic drag coefficient can be done through a method termed the equivalent flat plate area technique. This method involves estimation of an aerodynamic cleanliness factor, C_F, based on comparisons with historical data, such as what is depicted in Table 7.1 (McCormick 1995). (Other data sources include Loftin 1980, 1985; Sun et al. 2018.) The aerodynamic cleanliness factor is estimated by comparing a given aircraft configuration with the available data for similar configurations, and making a judgement call on the most appropriate value of C_F to use. This aerodynamic cleanliness factor is then multiplied by the wetted area of the aircraft, S_{wet}, to find the equivalent flat plate area, f:

$$f = C_F S_{\text{wet}} = C_{D_0} S. \tag{7.4}$$

(Note that the wetted area is all of the exposed surface area of all parts of the aircraft, including the surface areas of the wing, fuselage, empennage, etc. This can be readily calculated using computer-aided design packages or estimated via simple geometric approximations.) The idea of the equivalent flat plate area is that it is a measure of the total surface area of a flat plate, with a drag coefficient of 1.0, which would produce the same amount of drag as the aircraft. Thus, based on this notional equivalence, we can find the parasitic drag coefficient referenced to the wing planform area, S, as shown in Eq. (7.4).

A second method for estimating the parasitic drag coefficient is the component buildup method. This approach decomposes the overall drag of the aircraft into contributions from individual components (*e.g.*, wing, fuselage, etc.), which are individually estimated and summed up. The drag

Table 7.1 Overall skin friction coefficients for various aircraft.

C_F	Aircraft	Description
0.0100	Cessna 150	Single prop, high wing, fixed gear
0.0095	PA-28	Single prop, low wing, fixed gear
0.0070	B-17	Four props, World War II bomber
0.0067	PA-28R	Single prop, low wing, retractable gear
0.0066	C-47	Twin props, low wing, retractable gear
0.0060	P-40	Single prop, World War II fighter
0.0060	F-4C	Jet fighter, engines internal
0.0059	B-29	Four props, World War II bomber
0.0054	P-38	Twin props, twin-tail booms, World War II fighter
0.0050	Cessna 310	Twin props, low wing, retractable gear
0.0049	Beech V35	Single prop, low wing, retractable gear
0.0046	C-46	Twin props, low wing, retractable gear
0.0046	C-54	Four props, low wing, retractable gear
0.0042	Learjet 25	Twin jets, pod-mounted on fuselage, tip tanks
0.0044	CV 880	Four jets, pod-mounted under wing
0.0041	NT-33A	Training version of P-80 (see below)
0.0038	P-51F	Single prop, World War II fighter
0.0038	C-5A	Four jets, pod-mounted under wing, jumbo jet
0.0037	Jetstar	Four jets, pod-mounted on fuselage
0.0036	747	Four jets, pod-mounted under wing, jumbo jet
0.0033	P-80	Jet fighter, engines internal, tip tanks, low wing
0.0032	F-104	Jet fighter, engines internal, midwing
0.0031	A-7A	Jet fighter, engines internal, high wing

Source: McCormick (1995). © John Wiley & Sons.

contribution of each component is approximated based on historical data, which can be obtained from various sources (*e.g.*, Raymer 2018; Hoerner 1965; or Roskam 1971). This is potentially a slightly higher fidelity approach than the aerodynamic cleanliness factor method, since different components with differing drag characteristics may be combined into one design. However, the component buildup method neglects interaction effects between various components, which are known to be significant. For example, horseshoe vortices form at the wing-fuselage junction and represent a significant contribution to the overall drag. This type of drag is not captured by each component in isolation, but must be accounted for by estimating a factor to capture the interaction effects. Examples of the component drag buildup method are provided by Feagin and Morrison (1978), McCormick (1995), and Roskam (2008).

The Oswald efficiency factor, first defined by William B. Oswald (Oswald 1932), is essentially a correction factor for the induced drag term as a measure of the departure from the ideal elliptical lift distribution ($e = 1$). According to McCormick (1995), typical values of e are approximately 0.6 for low wing designs and 0.8 for high wing designs. The efficiency factor may be approximated by

$$e = (k\pi \text{AR} + 1 + \delta)^{-1}, \tag{7.5}$$

where k describes the sensitivity of the parasitic drag coefficient to changes in the square of lift coefficient,

$$k = \frac{dC_{D_0}}{dC_L^2},$$
(7.6)

which is usually small ($k \cong 0.009 - 0.012$). The induced drag factor, δ, may be estimated from the taper ratio, $\Lambda = c_{tip}/c_{root}$, and the wing aspect ratio in Figure 7.2. Other methods and resources for estimating the Oswald efficiency factor are provided by Hoerner (1965), Roskam (2008), Böhnke et al. (2011), and Niţă and Scholz (2012).

Beyond these basic approaches, there are numerous computational tools available that can provide low-order estimates of drag. One noteworthy resource is OpenVSP (Vehicle Sketch Pad), which is an open-source tool developed by NASA that combines a basic CAD tool that is custom tailored to aircraft layouts, with many built-in computational tools to predict the performance of the aircraft. OpenVSP has companion tools such as VSPAERO (a vortex lattice code) which uses flat plate correlations for estimating the skin friction drag and Trefftz-plane analysis for estimating the induced drag acting on the aircraft (for more details on the theory behind these approaches, see Drela 2014 or Katz and Plotkin 2001). OpenVSP and VSPAERO are described in detail by Fredericks et al. (2010), and are available from NASA (2020).

Now, let's explore how the drag polar impacts the power required for flight. In level flight, the thrust required to sustain flight is equal to the drag (assuming that α_T is small, leading to $T = D$ from Eq. (7.2)). Since drag is defined as

$$D = \frac{1}{2}\rho V^2 S C_D,$$
(7.7)

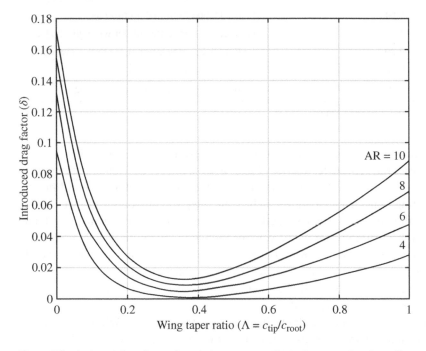

Figure 7.2 Induced drag factor for varying aspect ratio and taper ratio wings. Source: McCormick (1995). © John Wiley & Sons.

where V is the flight velocity, S is the wing area, and ρ is the air density, and power is equal to the product of thrust and airspeed,

$$P = TV, \tag{7.8}$$

we can write an expression for the power required for level flight,

$$P_R = \frac{1}{2}\rho V^3 S \left(C_{D_0} + \frac{C_L^2}{\pi \, \mathrm{AR} \, e} \right). \tag{7.9}$$

Since the lift equation is

$$L = \frac{1}{2}\rho V^2 S C_L, \tag{7.10}$$

and $L = W$ from Eq. (7.1), we can solve for lift coefficient and substitute into Eq. (7.9) to obtain

$$P_R = \left(\frac{1}{2} \rho S C_{D_0} \right) V^3 + \left(\frac{2W^2}{\rho \, S\pi \, \mathrm{AR} \, e} \right) V^{-1}. \tag{7.11}$$

The first term and second term are the parasite power and the induced power, respectively. Figure 7.3 shows the power-velocity curve for a typical light aircraft ($W = 3000$ lb, $S = 147$ ft^2, AR $= 10$, $C_{D_0} = 0.028$, and $e = 0.8$) in level flight at standard sea level conditions. The total power curve has a characteristic U-shape due to the decaying influence of the induced power and the increasing magnitude of the parasite power as the flight velocity increases.

Some performance parameters for level flight can be determined by C_{D0} and e (Anderson 1999). For example, the maximum lift-to-drag ratio is

$$(L/D)_{\mathrm{max}} = \left(\frac{\pi \, \mathrm{AR} \, e}{4C_{D_0}} \right)^{1/2}, \tag{7.12}$$

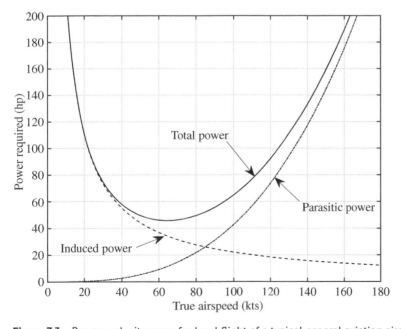

Figure 7.3 Power–velocity curve for level flight of a typical general aviation aircraft.

and the velocity for $(L/D)_{max}$ or the minimum thrust required is

$$V_{(L/D)_{max}} = \left(\frac{2}{\rho}(C_{D_0} \pi \, AR \, e)^{-1/2} \frac{W}{S} \right)^{1/2}. \tag{7.13}$$

Other flight performance parameters such as maximum speed, range, endurance, etc. are also related to C_{D_0} and e. The task of flight testing the level flight performance is to determine the parasite drag coefficient and the Oswald efficiency factor, which together define the drag polar for the aircraft.

7.1.2 The PIW–VIW Method

The main idea of the PIW–VIW method is to generalize Eq. (7.11) such that the effects of uncontrolled variables are factored out, and the equation is cast in a form for linear regression. In flight testing, the air density ρ in the Earth's atmosphere is a variable depending on the altitude, and the aircraft weight W decreases in time due to the fuel consumption. Figure 7.4 shows the effect of the altitude on the power-velocity curve for light aircraft, and Figure 7.5 shows the effect of the weight change on the power-velocity curve for light aircraft. Since both density and weight have a significant impact on the power required for flight, we must employ methods to factor out these effects. We'll develop methods to normalize the results by determining a generalized power parameter (PIW, or "power independent of weight") and a generalized velocity parameter (VIW, or "velocity independent of weight"). The objective of these normalized parameters is to present flight testing results for standardized density and weight conditions, which are typically sea level density and maximum takeoff weight, respectively.

The effects of air density are accounted for with the use of equivalent airspeed (V_e), which is related to the true airspeed (V) by

$$V_e = V \sqrt{\sigma}, \tag{7.14}$$

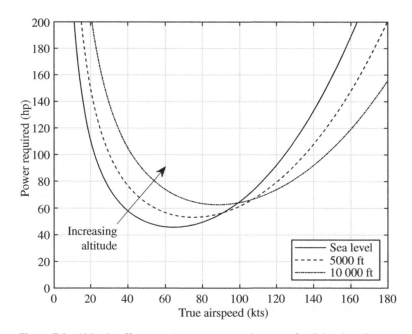

Figure 7.4 Altitude effects on the power–velocity curve for light aircraft.

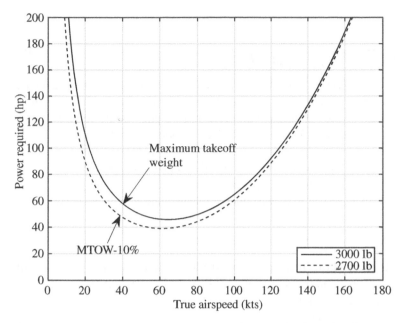

Figure 7.5 Weight effects on the power–velocity curve for light aircraft.

where

$$\sigma = \frac{\rho}{\rho_{\mathrm{SL}}} \tag{7.15}$$

is the density ratio (defined in Chapter 2) and ρ and ρ_{SL} are the air densities at altitude and sea level, respectively. Equivalent airspeed is directly proportional to dynamic pressure, and is related to the lift with an assumed density of ρ_{SL} in Eq. (7.10). (Further details about the definition of equivalent airspeed, true airspeed, and other airspeeds are provided in Chapter 8.)

Further, since the aircraft weight changes during flight, the velocity independent of weight is introduced, *i.e.*,

$$\mathrm{VIW} = \frac{V_e}{(W_T/W_S)^{1/2}}, \tag{7.16}$$

where W_T and W_S are the weight in the testing condition and at the standard condition, respectively. The standard weight is typically the maximum takeoff weight of the aircraft. The definition of VIW in Eq. (7.16) arises by setting Eq. (7.10) equal to Eq. (7.1), with $\alpha_T = 0$, and writing the expression twice – once for the test weight W_T with a corresponding V_e in the lift equation, and a second time for the standard weight W_S with a corresponding VIW in the lift equation. Then, forming a ratio of these expressions of the lift equation, we obtain the expression for VIW in Eq. (7.16).

Accordingly, the generalized power parameter is introduced as

$$\mathrm{PIW} = \frac{P_R \sqrt{\sigma}}{(W_T/W_S)^{3/2}}, \tag{7.17}$$

where PIW denotes the power independent of weight. This parameter is derived in a manner similar to that for VIW, where Eq. (7.9) is written twice (once with P_R, ρ, and V_i, and a second time with PIW, ρ_{SL}, and VIW), and ratioed. Equations (7.14–7.16) are substituted into this ratio and variables collected to find PIW as a function of P_R and standardized conditions (yielding Eq. (7.17)).

With these generalized variables, the power-velocity relation, Eq. (7.11), is rewritten as

$$\text{PIW} = \left(\frac{1}{2}\rho_{\text{SL}}SC_{D_0}\right)(\text{VIW})^3 + \left(\frac{2W_S^2}{\rho_{\text{SL}}\,S\pi\,\text{AR}\,e}\right)(\text{VIW})^{-1}. \tag{7.18}$$

Clearly, the advantage of Eq. (7.18) is that the PIW - VIW relation does not explicitly depend on the air density and testing weight at an altitude, but on ρ_{SL} and W_S which are known constants. Thus, data of the power and airspeed obtained at different altitudes and weights should collapse onto a single curve, irrespective of the altitude or fuel burned for that test point. By reducing the data via Eq. (7.18), we ensure that the aircraft is operating at a consistent lift coefficient, independent of the altitude or weight condition of the aircraft. This ensures that a given test point will correctly collapse to the right location on the drag polar in Eq. (7.3).

If we multiply each side of Eq. (7.18) by VIW, we obtain a linear relation between (PIW)(VIW) and $(\text{VIW})^4$, *i.e.*,

$$(\text{PIW})(\text{VIW}) = a\,(\text{VIW})^4 + b, \tag{7.19}$$

where the slope is $a = \rho_{\text{SL}}SC_{D_0}/2$ and the intercept is $b = 2W_S^2/(\rho_{\text{SL}}S\pi\,\text{AR}\,e)$. Linear regression of flight test data of the power required and the corresponding airspeed in level flight with Eq. (7.19) leads to determination of the drag polar. The parasitic drag coefficient and Oswald efficiency factor can be evaluated from the linear regression data by

$$C_{D_0} = \frac{2a}{\rho_{\text{SL}}S}, \tag{7.20}$$

and

$$e = \frac{2W_S^2}{b\pi\,\text{AR}\,\rho_{\text{SL}}S}. \tag{7.21}$$

Flight testing of a typical general aviation aircraft, with an internal combustion engine driving a propeller, centers on measurements of power required and airspeed to provide the necessary input to Eq. (7.19). Measurement of airspeed is straightforward in any aircraft, using the airspeed indicator (see Chapter 3). However, standard general aviation aircraft are not equipped to directly measure the power produced by the engine in flight. Instead, we must rely on documented performance characteristics of the engine and propeller (*e.g.*, through ground testing of the propulsion system), and measurements of key parameters in flight such as engine speed, manifold pressure, and fuel flow rate. We'll next turn our attention to the underlying theory of internal combustion engines and propeller aerodynamics, in order to have a basis for estimation of power required in flight.

7.1.3 Internal Combustion Engine Performance[1]

Most general aviation aircraft are driven by propellers powered by reciprocating engines. Internal combustion engines for general aviation aircraft are basically the same as those for ground vehicles, except that they are designed for high specific power output and reliability. Figure 7.6 shows a Teledyne Continental IO-360-ES engine, with Figure 7.7 providing a schematic drawing of the engine and its components. A brief discussion on power of internal combustion engines is given from a perspective of flight. Most reciprocating engines are based on a four-stroke cycle, as illustrated in Figure 7.8 (Ferguson and Kirkpatrick 2016). Each cylinder requires four strokes of its piston and

---

1 Additional details are available in an online supplement, "Basic Performance Prediction of Internal Combustion Engines."

Figure 7.6 Continental IO-360 engine. Source: Courtesy of Continental Aerospace technologies.

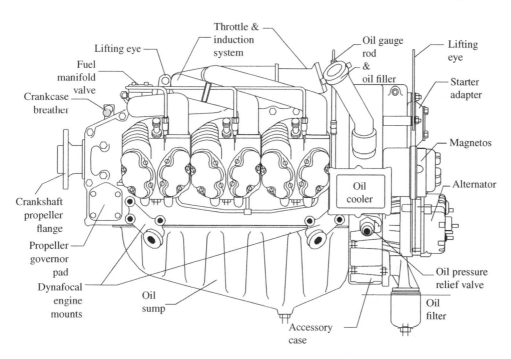

Figure 7.7 Side view of the 210-hp Continental Motors IO-360-ES. Source: Continental Motors Inc.

two revolutions of the crankshaft. Each of the cylinders of an aircraft engine has two spark plugs, each fired by an independent ignition system for redundancy. Each ignition system is based on an engine-driven magneto, which will continue firing the spark plugs as long as the engine is rotating (thus, no electrical system is required for sustaining engine operation!). The basic operation of a

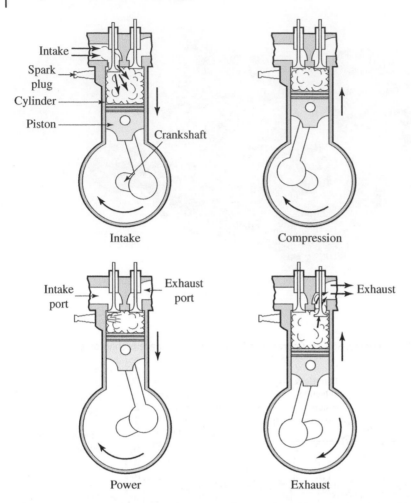

Figure 7.8 Four-stroke spark ignition cycle. Source: Ferguson and Kirkpatrick (2016). © John Wiley & Sons.

four-stroke engine is described as follows (see Heywood 2018; Ferguson and Kirkpatrick 2016; or FAA 2016 for further details).

First, in the intake stroke, the piston starts from its highest position (termed top dead-center, TDC) and moves down to its lowest position, bottom dead-center (BDC), by the end of the stroke. This piston movement increases the volume inside the cylinder. The inlet valve opens shortly before the beginning of the intake stroke, allowing a fresh charge of fuel-air mixture to enter the cylinder, and closes at the end of the stroke.

Second is the compression stroke, where the piston moves from BDC back up to TDC. During this stroke the piston does work on the contents of the cylinder, compressing the fuel-air mixture, which leads to an increase in the pressure of the mixture. Near the end of the compression stroke, the spark plug fires, leading to rapid combustion of the fuel-air mixture inside the cylinder.

The rapid energy release associated with combustion of the fuel-air mixture dramatically increases the pressure and temperature inside the cylinder, forcing the piston downward and imparting significant power to the crankshaft during the power stroke. The pressures inside the cylinder during the power stroke are much greater than during the compression stroke, leading to the work done on the piston in the power stroke being a factor of 5 greater than the work done by the piston in the compression stroke, and positive net work done by the engine (Heywood 2018).

Finally, the exhaust valve opens near the beginning of the exhaust stroke. The residual elevated pressure of the combustion products (relative to the exhaust manifold pressure) induces flow out of the cylinder. Simultaneously, the piston moves upward and sweeps the exhaust gases out. Shortly before the piston reaches TDC, the inlet valve opens as a precursor to the intake stroke. At the conclusion of the exhaust stroke, the exhaust valve closes, and the cycle repeats.

Estimation of the shaft power produced by an engine in flight can be done via several different approaches. One method involves installation of a torque meter on board the aircraft. Once a reliable in situ measurement of torque (τ) is made, the shaft power can be found from $P_{shaft} = \tau N$ (where N is engine speed, with units of s^{-1}). However, installation of a torque meter can be expensive and requires modification to the aircraft. This involves mounting the torque meter between the propeller and the shaft, which increases the distance between the propeller and the engine cowling (possibly changing the propeller efficiency characteristics). Furthermore, the presence of the torque meter may change the drag on the aircraft, and possibly the flow of cooling air into the engine cowling. Thus, the use of a torque meter involves tradeoffs that must be carefully considered in planning a flight test program.

Another method involves direct measurements of the fuel flow rate through the engine, which is directly related to shaft power by the specific fuel consumption. The shaft power available is then found through the definition of specific fuel consumption (sfc) and incorporating the measured fuel mass flow rate,

$$P_{shaft} = \frac{\dot{m}_f}{sfc}. \tag{7.22}$$

The difficulty is that specific fuel consumption varies with the mixture setting. However, Kimberlin (2003) provides a suggested procedure for accounting for this effect, where a precise measurement of exhaust gas temperature (EGT) on the engine is used to lean the mixture to the desired value.

Related to this method, some avionics displays provide a direct indication of percent power. This reading is based on calibrated sfc values and a measurement of fuel flow rate to provide a direct reading of power via Eq. (7.22). This estimate of power is often reported as a percentage of maximum installed power.

A fourth method involves the use of engine performance charts provided by the manufacturer (*e.g.*, Figure 7.9 which is reproduced from data for the IO-360-ES provided by Continental Motors 2011). The charts are composed of two portions: sea-level performance curves on the left side, and altitude performance curves on the right side. When working with this type of engine chart, measurements of the engine speed (in rpm), manifold pressure (in inches of mercury), pressure altitude, and outside air temperature are required. Use of the chart (Figure 7.9) proceeds along the following steps:

(1) Using manifold pressure and engine rpm, find the corresponding point on the altitude chart on the right side. Identify this point on the chart as point A.
(2) With the manifold pressure and engine rpm, find the appropriate point on sea level chart on the left side. Identify this point on the chart as point B.
(3) Read the value of engine power from point B, and mark that same power level on the y-axis (sea level) of the altitude chart on the right. Identify this point on the chart as point C.
(4) Draw a straight line connecting points C and A.
(5) Find the point where the line crosses the measured pressure altitude, and identify this point on the altitude chart as point D.
(6) Read the brake horsepower for point D, which is $P_{shaft, std}$

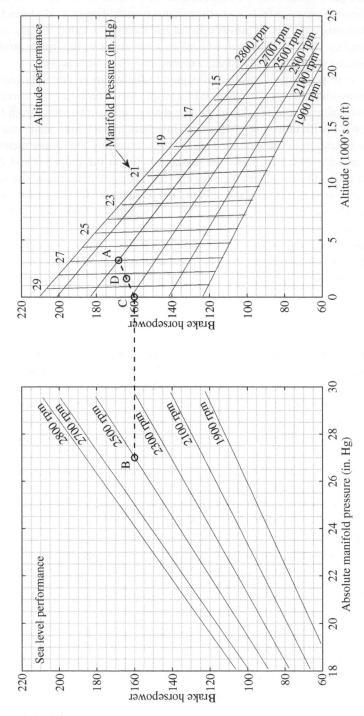

Figure 7.9 Engine performance charts for the Continental IO-360-ES. Source: Data from Continental Motors (2011).

(7) Adjust the brake horsepower for non-standard temperature by

$$P_{\text{shaft,test}} = P_{\text{shaft,std}} \sqrt{\frac{T_{\text{std}}}{T_{\text{test}}}}, \tag{7.23}$$

where T_{std} is the standard temperature at the given pressure altitude, and T_{test} is the actual temperature measured at that condition (both in absolute units).

The example illustrated in Figure 7.9 is for an engine speed of 2500 rpm and a manifold pressure of 27 inHg, which gives a sea level power of 160 hp (point C). Then, with a pressure altitude of 1620 ft, point D falls at 164.2 hp. At this altitude, the standard temperature is 512.89°R. But, with an actual temperature of 500°R, the power output is adjusted to 166.3 hp by Eq. (7.23).

7.1.4 Propeller Performance

For propeller-driven aircraft, the thrust for flight is provided by the combination of the reciprocating engine and propeller: the propeller transmits the power from the shaft output of the engine to the flow (recall from Eq. (7.8) that power is equal to the product of thrust and velocity). Not all of the power produced by the engine is actually imparted to the airflow by the propeller, since some of the shaft power is dissipated as aerodynamic drag acting on the propeller blades. We describe the relationship between engine shaft power – P_{shaft}, often also referred to as brake horsepower – and the power available for a flight condition by $P_A = \eta_{\text{prop}} P_{\text{shaft}}$, where η_{prop} is the propeller efficiency. If an aircraft is in steady, level flight then the power available is set equal to the power required such that

$$P_R = P_A = \eta_{\text{prop}} P_{\text{shaft}}. \tag{7.24}$$

Propeller efficiency will vary depending on the operating conditions for the aircraft, which dictate the local angle of attack that the blade experiences. When the airfoil at a local cross-section of the propeller blade is at a high angle of attack, the flow can be locally stalled and the efficiency drops dramatically. At the other extreme, if the local airfoil section is at a low angle of attack, the propeller will not be producing much lift but will still have substantial drag, so the efficiency is again low. Given the significance of angle of attack, we need to relate the blade pitch (mounting angle) to the flow angle of attack (relative to a local cross-section on the blade). We can do this by drawing a velocity triangle for a given spanwise location along the propeller blade, as shown in Figure 7.10.

Figure 7.10(a) depicts a front view of the propeller, looking perpendicular to the rotation plane of the propeller, where the freestream velocity of the aircraft is oriented into the page. Figure 7.10(b) shows a cross-section of the propeller blade at a specific spanwise location (r) along the blade. The rotational velocity, ωr, will vary significantly along the span of the blade, starting with low values near the hub, and increasing linearly with r out to the tip. The combination of the aircraft freestream velocity (V) and the local rotational velocity forms a velocity triangle, which defines the inflow angle φ between the local relative velocity vector, V_R, and the rotation plane of the propeller. But, the spinning propeller experiences an effective velocity, V_E, which includes the effects of an induced velocity, w, in addition to the relative velocity. The induced velocity, w, is analogous to the downwash experienced by a finite wing and is not known a priori. As illustrated in Figure 7.10(b), the downwash is comprised of an axial inflow component, w_a, and a tangential (swirling) component, w_t, in the plane of rotation. Vector addition of w and V_R results in the induced angle of attack, α_i. With the blade pitch angle, β, defined between the airfoil chord line and the rotational plane, the local angle of attack experienced by a blade section is $\alpha = \beta - \varphi - \alpha_i$.

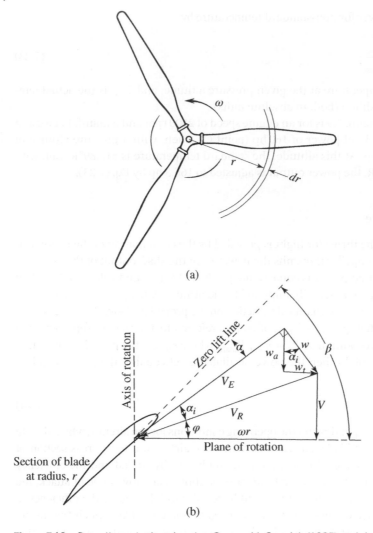

Figure 7.10 Propeller velocity triangles. Source: McCormick (1995). © John Wiley & Sons.

Since the rotational velocity changes with span, the local angle of attack (α) also varies across the span. In order to obtain optimum angle of attack at each spanwise location, the blade pitch angle (β) is twisted along the span, with high blade pitch at the root and low pitch at the tip. Typical twisting of a propeller blade is illustrated in Figure 7.11, which depicts a Hartzell 4-bladed propeller.

At this juncture, it's helpful to form a nondimensional parameter that describes the ratio of the freestream velocity and rotational speed of the propeller. The advance ratio,

$$J = \frac{V}{ND_{\text{prop}}}, \tag{7.25}$$

where N is revolutions per unit time and D_{prop} is the propeller diameter, describes how far forward the propeller moves for a given amount of rotation in the propeller plane. Since $\omega = 2\pi N$ and $R = D/2$, advance ratio can also be expressed as

$$J = \frac{\pi V}{\omega R}, \tag{7.26}$$

Figure 7.11 Side view of Hartzell propeller, illustrating blade twist along the span. Source: Photo courtesy of Hartzell Propeller Inc.

and can be related to the inflow angle at the tip by

$$\varphi_{\text{tip}} = \tan^{-1}(\lambda), \tag{7.27}$$

where $\lambda = J/\pi$ is the advance ratio coefficient. We'll also define nondimensional parameters for thrust (T) and power (P_{shaft}) as

$$C_T = \frac{T}{\rho N^2 D_{\text{prop}}^4}, \tag{7.28}$$

and

$$C_P = \frac{P_{\text{shaft}}}{\rho N^3 D_{\text{prop}}^5}, \tag{7.29}$$

where ρ is freestream density and N must have units of revolutions per second. (Note that there are various definitions of thrust and power coefficient in use; the definitions here are the most common for presenting thrust stand data. Be careful when consulting various data sources, to ensure which definitions of coefficients are being used!) With these definitions, it can be helpful to recognize that the product ND_{prop} is proportional to rotational velocity, and D_{prop}^2 is proportional to the swept area of the propeller disk. Also, since $\eta_{\text{prop}} = TV/P_{\text{shaft}}$, we can express the propeller efficiency as

$$\eta_{\text{prop}} = J\frac{C_T}{C_P}. \tag{7.30}$$

Now, bringing our discussion back to propeller efficiency, we can plot the efficiency factor as a function of advance ratio for a series of blade pitch angles, as shown in Figure 7.12(a) for a Hartzell 3-bladed propeller. The general trend of these curves is a linear increase in η_{prop} with J, a rolling off at a maximum efficiency, and then a rapid decrease in efficiency down to zero. Propeller efficiency tends to zero at $J = 0$, due to the definition of power ($P = TV$) and the lack of airspeed at low

(a)

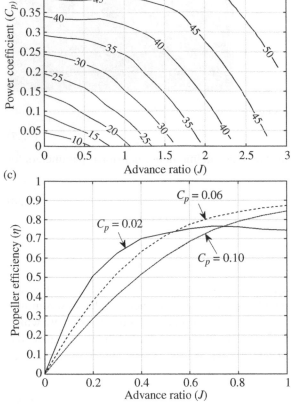

Figure 7.12 Propeller efficiency for a Hartzell 3-bladed, 74-in. diameter propeller at a tip Mach number of 0.5, including installation effects on a Cirrus SR20. (a) efficiency curves with varying blade pitch, (b) contours of constant blade pitch over varying power coefficient and advance ratio, and (c) efficiency curves with varying power coefficient. Source: Data from Hartzell Propeller.

(b)

(c)

advance ratio. Also, at low advance ratios, there will likely be large portions of separated flow on the propeller blade since low freestream velocity will lead to high local angle of attack (see the velocity triangle in Figure 7.10). At high advance ratio, on the other hand, the angle of attack at the blade section is decreasing to the point where the local angle becomes negative (in which case, the propeller is windmilling and adding drag to the aircraft!).

For the data shown in Figure 7.12(a), blade pitch (β) is the overall pitch of the propeller, set at the blade root. For a fixed-pitch propeller, the choice of that pitch angle cannot be adjusted in flight, resulting in off-optimal propeller efficiency over a wide range of advance ratios (flight speeds). This limitation led to the invention of the constant-speed propeller, which is found on many general aviation aircraft today. The constant-speed propeller automatically adjusts the blade pitch angle in order to maintain the desired operating condition and efficiency. Each curve shown in Figure 7.12(a) is a different blade pitch angle for this constant-speed propeller. Optimal efficiency ($\eta_{\text{prop}} = 0.85$ to 0.90) can be obtained over a much wider range of advance ratios using this technique.

Our focus in flight testing is to come up with an appropriate estimate of propeller efficiency at our desired flight test point. For our purposes, there are two main approaches that we can take to estimating the propeller efficiency, which we'll detail as follows. The first approach is to rely on empirical measurements or generalizations of data sets, whereas the second approach involves more accurate modeling but requires knowledge of the propeller blade geometry.

First, let's consider available measured data, such as what is shown in Figure 7.12 for an installed propeller-engine combination found on a Cirrus SR20. Unfortunately, Figure 7.12(a) is not very useful for us in practice, since we don't have a direct reading of propeller pitch angle (β) in the cockpit. Thus, we need to relate propeller efficiency to other parameters which we can measure – here, we'll express propeller efficiency as a function of advance ratio for several different power coefficients. In order to establish this relationship, Figure 7.12(b) relates power coefficient to pitch angle across a range of advance ratios. For a given contour of blade pitch angle, the power produced by the propeller reduces as advance ratio increases. Combining the data of Figure 7.12(a) and (b), we can plot propeller efficiency as a function of advance ratio for a range of power coefficients.

Figure 7.12(c) allows us to determine η_{prop} from measurements that we can make in flight. Advance ratio, Eq. (7.25), requires measurement of airspeed, propeller diameter, and engine speed. Power coefficient, Eq. (7.29), also requires measurements of engine shaft power (see Section 7.1.3) and local density. By interpolating between power coefficient curves in Figure 7.12(c), we can find η_{prop} for any desired flight condition. (Note that the data shown in Figure 7.12 is for a tip Mach number of 0.5. Higher tip Mach conditions would likely require separate data sets in order to capture compressibility effects.)

Beyond the data shown in Figure 7.12, propeller data can also be obtained from other sources. Hartman and Biermann (1938) provide efficiency data for several propellers based on the Clark-Y and R.A.F. 6 airfoils with 2, 3, or 4 blades; while Gilman (1953) and Lowry (1999) provide generalized propeller efficiency charts that may be extended to many types of aircraft. McCrink and Gregory (2017) provide data on small propellers used on unmanned aircraft, as measured in a wind tunnel.

The second approach, based on higher-fidelity analysis for prediction of blade performance, requires exact knowledge of the propeller blade geometry (i.e., airfoil section, chord, and pitch angle as a function of span, as well as the propeller diameter and number of blades). This theoretical approach is termed blade element momentum theory (BEMT), which segments the propeller blade into two-dimensional slices and uses 2D airfoil theory to predict and sum up the performance of each spanwise airfoil section. However, the local velocity experienced at a given blade section is not known a priori, due to the effects of inflow and swirl. These unknown velocity components (w_a and w_t shown in Figure 7.10) are found by coupling the 2D analysis with 1D momentum theory applied to the streamtube of flow through the propeller. The coupling of the 2D annular analysis with 1D streamtube analysis allows us to close the equations, calculate the inflow as a function of span, and determine the sectional forces and moments. These sectional force contributions are integrated across the span of the blade in order to determine the total thrust and torque acting on the propeller. Further details on the theory behind BEMT are provided by McCormick (1995, 1999), Leishman (2016), or McCrink and Gregory (2017). Other treatments of propeller theory are provided by Glauert (1935), Goldstein (1929), Theodorsen (1948), or Wald (2006).

In order for BEMT to provide reasonably accurate predictions of the propeller efficiency, a number of correction factors are needed. Notably, BEMT does not account for tip effects (Shen et al. 2005), which involves pressure relief at the blade tips (loading diminishing to zero). Similarly, compressibility effects are often important, especially with the high subsonic flows near the blade tips. Finally, since BEMT is applied to 2D annular sections of the propeller blades, the analysis presumes that there is no radial flow. However, this assumption can be invalidated by outward radial pumping of flow within the thick boundary layer or any separated regions. McCrink and Gregory (2017) document correction factors for each of these physical effects, showing much improved predictions of the propeller efficiency.

Let's now summarize our discussion of propeller efficiency. We've seen that the efficiency varies as a function of advance ratio, which depends on flight conditions. The process of finding propeller efficiency in flight testing work involves calculating advance ratio in Eq. (7.25) and power coefficient from Eq. (7.29), and reading the efficiency from Figure 7.12(c) or other data sources. Thus, the required input data are an estimate of engine power from the methods detailed in Section 7.1.3, propeller diameter, ambient pressure and temperature, engine speed (rpm), and true airspeed.

7.2 Flight Testing Procedures

We've now established sufficient theory to turn our attention to flight testing procedures required for determining the drag polar. Before we discuss the details of flight testing, it's helpful to provide an interim summary that recapitulates the key points that will be applied to flight tests. Our goal in this flight test is to determine the drag polar – specifically, to find C_{D_0} and e in Eq. (7.3). We will do this through the PIW–VIW method, which will take in-flight measurements of power and velocity, correct these measured values to standardized values, and plot them as $(PIW)(VIW)$ versus $(VIW)^4$ from Eq. (7.19). From this plot, we can apply a linear fit to find C_{D_0} from the slope (Eq. (7.20)) and e from the y-intercept (Eq. (7.21)).

Prior to flight, the aircraft weight must be recorded, including the operating empty weight, fuel weight, and pilot & passenger weights (see Chapter 6 for details). Also, the propeller diameter must be identified from the manufacturer data, or directly measured. Flight testing is performed such that a number of stabilized flight conditions are flown throughout the velocity range of the aircraft (from near stall up to maximum speed). The best practice is to spread the test points throughout the velocity range, with sufficient range to cover the nonlinear portion of the drag curve and the back side of the power curve. All of the stabilized flight conditions should be flown at the same pressure altitude, but the test can be repeated at other altitudes to ensure that the data collapses onto a single trend (assuming negligible changes in viscous effects at different Reynolds numbers). Before commencing data acquisition, the aircraft should be in trimmed flight for cruise conditions, with the mixture leaned to an appropriate setting. The test is begun at maximum velocity with full throttle; each successive test point is set by pulling the throttle back to reduce the manifold pressure in increments of approximately 1 inHg, and pitching the aircraft in order to maintain constant altitude. The aircraft should be allowed to stabilize in the new flight configuration before each data point is recorded. Data to record at each set point includes:

- Pressure altitude (read from the altimeter, set to a reference pressure of 29.92 inHg)
- Outside air temperature (if available, *e.g.*, from the primary flight display)

- Indicated airspeed (from the airspeed indicator)
- True airspeed (if available, *e.g.*, from the glass panel avionics)
- The following engine parameters (from an engine monitor, or the multifunction flight display):
  o Manifold pressure (inches of Hg)
  o Engine speed (rpm)
  o Percent power (%)
  o Fuel burned (gal)
  o Fuel flow rate (gal/hour)
  o Exhaust gas temperature ($°F$)

Data analysis proceeds along the following lines. First, the aircraft test weight must be determined for each test point. This can be found by subtracting off the weight of fuel burned from the takeoff weight. Following this, the ratio of test weight to standard weight, W_T/W_S, can be calculated. (Typically the maximum takeoff weight is used as the standard weight.)

Density ratio (Eq. (7.15)) is found from the ideal gas law and measurements of pressure altitude and outside air temperature. Equivalent airspeed can be found in one of two ways, based on available data. If a reading of outside air temperature is available, then the avionics likely reports a reading of true airspeed. Equivalent airspeed is found from true airspeed by Eq. (7.14) and the measured value of density ratio. However, if temperature and true airspeed are not available, then one may assume that indicated airspeed is equal to equivalent airspeed (see Chapter 8 for further details on airspeeds and calibration of the airspeed indicator). Once equivalent airspeed is known, the weight-corrected airspeed (VIW) is found from Eq. (7.16).

Engine shaft power is estimated from one of the methods discussed in Section 7.1.3 – typically, this could be a direct reading of percent power (relative to maximum installed power); or a measurement of engine speed, manifold pressure, pressure altitude, and outside air temperature with an engine performance chart (Figure 7.9). Propeller efficiency is found by a performance chart such as that shown in Figure 7.12. Measurements of engine speed, propeller diameter, and true airspeed are needed to find the advance ratio (the x-axis of Figure 7.12), and shaft power, density, engine speed, and propeller diameter are needed to find the power coefficient (Eq. (7.29)). Once an efficiency value is read from Figure 7.12(c), the power required for that flight condition is determined from Eq. (7.24). Finally, the PIW value is found from power required by Eq. (7.17).

Once PIW and VIW are found for the full range of velocity set points, (PIW)(VIW) versus (VIW)^4 is plotted for Eq. (7.19) and a linear fit is applied to the data. The slope of the best fit line provides the parasitic drag coefficient from Eq. (7.20), while the Oswald efficiency factor is found from the y-intercept and Eq. (7.21).

7.3 Flight Test Example: Cirrus SR20

The Cirrus SR20 (sixth generation) is a single engine monoplane that was used as a flight testing platform for this example. The aircraft is powered by a 74 in. constant speed Hartzell propeller, driven by a Lycoming IO-390-C3B6 4-cylinder horizontally-opposed piston engine. This engine is rated for 215 brake horsepower at 2700 rpm at sea level, and yields a cruise speed range of approximately 115–155 kts. The aircraft has a maximum takeoff weight of 3150 lb and a total fuel capacity of 58.5 gal. The wing span is 38.3 ft and wing area is 144.9 ft^2, giving an aspect ratio of 10.12. The Cirrus Perspective$^+$ primary flight display and multifunction flight display provide digital readouts of the aircraft and engine performance and operating information. Additionally, engine operation data such as cylinder head temperatures, exhaust gas temperatures, and fuel flow rates can be downloaded from the on-board recorder for analysis.

The aircraft was flown at a pressure altitude of 3000 ft with the autopilot engaged for altitude and heading hold. The aircraft was flown at a series of stabilized indicated airspeeds ranging from 81 to 150 kts. Pressure altitude, outside air temperature, indicated airspeed, true airspeed, percent power, fuel flow rate, fuel burned, engine rpm, exhaust gas temperatures, and manifold pressure were recorded at each condition. Once the parameters for each set point were recorded, the power setting was changed, and sufficient time elapsed to allow the aircraft to stabilize. This process was followed for each test point, while holding altitude constant.

Table 7.2 lists the basic pre-flight data needed for level flight testing. Relevant flight data extracted from the avionics system are shown in Table 7.3. Then, the derived quantities from the flight data are listed in Table 7.4, where the standard weight W_S was 3150 lb (maximum takeoff weight), and the test weight W_T was calculated based on the fuel consumption. True airspeed was found using the methods described in Chapter 8 (in fact, the same data set is used here), and VIW was calculated based on the true airspeed and the density ratio. (Note that a direct readout of true airspeed was available on the glass panel avionics display.) Shaft power was estimated from the percent power reading on the avionics and the rated power of the engine. The power required is $P_R = \eta_{\text{prop}} P_{\text{shaft}}$, with propeller efficiency found from the data in Figure 7.12, and PIW is given by Eq. (7.17). Figure 7.13 shows the linear relationship between (PIW)(VIW) and (VIW)^4 as

Table 7.2 Basic flight test data.

Aircraft parameter	Value
Weight at startup	2731 lb
Total fuel volume	58.5 gal
Total fuel weight	351 lb
Maximum takeoff weight W_S	3150 lb
Pressure altitude	3000 ft
Maximum rated power	215 hp
Propeller diameter	74 in.
Wing span	38.3 ft
Wing area	144.9 ft^2
Aspect ratio	10.12

Table 7.3 Flight data from Cirrus SR20 level flight.

Fuel burned (gal)	IAS (kts)	OAT (°C)	TAS (kts)	Manifold pressure (in. Hg)	Engine speed (rpm)	Percent power	Fuel flow rate (gal/h)
3.3	81	−4	83	14	2480	32	8.2
4.4	91	−4	93	15	2530	34	8.6
5.6	100	−4	103	15	2530	37	8.8
6.7	110	−4	114	17	2540	43	9.4
7.9	120	−4	124	19	2540	52	10.4
9.6	130	−4	133	20	2530	60	11.5
11.1	141	−4	145	24	2540	75	13.0
13.1	150	−5	153	27	2540	88	15.0

Table 7.4 Derived quantities from flight test data.

W_T (lb)	W_T/W_S	V_e (ft/s)	VIW (ft/s)	J	C_p	η	P_R (ft-lb/s)	σ	PIW (ft-lb/s)
2711	0.8607	137.5	148.2	0.55	0.026	0.790	29,900	0.9599	36,685
2705	0.8586	154.0	166.2	0.60	0.026	0.805	32,359	0.9599	39,849
2697	0.8563	170.7	184.4	0.67	0.029	0.816	35,693	0.9599	44,130
2691	0.8542	187.8	203.2	0.73	0.033	0.828	42,100	0.9599	52,243
2684	0.8519	204.3	221.3	0.80	0.040	0.850	52,290	0.9599	65,149
2673	0.8487	220.3	239.1	0.86	0.047	0.859	60,968	0.9599	76,396
2664	0.8458	239.0	259.9	0.93	0.057	0.867	76,879	0.9599	96,823
2652	0.8420	253.3	276.1	0.99	0.067	0.869	90,396	0.9634	114,833

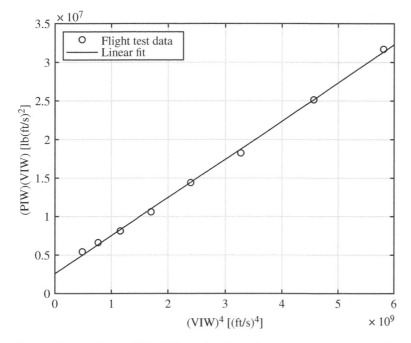

Figure 7.13 Flight test PIW–VIW data for determining the drag polar of the Cirrus SR20. A linear fit to the data indicates $C_{D_0} = 0.0287$ and $e = 0.7016$.

indicated by Eq. (7.19) for the Cirrus SR20. A linear fit to the data gives a parasite drag coefficient of $C_{D_0} = 0.0287$ and an Oswald efficiency factor of $e = 0.7016$ from the slope and intercept, and Eqs. (7.20, 7.21), respectively.

Nomenclature

a	slope of PIW–VIW curve, $\rho_{SL} S C_{D_0}/2$
AR	aspect ratio
b	intercept of PIW–VIW curve, $2W_S^2/(\rho_{SL} S \pi \, AR \, e)$

c_{root}	root chord
c_{tip}	tip chord
C_D	drag coefficient
C_{D_i}	induced drag coefficient
C_{D_0}	parasite drag coefficient
C_F	aerodynamic cleanliness factor
C_L	lift coefficient
C_P	power coefficient, $P_{shaft}/\rho N^3 D_{prop}^5$
C_T	thrust coefficient, $T/\rho N^2 D_{prop}^4$
D	drag
D_{prop}	propeller diameter
D_i	induced drag
D_0	parasite drag
e	Oswald efficiency factor
f	equivalent flat plate area
J	propeller advance ratio, V/ND
k	sensitivity of parasitic drag coefficient to changes in lift
L	lift
$\dot{m}_f$	mass flow rate of fuel into the cylinder
N	crankshaft rotational speed
P	power
P_R	power required
P_{shaft}	shaft power of an engine
r	spanwise radius along propeller blade
R	propeller radius
S	wing area
S_{wet}	wetted area of an aircraft
T	thrust
T_{std}	standard temperature
T_{test}	test temperature
V	velocity (true airspeed)
V_e	equivalent airspeed
V_E	effective velocity at propeller
V_R	relative velocity at propeller
w	induced velocity at propeller
w_a	axial component of induced velocity
w_t	tangential (swirling) component of induced velocity
W	aircraft weight
W_S	standard weight (typically maximum takeoff weight)
W_T	test weight
α	angle of attack
α_i	induced angle of attack
α_T	thrust angle
β	propeller blade pitch angle
δ	induced drag factor
φ	propeller inflow angle
η_{prop}	propeller efficiency
λ	advance ratio coefficient, J/π

Λ	wing taper ratio
ρ	air density
ρ_{SL}	standard sea level density
σ	density ratio, ρ/ρ_{SL}
τ	engine torque
ω	rotational speed

Acronyms and Abbreviations

BDC	bottom dead-center of a piston in a cylinder
BEMT	blade element momentum theory
bhp	brake horsepower
EGT	exhaust gas temperature
PIW	power independent of weight (standardized power)
rpm	revolutions per minute
sfc	specific fuel consumption
std	standard
tip	blade tip
TDC	top dead-center of a piston in a cylinder
VIW	velocity independent of weight (standardized velocity)

References

Anderson, J.D. (1999). *Aircraft Performance and Design*. New York: McGraw-Hill.

Böhnke, D., Jepsen, J., Pfeiffer, T., et al. (2011). An integrated method for determination of the Oswald factor in a multi-fidelity design environment. Proceedings of the third Council of European Aerospace Societies (CEAS) Air & Space Conference and the XXI Italian Association of Aeronautics and Astronautics (AIDAA) Congress, Venice, Italy. https://elib.dlr.de/74495/.

Continental Motors (2011). *IO-360 Series E, vngine Maintenance and Operator's Manual*, Publication Number X30617. Mobile, AL: Continental Motors, Inc.

Drela, M. (2014). *Flight Vehicle Aerodynamics*. Cambridge, MA: MIT Press.

Feagin, R. C. and Morrison, Jr.,, W. D. (1978). Delta Method, An Empirical Drag Buildup Technique. NASA Contractor Report 151971, http://hdl.handle.net/2060/19790009630.

Federal Aviation Administration. (2016). *Pilot's Handbook of Aeronautical Knowledge*, FAA-H-8083-25B, United States Department of Transportation, Federal Aviation Administration https://www.faa.gov/regulations_policies/handbooks_manuals/aviation/phak/media/pilot_handbook.pdf.

Ferguson, C.R. and Kirkpatrick, A.T. (2016). Chapter 1: Introduction to internal combustion engines. In: *Internal Combustion Engines: Applied Thermosciences*, 3e, 1–30. Chichester, UK: John Wiley.

Fredericks, W.J., Antcliff, K.R., Costa, G. et al. (2010). *Aircraft Conceptual Design Using Vehicle Sketch Pad*. Orlando, FL: AIAA 2010-658, Proceedings of the 48th AIAA Aerospace Sciences Meeting http://dx.doi.org/10.2514/6.2010-658.

Gilman, Jr.,, J. (1953). Propeller-Performance Charts for Transport Airplanes. NACA Technical Note 2966, http://hdl.handle.net/2060/19930084064.

Glauert, H. (1935). Airplane propellers. In: *Aerodynamic Theory* (ed. W.F. Durand), 169–360. Berlin: Julius Springer https://doi.org/10.1007/978-3-642-91487-4_3.

Goldstein, S. (1929). On the vortex theory of screw propellers. *Proceedings of the Royal Society A* 123 (792): 440–465. https://doi.org/10.1098/rspa.1929.0078.

Gur, O., Mason, W.H., and Schetz, J.A. (2010). Full-configuration drag estimation. *Journal of Aircraft* 47 (4): 1356–1367. https://doi.org/10.2514/1.47557.

Hartman, E. P. and Biermann, D. (1938). The Aerodynamic Characteristics of Full-Scale Propellers Having 2, 3, and 4 Blades of Clark Y and R.A.F. 6 Airfoil Sections. NACA Technical Report 640, http://hdl.handle.net/2060/19930091715.

Heywood, J.B. (2018). Chapter 1: Engine types and their operation. In: *Internal Combustion Engine Fundamentals*, 2e. New York: McGraw-Hill.

Hoerner, S.F. (1965). *Fluid Dynamic Drag*. Brick Town, NJ: Hoerner Fluid Dynamics https://hoernerfluiddynamics.com/.

Katz, J. and Plotkin, A. (2001). *Low-Speed Aerodynamics*, 2e. Cambridge, UK: Cambridge University Press.

Kimberlin, R.D. (2003). Chapter 5: Determination of engine power in flight. In: *Flight Testing of Fixed-Wing Aircraft*, 55–58. Reston, VA: American Institute of Aeronautics and Astronautics.

Leishman, J.G. (2016). Chapter 3: Blade element analysis. In: *Principles of Helicopter Aerodynamics*, 2e, 125–152. Cambridge, UK: Cambridge University Press.

Loftin, L.K. Jr., (1980). *Subsonic Aircraft: Evolution and the Matching of Size to Performance*. NASA Reference Publication 1060 http://hdl.handle.net/2060/19800020744.

Loftin, Jr.,, L. K.. (1985). Quest for Performance: the Evolution of Modern Aircraft. NASA-SP-468, http://hdl.handle.net/2060/19850023776.

Lowry, J.T. (1999). *Performance of Light Aircraft*, 291–319. Reston, VA: American Institute of Aeronautics and Astronautics.

McCormick, B.W. (1995). Chapter 6: The production of thrust. In: *Aerodynamics, Aeronautics, and Flight Mechanics*, 2e. New York: John Wiley & Sons.

McCormick, B.W. (1999). *Aerodynamics of V/STOL Flight*. New York: Dover Chapter 4.

McCrink, M.H. and Gregory, J.W. (2017). Blade element momentum modeling of low-Reynolds electric propulsion systems. *Journal of Aircraft* 54 (1): 163–176. https://doi.org/10.2514/1.C033622.

NASA. (2020). *OpenVSP*. http://openvsp.org/v.3.21.2 (accessed 23 August 2020).

Niță, M. and Scholz, D. (2012). Estimating the Oswald factor from basic aircraft geometrical parameters. Deutscher Luft- und Raumfahrtkongress, paper number 281424, http://nbn-resolving.org/urn:nbn:de:101:1-201212176728.

Oswald, W.B. (1932). General formulas and charts for the calculation of airplane performance. NACA Contractor Report 408, http://hdl.handle.net/2060/19930091482.

Raymer, D. (2018). *Aircraft Design: A Conceptual Approach*, 6e. Reston, VA: American Institute of Aeronautics and Astronautics.

Roskam, J. (1971). *Methods for Estimating Drag Polars of Subsonic Airplanes*. Lawrence, KS: DARcorporation.

Roskam, J. (2008). *Airplane Design, Part VI: Preliminary Calculation of Aerodynamic*. Thrust and Power Characteristics, Lawrence, KS: DARcorporation.

Shen, W.Z., Mikkelsen, R., and Sørensen, J.N. (2005). Tip loss corrections for wind turbine computations. *Wind Energy* 8 (4): 457–475. http://dx.doi.org/10.1002/we.153.

Sloof, J.W. (1986). Aircraft drag prediction and reduction: computational drag analyses and minimization; mission impossible? AGARD Report No. 723, Addendum 1.

Smetana, F. O., Summey, D. C., Smith, N. S., and Carden, R. K. (1975). Light aircraft lift, drag, and moment prediction – a review and analysis. NASA Contractor Report 2523, http://hdl.handle.net/2060/19750016605.

Sun, J., Hoekstra, J. M., and Ellerbroek, J. (2018). Aircraft drag polar estimation based on a stochastic hierarchical model. Proceedings of the Eighth SESAR Innovation Days. http://resolver.tudelft.nl/uuid:9977a616-4506-4acb-9d27-b9e27f9ddb8d.

Theodorsen, T. (1948). *Theory of Propellers*. New York: McGraw-Hill Book Company.

Wald, Q.R. (2006). The aerodynamics of propellers. *Progress in Aerospace Sciences* 42: 85–128.

8

Airspeed Calibration

In this chapter, we will discuss the flight testing methods for airspeed calibration. Not only are accurate measurements of airspeed important for safety of flight, but flight testing professionals need high-quality airspeed data for evaluation of all performance characteristics of the aircraft. However, an aircraft's pitot-static system for airspeed measurement is affected by several factors that introduce errors in the measurement. We'll discuss these error sources, with particular attention given to error from the location of the static port on the aircraft (referred to as *position error*). We'll start with a discussion of the effects of the locations of pitot probes and static ports on airspeed measurements and define the various airspeeds that are used in aviation. In order to account for these error sources, careful flight test calibrations are performed. We'll describe several airspeed calibration methods, including tower fly-by, radar, chase aircraft, trailing bomb, ground course visual sighting, and global positioning system (GPS) methods. In particular, the GPS method is discussed in detail, with detailed testing procedures and examples provided.

8.1 Theory

The basic functioning of the pitot-static system has already been described in Chapter 3 (see Section 3.1.2 on pressure-based instrumentation). The pitot-static system basically consists of pitot-static probe, airspeed indicator (ASI), and altimeter, as shown in Figure 3.4. The pitot-static probe is mounted facing forward, and the static pressure is detected at static ports on one or both sides of the aircraft (sometimes static pressure measurement is integrated into the pitot probe). Total pressure is measured at the tip of the pitot probe, while the static port measures static pressure. The pressure measured on the side of the fuselage is not necessarily the same as freestream static pressure, since the flow accelerates or decelerates at various locations along the fuselage (also leading to changes in static pressure along the fuselage, as described by the Bernoulli equation). As the airspeed (and angle of attack) of the aircraft changes, the pressure distribution around the aircraft will also change. Thus, it can be difficult to identify a single point (or small selection of points) on the fuselage where the local pressure is the same as freestream static pressure and remains so across a wide range of airspeeds. Thus, the static pressure measurement usually does not provide a measurement of the true freestream static pressure across a range of all angles of attack (airspeeds). This error is known as position error, which varies for different aircraft and airspeeds. Our primary focus in this chapter will be on describing flight testing methods used to perform a calibration of the airspeed system to account for position error and other error sources.

We'll begin with a general set of definitions of different kinds of airspeeds, followed by more detailed discussion on several of these. First, the true airspeed (TAS) is the actual velocity of the aircraft relative to the surrounding air (thus, it is "true") and is the airspeed type that engineers

Introduction to Flight Testing, First Edition. James W. Gregory and Tianshu Liu.
© 2021 John Wiley & Sons Ltd. Published 2021 by John Wiley & Sons Ltd.
Companion website: https://www.wiley.com/go/flighttesting

most often use for performance calculations. TAS is related to the ground speed (GS) and wind velocity by

$$V = V_{GS} - V_{wind}, \tag{8.1}$$

where V_{GS} is the velocity of the aircraft relative to the ground (typically measured by GPS) and V_{wind} is the velocity of the air relative to the ground. (Note the frames of reference and sign convention used in the equation; positive V_{wind} is a tailwind.) Next, we have the equivalent airspeed (EAS, or V_e), which is related to TAS by assuming standard sea-level density instead of actual density in the calculation of V_e. Then, calibrated airspeed (CAS, or V_{cal}) is related to EAS by also assuming standard sea-level pressure and speed of sound (*i.e.*, temperature). EAS and CAS are often described as being related to one another by compressibility effects (EAS being CAS corrected for compressibility), but this is only partially true – it is best to think of CAS as assuming standard pressure and temperature, along with the standard density assumption of EAS. Finally, the indicated airspeed (IAS, or V_i) is simply defined as the value of airspeed that is directly read from the ASI. This reading is subjected to a range of error sources (discussed in Section 8.2), which differentiate it from CAS. The central focus of this chapter is to define the relationship between IAS and CAS through flight test calibration. We'll see that the leading calibration technique involves direct measurement of IAS, and by measurements of GS converted to CAS via TAS and EAS.

Now, why do we need all of these different kinds of airspeeds? Each one has a specific rationale behind it. Ground speed is important for determining how long it will take to fly between the flight origin and destination points (and thus, the fuel required). TAS is important for engineers doing analysis of aircraft aerodynamics and performance, and for pilots in determining ground speed. Equivalent airspeed is useful since most aircraft performance limits such as stall speeds, maximum speeds, flutter boundary speeds, gear or flap extension speeds, or any other limit are based on dynamic pressure. CAS creates a simpler calibration for the instrument, leading to less mechanical complexity for traditional ASIs. And indicated airspeed is differentiated from CAS due to the presence of position error. A summary of these airspeed definitions is provided in Table 8.1, and we'll discuss each of these in greater detail as follows. (As a brief side note, the abbreviations KIAS, KCAS, KEAS, and KTAS are often used in aviation and flight testing, where the airspeed definition and the unit system are merged into one acronym. The "K" refers to kts, or nautical miles per hour, so 100 KIAS is an indicated airspeed of 100 kts.) We'll now work our way through each type of airspeed in detail.

Table 8.1 Speed definitions.

Speed	Symbol	Definition	Description
GS	V_{GS}	Ground speed	Aircraft speed relative to the ground
WS	V_{wind}	Wind speed	Velocity of the air relative to the ground
TAS	V	True airspeed	Actual velocity of the aircraft relative to the air. Most often used in engineering analysis
EAS	V_e	Equivalent airspeed	Related to TAS by assuming standard, sea-level density. Most often used for analysis of loads or flutter
CAS	V_{cal}	Calibrated airspeed	Related to EAS by assuming standard, sea-level pressure and temperature
IAS	V_i	Indicated airspeed	Speed read directly from the airspeed indicator, related to CAS through a calibration of errors such as position error

8.1.1 True Airspeed

Dynamic pressure is defined as:

$$q = \frac{1}{2}\rho V^2,$$ (8.2)

where ρ is density and V is the local velocity (TAS). For an incompressible flow, the dynamic pressure is equal to the difference between total pressure and freestream static pressure. Solving for velocity results in a simple expression that is a function of only dynamic pressure and density:

$$V = \sqrt{\frac{2q}{\rho}}.$$ (8.3)

However, this definition of dynamic pressure ignores the effects of compressibility. (Air is always compressible! We sometimes neglect compressibility effects at low Mach numbers, but compressibility effects are always present.) Thus, ASIs are more accurately calibrated based on isentropic flow theory.

The pressure difference measured by a conventional pitot-static system in subsonic flow is termed the impact pressure, defined as:

$$q_c = p_T - p = p\left(\frac{p_T}{p} - 1\right),$$ (8.4)

where p_T is total pressure and p is static pressure. (Note that impact pressure is distinct from dynamic pressure, as it includes the effect of air compression.) Using the isentropic flow relation:

$$\frac{p_T}{p} = \left(1 + \frac{\gamma - 1}{2}M^2\right)^{\frac{\gamma}{\gamma-1}},$$ (8.5)

we have

$$\frac{q_c}{p} = \left[\left(1 + \frac{\gamma - 1}{2}M^2\right)^{\gamma/(\gamma-1)} - 1\right] = \left[\left(1 + \frac{\gamma - 1}{2}\frac{V^2}{\gamma RT}\right)^{\gamma/(\gamma-1)} - 1\right],$$ (8.6)

where γ is the specific heat ratio (1.4 for air), M is Mach number, R is the gas constant for air, and T is temperature. Solving Eq. (8.6) for the velocity (TAS), we have

$$V = \sqrt{\frac{2\gamma p}{(\gamma - 1)\rho}\left[\left(\frac{q_c}{p} + 1\right)^{(\gamma-1)/\gamma} - 1\right]}.$$ (8.7)

8.1.2 Equivalent Airspeed

Based on TAS, the equivalent airspeed (EAS) is given by

$$V_e = V\sqrt{\sigma},$$ (8.8)

where $\sigma = \rho/\rho_{SL}$ is the density ratio (see Chapter 2) and ρ_{SL} is the air density at the sea level. Combining Eqs. (8.7) and (8.8), the equivalent airspeed can also be expressed as:

$$V_e = \sqrt{\frac{2\gamma p}{(\gamma - 1)\rho_{SL}}\left[\left(\frac{q_c}{p} + 1\right)^{(\gamma-1)/\gamma} - 1\right]},$$ (8.9)

where standard sea-level density is assumed instead of actual density. The EAS is directly proportional to dynamic pressure, which appears in all force and moment equations and therefore commonly correlates directly to structural loads on the airframe. Thus, EAS is often used in structural load analysis and flutter flight testing. As flight altitude increases, EAS will be lower than TAS, since the fixed value of density is in the denominator of Eq. (8.9) and density decreases with altitude. Conveniently for pilots, stall speeds and other critical flight speeds maintain fixed EAS values that are independent of altitude.

8.1.3 Calibrated Airspeed

For a measured pressure difference q_c, calculation of the airspeed V by Eq. (8.7) depends on the local static pressure and density, both of which vary with altitude and atmospheric conditions. In order to define a consistent calibration for airspeed, CAS is introduced with the assumption of standard sea-level pressure and density. Most ASIs are calibrated based on the equation for CAS, defined as:

$$V_{cal} = \sqrt{\frac{2\gamma p_{SL}}{(\gamma - 1)\rho_{SL}}\left[\left(\frac{q_c}{p_{SL}} + 1\right)^{(\gamma-1)/\gamma} - 1\right]}. \tag{8.10}$$

Equations (8.6) and (8.7) are based on isentropic flow theory, which accounts for compressibility effects. However, when standard, sea-level values of pressure and density are assumed for CAS in Eq. (8.10), we essentially recover the incompressible Bernoulli equation. We'll now demonstrate that Eq. (8.10) is equivalent to Eq. (8.3) with an assumption of standard, sea-level density. Starting with Eq. (8.5), we can perform a binomial expansion to obtain

$$\frac{p_T}{p} = 1 + \frac{\gamma}{2}M^2 + \frac{\gamma}{8}M^4 + \cdots. \tag{8.11}$$

Substituting Eq. (8.11) into Eq. (8.4), we obtain an alternate form of Eq. (8.6) as:

$$\frac{q_c}{p} = \left[\left(1 + \frac{\gamma - 1}{2}M^2\right)^{\gamma/(\gamma-1)} - 1\right] = \frac{\gamma}{2}M^2 + \frac{\gamma}{8}M^4 + \cdots. \tag{8.12}$$

Now, if we assume that $M \ll 1$ for incompressible flow ($q_c \approx q$), we can neglect all higher order terms such that

$$\frac{q}{p} \approx \frac{\gamma}{2}M^2. \tag{8.13}$$

Recognizing that $M = V/a$ and $a^2 = \gamma RT$, assuming standard sea-level conditions commensurate with the definition of CAS, and solving for CAS we have

$$V_{cal} \approx \sqrt{\frac{2q}{\rho_{SL}}}, \tag{8.14}$$

which is identical to Eq. (8.3) evaluated at standard, sea-level conditions. If the velocity and altitude of an aircraft are low, then Eq. (8.14) can be a reasonable approximation for Eq. (8.10) – this would be an application of an incompressible assumption, but this is not required for converting between EAS and CAS.

It's important to point out that Eq. (8.10) is the theory that is used to calibrate the ASI (Gracey, 1980). As pointed out by Schoolfield (1942), the difference between EAS and CAS is not really due to neglecting compressibility effects; rather, the difference is simply due to the assumption of standard, sea-level pressure (in addition to the assumption of standard, sea-level density inherent to both EAS and CAS).

The relationship between EAS and CAS can be defined by simply forming a ratio between Eqs. (8.9) and (8.10) as:

$$\frac{V_e}{V_{\text{cal}}} = \sqrt{\frac{p}{p_{\text{SL}}}} \frac{\sqrt{\left[\left(\frac{q_c}{p}+1\right)^{(\gamma-1)/\gamma}-1\right]}}{\sqrt{\left[\left(\frac{q_c}{p_{\text{SL}}}+1\right)^{(\gamma-1)/\gamma}-1\right]}}, \tag{8.15}$$

where actual pressure is used for EAS and standard sea-level pressure is assumed for CAS. To give more compact expressions for V_e and V_{cal}, we introduce a function defined as:

$$f(x) = \sqrt{\frac{\gamma}{\gamma-1}} \sqrt{\frac{(x+1)^{(\gamma-1)/\gamma}-1}{x}}, \tag{8.16}$$

which is sometimes called the compressibility factor (Schoolfield 1942; Gracey 1980). This function encapsulates the error inherent to the assumption of standard sea-level pressure and density in the definition of CAS. Based on $f(x)$, Eqs. (8.9) and (8.10) can be written as:

$$V_e = \sqrt{\frac{2q_c}{\rho_{\text{SL}}}} f(\delta^{-1} q_c / p_{\text{SL}}), \tag{8.17}$$

$$V_{\text{cal}} = \sqrt{\frac{2q_c}{\rho_{\text{SL}}}} f(q_c / p_{\text{SL}}), \tag{8.18}$$

where $\delta(h) = p(h)/p_{\text{SL}}$ is the pressure ratio depending on the altitude (see Chapter 2). Figure 8.1 shows the behavior of $f(x)$. Since x is proportional to the dynamic pressure, we see that the significance of $f(x)$ increases with airspeed. Now, to convert between calibrated and equivalent airspeeds, we form a ratio between V_e and V_{cal} as:

$$\frac{V_e}{V_{\text{cal}}} = \frac{f(q_c/p)}{f(q_c/p_{\text{SL}})} = \frac{f(\delta^{-1} q_c / p_{\text{SL}})}{f(q_c / p_{\text{SL}})}. \tag{8.19}$$

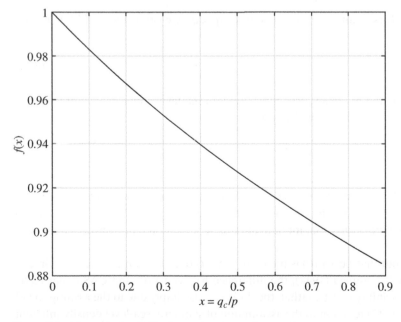

Figure 8.1 The compressibility factor for airspeed correction.

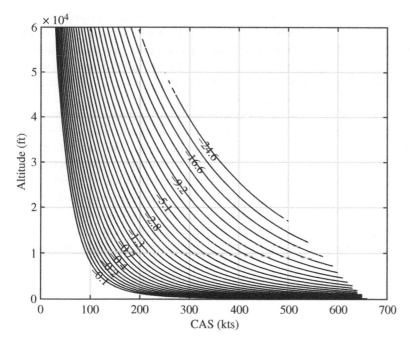

Figure 8.2 Correction (in kts) between EAS and CAS for a range of airspeeds and altitudes (subsonic conditions only).

To provide an indication of the significance of this correction, ΔV values, defined as:

$$\Delta V = V_e - V_{cal} = V_{cal}\left(\frac{V_e}{V_{cal}} - 1\right), \tag{8.20}$$

are plotted in Figure 8.2 across a range of CASs and altitudes (contours are spaced logarithmically, and only the subsonic regions are shown). The correction factor is minimal (a small fraction of a knot) at most airspeeds and altitudes where light aircraft operate. Thus, CAS is often a good proxy for EAS, where simplicity is gained using CAS with very little error. However, as airspeed and/or altitude increase, the significance of the correction factor grows – high-altitude, high-speed flight vehicles such as the Space Shuttle Orbiter or the SR-71 Blackbird had dedicated equivalent ASIs since specific knowledge of EAS was critical in those flight regimes (Corda 2017).

8.1.4 Indicated Airspeed

Indicated airspeed is the speed that is read directly from the ASI. It differs from CAS by various error sources, including instrument error, time lags in the pressure system, and position error. The relationship between CAS and IAS is defined by calibration of these error sources. Instrument error is usually very low (a fraction of a knot), but position error can be substantial. We'll discuss these error sources in Section 8.2, with a discussion of airspeed calibration methods in the following section.

8.1.5 Summary

We'll now summarize the various airspeeds and conversion factors. Depending on the starting information and the desired end state, one can work through the conversions in two different directions. For example, to determine CAS for flight test calibration of the ASI, one might start with a measurement of ground speed. If wind speed is known, then the TAS can be found from Eq. (8.1). Then, EAS is found by applying Eq. (8.8), which assumes standard sea-level density. EAS is then converted to CAS by (8.15). Conversely, once flight testing is complete, a pilot may wish to find the TAS from the indicated airspeed. To do this, IAS is converted to CAS through the calibration that is defined in flight testing (methods described as follows). Once CAS is known, it is converted to EAS by Eq. (8.15), and EAS is converted to TAS by Eq. (8.8).

Perhaps surprisingly, equivalent airspeed is the preferred indication in the cockpit instead of TAS. Why would a pilot care more about an airspeed defined based on an assumed value of density, rather than the true speed at which the aircraft is flying? The answer lies in the fact that equivalent airspeed is directly proportional to dynamic pressure. This allows for fixed definition of flight envelope limits that do not change with altitude if expressed as equivalent airspeed. Thus, critical flight speeds such as stall speed, flap extension speed, landing gear extension speed, corner speed, and never-exceed speed will not change with altitude when expressed as equivalent airspeed, even though the corresponding TAS will change. And, since calibrated and indicated airspeeds are approximately equal to equivalent airspeed, the ASI in the cockpit provides a reliable instrument for safety of flight.

8.2 Measurement Errors

8.2.1 Instrument Error

Instrument error in the pitot-static system is due to a difference between the indicated airspeed and the airspeed corresponding to the actual pressure difference (impact pressure) applied to the ASI. As shown in Figures 3.3 and 3.4, the total pressure is sensed by an opening in the forward-facing tube (pitot tube) and the static pressure is sensed by orifices in the side of the static pressure tube or by a set of holes in the side of an aircraft fuselage. The pressures that are sensed by the pitot tube and static pressure orifices are conveyed through tubing to a calibrated differential pressure sensor. The pressure difference (the impact pressure) $q_c = p_T - p$ is measured using a pressure-sensing device that is generally in the form of capsules, diaphragms, or bellows, as shown in Figure 8.3. The readings of the instrument may be affected by errors due to imperfections in the instrument mechanism. The instrument error depends on the elastic properties of the pressure capsule (scale error, hysteresis, and drift) and the effects of temperature, acceleration, and friction on the linkage mechanism. The scale error is the difference, for a given applied pressure, between the value indicated by the instrument and the correct value corresponding to the applied pressure. ASIs can be calibrated statically and dynamically in laboratories. Typically, the static calibration is sufficient for most flight testing measurements. Laboratory calibration data give the differences between the instrument-corrected values and gauge readings, which are expressed as $\Delta V_{ic} = V_{ic} - V_i$ for airspeed and $\Delta h_{ic} = h_{ic} - h_i$ for altitude, where the subscripts "ic" and "i" denote the instrument-corrected and actual gauge reading, respectively. The calibration results can be applied in the form of charts and computer schemes for corrections of the instrument errors.

8.2.2 System Lag

The pressure-sensing system is subject to an error due to a time lag in transmitting the pressure from the point of measurement to the sensor through tubing. This error is significant when the rate of pressure change is high in the case of a rapid climb or decent. For a given rate of change of

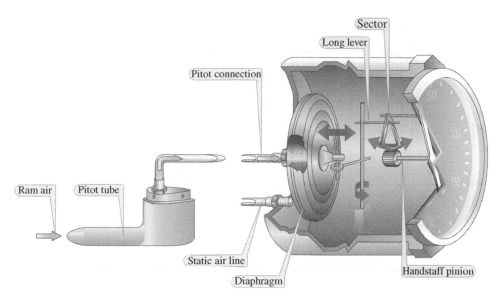

Figure 8.3 Cutaway diagram of the airspeed indicator, illustrating how the deflection of a sealed diaphragm transduces pressure to a reading of airspeed. Source: Federal Aviation Administration (2012).

pressure, the lag error depends primarily on the length and diameter of the tubing and the internal volume of the instrument.

The lag error is related to the two types of sources: acoustic lag and pressure lag. Since pressure waves propagate at the speed of sound along the tubing, the magnitude of the acoustic lag depends only on the speed of sound and the length of the tubing. Thus, an estimate for the acoustic lag time is $\tau = L/a$, where L is the tubing length and a is the speed of sound. For the tubing length of most aircraft, the error associated with the acoustic lag is not significant.

The pressure lag is related to the pressure drop between the two ends of the tubing. When air in the tubing between a pressure source and an instrument is flowing, the response of the pressure at the instrument can be delayed relative to the pressure at the source, leading to an inaccurate indication of the instrument. According to Gracey (1980), the pressure drop is proportional to the rate of change of pressure, i.e., $\Delta p = \lambda dp/dt$, where λ is a lag constant defined as $\lambda = 128\mu LC/\pi d^4 p$ (assuming laminar flow in the tubing). Here, L and d are the length and internal diameter of the tubing, respectively, C is the internal volume of the instrument, p is the pressure, and μ is the viscosity of air. When the value of λ for an instrument system is known, the errors in airspeed and altitude associated with a given rate of climb or descent of an aircraft (described by dp/dt) can be determined.

8.2.3 Position Error

Position error results from a difference between the local sensed value and the true value of pressure and is the key error source underlying the difference between IAS and CAS. Since the pitot-static system is measuring impact pressure (Eq. (8.4)), the two pressures to be measured are total pressure and freestream static pressure. As discussed in Chapter 3, a pitot probe measures the total pressure, and a static pressure port on the side of the pitot probe or on the side of the fuselage measures the static pressure (see Figure 3.3(b)). Position error is so-named since the error is directly related to placement of the pitot probe and static pressure port on the aircraft. For total pressure measurement, the leading source of position error is a misalignment between the pitot probe axis

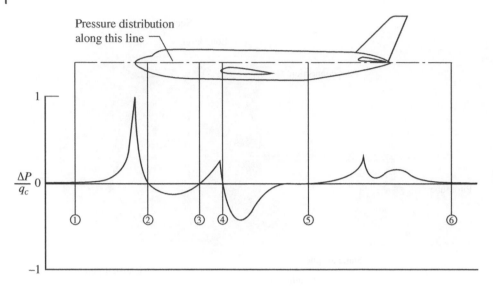

Figure 8.4 The pitot-static probe mounted on an airplane and the surface pressure distribution along the fuselage reference line (FRL). Source: Haering (1995).

and the flow direction, resulting in lower values of measured total pressure. However, Gracey (1980) showed that careful design of the probe head can result in accurate measurements across a reasonably wide range of angles ($\pm 20°$ or more). Also, the pitot probe can be placed in a location away from the fuselage, where it is out of the wake of the fuselage under strong crosswind conditions. Thus, errors in the measured total pressure are usually negligible across a wide range of angle of attack and sideslip. However, accurate measurements of static pressure are more difficult; we'll elaborate on these challenges as follows.

The main challenge with measurement of freestream static pressure is to identify a suitable location on the aircraft where the local static pressure value (p) is the same as the freestream value (p_∞). Furthermore, this desired equivalency must hold for a wide range of angles of attack and sideslip. Figure 8.4 illustrates a typical pressure field along the centerline of an aircraft – local pressure values are either above or below the freestream static value due to local acceleration or deceleration of the flow relative to the freestream velocity. Points 1–6 are identified on the pressure distribution as candidates for location of a static pressure port, since the local pressure is equal to freestream static pressure in each of these locations. However, as the angle of attack or sideslip of the aircraft change, the flow streamlines adjust and the pressure distribution will change.

Aircraft designers will thoughtfully place the static pressure port at the location predicted to have the least amount of pressure change across the operating envelope of the aircraft, but there will not be an ideal location. The local geometry of the static pressure port can also be carefully designed in order to minimize sensitivity to these changes in the pressure field. For example, Figure 3.3(b) shows a backward-facing step immediately in front of the pressure port. This creates a small separation bubble over the pressure port, where the local pressure is closer to freestream static pressure over a wider range of angle of attack and sideslip. Also, in order to reduce sensitivity to changes in sideslip angle, static pressure ports can be positioned on both sides of the aircraft and physically manifolded into one measurement.

Despite these design efforts, there is typically a residual error remaining in the measurement of freestream static pressure. This pressure difference, defined as the position error $\Delta p_{\text{pc}} = p - p_\infty$, has an impact on measurement of both altitude and airspeed, since the static pressure measurement is inherent to both. Thoughtful placement of the static pressure port can minimize the error at the airspeed associated with the most critical phases of flight (*e.g.*, approach speed), and the resulting errors can be in a direction that is conservative (*e.g.*, the ASI provides a lower-than-actual reading at low speeds, leading the pilot to fly at slightly higher speeds away from stall).

Position error is also a function of the source of the static pressure measurement. Since static pressure measurement is so critical for determining airspeed and altitude, aircraft usually have a backup static pressure source. The primary static pressure port(s) can possibly get clogged by oil, debris, moisture, or even ice. The backup system is usually a pressure measurement from within the cabin of an unpressurized aircraft. This approach to a backup has the advantage that it is not sensitive to the external environmental concerns; however, cabin pressure as a static source will have higher position error. In some aircraft, the cabin pressure is higher than the true freestream static pressure due to ram air effects (*e.g.*, see Gregory and McCrink 2016).

Since position error affects cockpit indications of both altitude and airspeed, we'll develop a relationship between the two. The error in a velocity measurement will be related to an error in measured pressure, and this pressure error will be related to an error in the indicated altitude. The airspeed position error is given by $\Delta V_{pc} = V_{cal} - V_{ic}$, where the subscript "pc" denotes the position error, and "cal" and "ic" denote the measurements of CAS and indicated airspeed corrected for instrument error, respectively. We denote the measured impact pressure (with the position error) as $q_{c,ic} = p_T - p$, and the true impact pressure (without the position error) as $q_c = p_T - p_\infty$, where the difference between the two results from differences between p (measured static pressure) and p_∞ (freestream static pressure). Thus, the position pressure error is given by

$$\Delta p_{pc} = p - p_\infty = q_c - q_{c,ic} = \Delta q_{c,ic}, \tag{8.21}$$

where zero error in the measurement of p_T is assumed. Since $q_{c,ic} = p_T - p$, and the error in total pressure measurement is negligible, a differential change in the instrument-corrected impact pressure is equivalent to a differential change in static pressure:

$$dq_{c,ic} = -dp. \tag{8.22}$$

From Eq. (8.6), at sea level we have

$$\frac{q_{c,ic}}{p_{SL}} = \left(1 + \frac{\gamma - 1}{2} \frac{V_{ic}^2}{a_{SL}^2}\right)^{\frac{\gamma}{\gamma-1}} - 1, \tag{8.23}$$

where a_{SL} is the speed of sound at the sea level. Differentiating Eq. (8.23) with respect to V_{ic} and incorporating Eq. (8.22) yields

$$\frac{dq_{c,ic}}{dV_{ic}} = -\frac{dp}{dV_{ic}} = \frac{\gamma p_{SL} V_{ic}}{a_{SL}^2}\left(1 + \frac{\gamma - 1}{2} \frac{V_{ic}^2}{a_{SL}^2}\right)^{\frac{1}{\gamma-1}}. \tag{8.24}$$

for subsonic conditions. (Note that a similar estimate for supersonic flow is given by Ward et al. (2006).) We can approximate the differentials by finite differences, where $dV_{ic} \approx V_{ic} - V_{cal} = -\Delta V_{pc}$ and $dp \approx p - p_\infty = \Delta p_{pc}$. Thus, Eq. (8.24) becomes

$$\frac{\Delta p_{pc}}{\Delta V_{ic}} \approx \frac{\gamma p_{SL} V_{ic}}{a_{SL}^2}\left(1 + \frac{\gamma - 1}{2} \frac{V_{ic}^2}{a_{SL}^2}\right)^{\frac{1}{\gamma-1}}. \tag{8.25}$$

In addition, altitude measurement is affected by the position error, and the altimeter position error is given by $\Delta h_{pc} = h_{cal} - h_{ic}$. Based on the hydrostatic equation (see Chapter 2), an estimate of the pressure position error is

$$\frac{\Delta p_{pc}}{\Delta h_{pc}} \approx \rho_{SL} \sigma g_0. \tag{8.26}$$

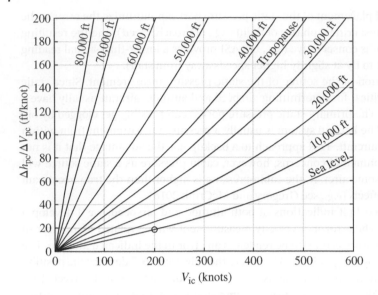

Figure 8.5 Relationship between altitude error and velocity error as a result of position error. Source: Herrington et al. (1966).

Equation (8.26) indicates that the position pressure error Δp_{pc} is proportional to Δh_{pc}, which is used in several airspeed calibration methods (the tower fly-by, tracking radar and ground camera methods). Since both the airspeed system and altimeter are connected to the same static pressure system in most aircraft, it is possible to relate the altimeter position error Δh_{pc} to the airspeed position error ΔV_{pc}. From Eqs. (8.25) and (8.26), we have

$$\frac{\Delta h_{pc}}{\Delta V_{pc}} \approx \frac{V_{ic}}{\sigma g_0}\left(1 + \frac{\gamma - 1}{2}\frac{V_{ic}^2}{a_{SL}^2}\right)^{1/(\gamma-1)} \qquad (8.27)$$

for subsonic conditions. Based on the above theory, the ASI and altimeter position errors are directly related, and thus the pressure position error can be determined using either ΔV_{pc} or Δh_{pc}. Figure 8.5 illustrates the relationship described by Eq. (8.27), where $\Delta h_{pc} = h_{cal} - h_{ic}$ in units of feet is related to $\Delta V_{pc} = V_{cal} - V_{ic}$ in units of kts. For example, if an aircraft is cruising at sea level at an instrument-corrected indicated airspeed of 200 kts, then $\Delta h_{pc}/\Delta V_{pc}$ is 18.5 ft/knot (see circle on Figure 8.5). If an aircraft has a known ΔV_{pc} of 4 kts (*i.e.*, CAS is 4 kts higher than indicated airspeed) from an airspeed calibration, then the altimeter will read 74 ft lower than actual when at sea level.

8.3 Airspeed Calibration Methods

Airspeed calibration methods are designed to obtain the relationship between the CAS (V_{cal}) and the instrument-corrected indicated airspeed (V_{ic}) mainly for deriving the position error. The static pressure error Δp_{pc} can be estimated from the proportional relation between ΔV_{pc} and Δp_{pc} in Eq. (8.25). The position error of the static pressure system can be difficult to determine since the pressure field influenced by aircraft's body is complicated. There are various calibration methods that have been devised for the determination of this error. Gracey (1980) has described some classical airspeed calibration methods, including trailing bomb, trailing cone, trailing anemometer, pacer aircraft, tower fly-by method, speed course method, tracking radar, and ground camera.

Figure 8.6 Effect of a wing/body on measurement of the static pressure in the incoming flow: (a) pitot-static probe near a wing and (b) pitot-static probe on the air data boom.

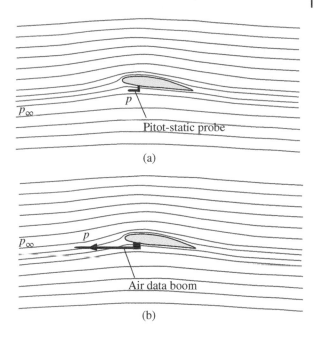

These methods could be classified based on the physical quantities derived from test flights such as the freestream static pressure methods and true-airspeed methods (Gracey 1980). Here, these methods are discussed briefly based on whether instrumentation for airspeed calibration is in the air or on the ground. Since most methods require some specialized equipment, they are not suitable for an introductory course of flight testing. Therefore, the GPS-based method as a simple modern method is described separately in more detail.

8.3.1 Boom-Mounted Probes

In flight testing, as shown in Figure 8.6, a pitot-static probe can be integrated in an air data boom extended upstream of the wing, where the measured static pressure represents the undisturbed pressure of the atmosphere at the altitude that the aircraft is flying. As shown in Figure 8.7, an air data boom installed at the nose of an X-15 experimental aircraft, allowing measurements of both the TAS and freestream flow angles, since the measurements are not influenced by the aircraft body. However, installation of an air data boom is not feasible unless an aircraft is experimental.

8.3.2 Trailing Devices and Pacer Aircraft

The methods of trailing bomb, trailing cone, trailing anemometer, and pacer aircraft utilize the reference static pressure sources accompanying the test aircraft in flight. Figure 8.8 illustrates the airspeed calibration methods of trailing bomb, trailing cone, and pacer aircraft. With the trailing bomb method, the static pressure p' measured by the aircraft is compared directly with the static pressure p measured by orifices on a bomb-shaped body suspended on a long tubing cable below the aircraft. There are two types of bombs, where the orifices are either on the body of the bomb or on the static pressure tube ahead of the bomb. The static pressure measurement on a trailing bomb is also affected by its body, and thus the static pressure on the bomb is not exactly the freestream static pressure p_∞. However, this error can be determined by calibration in a wind tunnel since the full-scale bomb is small enough for wind tunnel testing, such that corrections for the error can

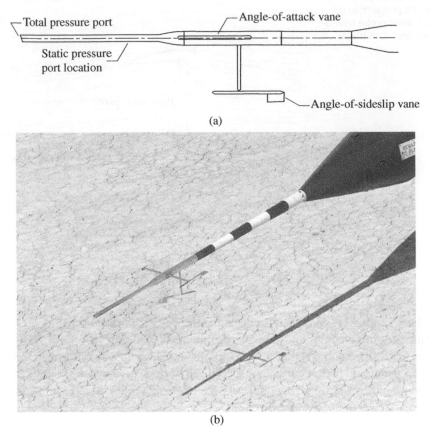

(a)

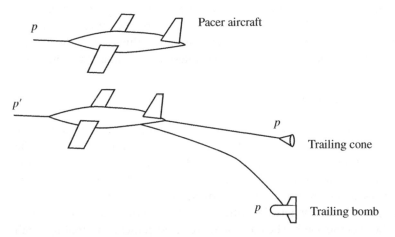

(b)

Figure 8.7 Air data boom (a) diagram and (b) installation on the nose of an X-15. Sources: (a) Haering (1995) and (b) NASA (2014).

Figure 8.8 Illustration of airspeed calibration methods of trailing bomb, trailing cone, and pacer aircraft. Source: Adapted from Gracey (1980).

be applied. Airspeed calibrations with a trailing bomb can be conducted through a wide range of altitudes and through a speed range from the stall speed to the maximum speed at which the bomb can be towed. The limiting speed is determined by the speed at which the suspension tubing develops unstable oscillations. The testing arrangements for a trailing light-weight cone and a trailing anemometer are similar to that for a trailing bomb.

With the pacer aircraft method, the freestream static pressure is derived from the calibrated static pressure installation of a pacer aircraft flying alongside the test aircraft being calibrated. The pacer aircraft maintains a steady flight, while the test aircraft flies in a tight formation on the pacer. When there is no relative movement, data are read. The difference $\Delta h = h' - h$ between the altimeter indication h' in the test aircraft and the corrected altimeter indication h in the pacer aircraft is measured. The pressures p' and p corresponding to the values of h' and h can be found. The difference between p' and p is then the position error Δp_{pc} for the test aircraft. Furthermore, a straightforward comparison between the CAS of the pacer aircraft and the instrument-corrected airspeed gives ΔV_{pc} such that Δp_{pc} can be determined by Eq. (8.27).

8.3.3 Ground-Based Methods

Ground-based airspeed calibration methods utilize equipment on the ground to record the flight time, position, distance, and altitude. These methods include the speed course, tower fly-by, tracking radar, and ground camera methods.

In the speed course method, the test aircraft is flown over a measured course on the ground at low altitudes, and the ground speed of the aircraft is determined by measuring the time for the aircraft to fly over the known distance between two landmarks. The wind speed and direction can be measured by a wind speed indicator. Flight along the course is done in opposite directions, such that the averaged ground speed will minimize wind effects. When the flight is in very calm air, the TAS is directly derived from measurements of the ground speed of the aircraft. The accuracy of this method is largely dependent on the accuracy of the time measurements of the speed run, the constancy of the wind speed, and the constancy of the airspeed throughout the speed run.

In the tower fly-by method, the aircraft flies at a constant airspeed and altitude by a tower of known height that has an observer and sensitive barometer on the top. The altitude of the aircraft relative to the tower can be determined by triangulation or with a theodolite. This altitude is compared with the value that is indicated by the aircraft's altimeter to determine the position error Δh_{pc}. This position error is then related to ΔV_{pc} by Eq. (8.27). The altitude of the test aircraft can be also determined using tracking radar and ground camera methods,[1] and the data reduction procedures are similar to those in the tower fly-by method.

8.3.4 Global Positioning System Method

The GPS provides accurate position and ground velocity of a flying aircraft (see Chapter 3). Thus, GPS-based methods are particularly attractive for determining the ground speed due to the simplicity and widespread availability of GPS receivers on board aircraft and integrated into data acquisition systems. In order to determine the TAS and wind velocity (both the magnitude and direction) from GPS measurements of ground speed, at least three flight tracks with sufficiently different headings are required, *with the aircraft flown at a constant indicated airspeed*. Three separate ground speed measurements are needed in order to form a system of three equations and three unknowns. The set of equations is based on Eq. (8.1), and the unknown parameters are wind speed magnitude, wind direction and the magnitude of the TAS. Ideally, the three headings for airspeed calibration

---

1 Details on a leading ground camera method, videogrammetry, are provided in an online supplement 3.

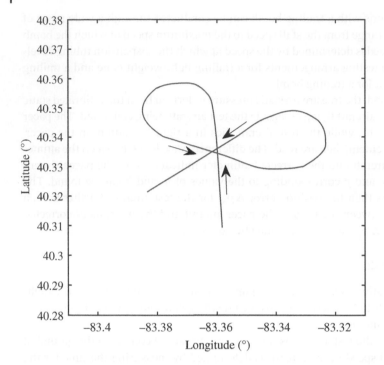

Figure 8.9 Sample ground track (a "cloverleaf") for GPS-based determination of TAS.

should equally divide a full 360°, with 120° separation between each neighboring heading. On each heading, after the aircraft has stabilized, the ground track and ground speed are recorded with the onboard GPS unit. When the GPS ground speed and the ground course for each track are known, three GPS velocity vectors for the three tracks can be reconstructed. A typical ground track for flight test measurement of a single value of TAS is shown in Figure 8.9. Three headings are flown, each separated by 120°, with ground speed and course measured on each leg.

We'll now develop the system of equations that can be solved to find the TAS from a set of three ground speed vectors. First let's consider the TAS of an aircraft represented as a vector, where the magnitude of the vector is the value of TAS, and the direction of the vector is the heading of the aircraft (the direction where the nose is pointed). We can take this vector and decompose it into components in a global coordinate system such that the TAS is

$$V = [V_x^2 + V_y^2]^{1/2}, \tag{8.28}$$

where V_x and V_y are the x- and y-components of TAS in this coordinate system (positive x is toward East, and positive y is toward North). Then, we take Eq. (8.1), which represents the relationship between ground speed, wind speed, and TAS and write it for each component of the TAS as:

$$V_x = V_{GS,x} - V_{wind,x}$$
$$V_y = V_{GS,y} - V_{wind,y}. \tag{8.29}$$

Substituting Eq. (8.29) into (8.28), and writing the result three separate times (once for each leg flown of a cloverleaf) we have

$$V = [(V_{GS,x,1} - V_{wind,x})^2 + (V_{GS,y,1} - V_{wind,y})^2]^{1/2}$$
$$V = [(V_{GS,x,2} - V_{wind,x})^2 + (V_{GS,y,2} - V_{wind,y})^2]^{1/2}$$
$$V = [(V_{GS,x,3} - V_{wind,x})^2 + (V_{GS,y,3} - V_{wind,y})^2]^{1/2}, \tag{8.30}$$

where $(V_{GS,x,k}, V_{GS,y,k})$ are the x- and y-components for the kth track ($k = 1, 2, 3$) of GPS ground speed, $(V_{wind,x}, V_{wind,y})$ are the wind velocity components (assumed constant), and V is the TAS. It is important to recognize that the winds must remain constant and the indicated airspeed must be held constant throughout the entire flight maneuver. In other words, the physical conditions must be set such that V and V_{wind} must be the same in all three equations in Eq. (8.30). This gives us a system of three equations and three unknowns (V, $V_{wind,x}$, $V_{wind,y}$), which can be solved numerically (*e.g.*, with `fsolve` in MATLAB).

We can also think of the system of equations in graphical terms. Figure 8.10 shows the vectors associated with the three ground tracks and illustrates how they are related to TAS and the wind vector. The GPS vectors are defined by the magnitude (ground speed) and direction (ground course) from each leg of the flight. Graphically, the end points of the three GPS ground velocity vectors are superimposed at a point P, and then the start points of the vectors must be located on a circle whose center is defined as the origin of a local coordinate system and radius represents the TAS magnitude. The vector from the center of the circle to the point P is the wind velocity vector. Since the coordinates of the start points of the GPS ground velocity vectors located on the circle are known, the vectors associated with the TAS and wind speed can be found. This can be done in a computer-aided drafting (CAD) software, or even on graph paper with a compass and ruler, where V_{wind} and V are directly measured. A diagram such as Figure 8.10 can be constructed by identifying a point P on the diagram, and then plotting the three GS vectors to terminate at P (vector length is proportional to the GPS-measured speed, and the orientation is determined by the GPS-measured ground course). A circle is then circumscribed around the tails of the three GS vectors. The radius of this circle is the TAS, and a vector from the origin of the circle to P is the wind vector.

8.4 Flight Testing Procedures

We've now seen that position error is the key source of error determining the difference between IAS and CAS. The goal of the airspeed calibration flight test is to define the relationship between the two. IAS will be found by direct measurement from the ASI. CAS, on the other hand, is found indirectly from TAS (measured by a series of ground speed recordings), and then by converting to EAS and then CAS. We'll focus our attention on the GPS method for finding TAS from a set of

Figure 8.10 Vector diagram for the GPS ground velocity, true airspeed, and wind velocity.

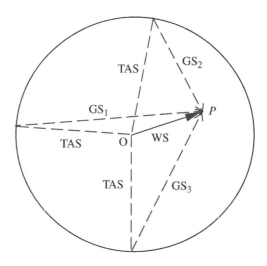

three ground speed measurements at the same flight condition. Ideally, the flight test should be performed under calm conditions, or at least when the winds are steady.

The airspeed calibration should be performed across a range of airspeeds that cover the operating envelope of the aircraft, with suitable spacing to capture any nonlinear effects which may be present. For example, a range of indicated airspeeds could be tested from near stall speed up to maximum speed, in increments of 10 kts. At each airspeed test point, a cloverleaf should be flown (see Figure 8.9) with three headings separated by 120°. The best practice is to make the turns the "long way" around (turning through 240° to the next heading), such that the three ground courses intersect within the same air mass. Throughout a single cloverleaf maneuver, the indicated airspeed must be maintained constant (this is easily done by not adjusting the throttle setting during a single cloverleaf maneuver). Each ground speed leg of the cloverleaf should be maintained for a duration sufficient for the aircraft to stabilize and to record enough data for averaging (15–30 seconds, typically). On each heading, each of the following parameters should be recorded after flight has stabilized:

- Ground speed
- Ground course
- Indicated airspeed
- Heading
- Fuel burned
- Pressure altitude (with the altimeter set to a reference pressure of 29.92 inHg)
- Outside air temperature (OAT)

(Note that there are strong similarities between the data set and methods required here, and the data acquired for level flight testing described in Chapter 7. For planning of efficient flight test programs, a flight testing team may wish to combine the two flight tests into one measurement campaign.) After a set of three headings is flown, the throttle is adjusted until a new desired IAS set point is achieved in steady flight, and another cloverleaf maneuver is flown. This process is repeated for each indicated airspeed in the test matrix.

Factors potentially affecting the calibration of the ASI include flap setting, and whether an alternate source of static pressure is selected. The flap setting affects the trim condition of the aircraft, leading to a different angle of attack and a different pressure field that corresponds to a given flight speed. Thus, the position error will be different for each flap setting and must be calibrated. Similarly, when the alternate static source is selected, the pressure being measured (*e.g.* cabin pressure) will likely have increased deviation from the true freestream value, leading to increased position error and requiring calibration for this condition.

Post-processing of the data is done as follows. For each airspeed setting within the test matrix, the set of three ground speed readings are used with Eq. (8.30) to determine the TAS. TAS is then converted to equivalent airspeed by the density ratio with Eq. (8.8). Finally, EAS is converted to CAS by Eq. (8.15) – this correction is usually minimal for light aircraft in relatively low-altitude, low-speed flight. These readings of CAS are then plotted against the direct measurements of indicated airspeed to form the calibration. Since position error is usually small, the relationship between IAS and CAS should be close to 1:1.

8.5 Flight Test Example: Cirrus SR20

Tests were conducted on a Cirrus SR20 (sixth-generation) aircraft, in which three nominal magnetic headings of 360°, 120°, and 240° were flown at each airspeed. On each heading, after the aircraft

Table 8.2 Flight data obtained at a pressure altitude of 3000 ft.

IAS (kts)	OAT (°C)	Nominal heading 360°		Nominal heading 120°		Nominal heading 240°	
		GS (kts)	Track (°)	GS (kts)	Track (°)	GS (kts)	Track (°)
81	−4	82	005	90	119	78	235
91	−4	92	005	100	119	88	236
100	−4	102	004	110	119	98	237
110	−4	113	004	120	119	108	238
120	−4	123	004	130	119	118	238
130	−4	134	004	140	119	126	239
141	−4	145	005	152	121	137	239
150	−5	155	005	160	119	144	240

had stabilized, the ground track and ground speed were recorded utilizing the aircraft's GPS unit (as displayed on the multifunction display). Three GPS ground speed vectors were obtained for each power setting over a matrix of indicated airspeeds ranging from 81 up to 150 kts (see Table 8.2) at a pressure altitude of 3000 ft. Following the data analysis procedure described in Section 8.3, the reduced data is shown in Table 8.3, where TAS and the wind vector results from solving the system of Eq. (8.30) for the ground speeds. Equivalent airspeed and CAS are also shown in Table 8.3 for these flight test conditions at a pressure altitude of 3000 ft and OAT values shown in Table 8.2. The resulting values of CAS are plotted against the indicated airspeed values in Figure 8.11 to define the calibration. A linear fit to this calibration data has a slope of 0.998 and is offset from the ideal case (IAS = CAS) by about 0.5 knot. The offset is likely due to experimental uncertainty in the measurement of OAT. The slope is very close to the ideal case since the values for IAS reported by the avionics system have likely been already corrected for position error. The Pilot's Operating Handbook (POH)-derived calibration between IAS and CAS, with a slope of 1.045, most likely only applies to the backup (analog) ASI.

Table 8.3 Airspeeds and wind speeds for airspeed calibration at a pressure altitude of 3000 ft.

IAS (kts)	TAS (kts)	EAS (kts)	CAS (kts)	Wind	
				Speed (kts)	Direction (°)
81	83.1	81.4	81.5	7.2	281.6
91	93.1	91.3	91.3	7.2	282.0
100	103.2	101.1	101.1	7.1	281.7
110	113.5	111.2	111.3	7.0	276.7
120	123.5	121.0	121.1	7.0	276.8
130	133.2	130.5	130.6	8.1	266.8
141	144.5	141.6	141.7	8.7	270.3
150	152.9	150.1	150.2	9.5	260.6

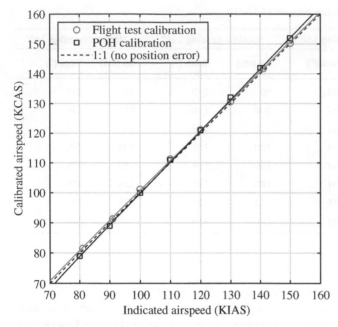

Figure 8.11 Airspeed calibration flight test results for the SR20G6, compared with values from the Pilot's Operating Handbook (POH).

Nomenclature

a	speed of sound, $\sqrt{\gamma RT}$
C	internal volume of the instrument
d	tubing internal diameter
$f(x)$	compressibility factor
g_0	gravitational acceleration at sea level
h	altitude
k	index value
L	tubing length
M	mach number
p	static pressure
p_T	total pressure
q	dynamic pressure
q_c	impact pressure
R	gas constant for air
t	time
T	temperature
V	true airspeed
V_{cal}	calibrated airspeed
V_e	equivalent airspeed
V_{GS}	ground speed
V_i	indicated airspeed
V_{wind}	wind speed
x	dummy variable

δ pressure ratio, p/p_{SL}

Δh_{pc} altimeter position error, $h_{cal} - h_{ic}$

Δp_{pc} pressure position error, $p - p_{\infty}$

ΔV_{pc} velocity position error, $V_{cal} - V_{ic}$

γ specific heat ratio

λ lag constant

μ viscosity of air

ρ density

σ density ratio, ρ/ρ_{SL}

τ acoustic lag time

Subscripts

cal	calibrated
i	indicated
ic	instrument corrected
pc	position error correction
SL	sea level
∞	freestream value

Acronyms and Abbreviations

ASI	airspeed indicator
CAD	computer-aided drafting
CAS	calibrated airspeed
EAS	equivalent airspeed
GPS	global positioning system
GS	ground speed
IAS	indicated airspeed
KEAS	knots equivalent airspeed
KIAS	knots indicated airspeed
OAT	outside air temperature
TAS	true airspeed
WS	wind speed

References

Corda, S. (2017). *Introduction to Aerospace Engineering with a Flight Test Perspective*, 673–674. Chichester, West Sussex, UK: Wiley.

Federal Aviation Administration (2012). *Instrument Flying Handbook*, FAA-H-8083-15B, United States Department of Transportation, Federal Aviation Administration, Airman Testing Standards Branch, AFS-630, P.O. Box 25082, Oklahoma City, OK 73125. https://www.faa.gov/regulations_policies/handbooks_manuals/aviation/media/FAA-H-8083-15B.pdf

Gracey, W. (1980). Measurement of Aircraft Speed and Altitude. *NASA Technical Report*, NASA-RP-1046. http://hdl.handle.net/2060/19800015804.

Gregory, J.W. and McCrink, M.H. (2016). *Accuracy of Smartphone-Based Barometry for Altitude Determination in Aircraft Flight Testing*. AIAA Paper 2016-0270, AIAA Atmospheric Flight Mechanics Conference, San Diego, CA, January 4–8, 2016. https://doi.org/10.2514/6.2016-0270

Haering, Jr., E.A. (1995). Airdata Measurement and Calibration. *NASA Technical Memorandum*, NASA-TM-104316. http://hdl.handle.net/2060/19990063780.

Herrington, R.M., Shoemacher, P.E., Bartlett, E.P., and Dunlap, E.W. (1966). *Flight Test Engineering Handbook*. USAF Technical Report 6273, Edwards AFB, CA: US Air Force Flight Test Center. Defense Technical Information Center Accession Number AD0636392, https://apps.dtic.mil/docs/citations/AD0636392.

NASA. (2014). X-15 #1 Rocket-Powered Aircraft. https://www.nasa.gov/centers/dryden/multimedia/imagegallery/X-15/E-5251.html

Schoolfield, W.C. (1942). A simple method of applying the compressibility correction in the determination of true air speed. *Journal of the Aeronautical Sciences* 9 (12): 457–464. https://doi.org/10.2514/8.10935.

Ward, D., Strganac, T., and Niewoehner, R. (2006). *Introduction to Flight Test Engineering*, 3e, vol. 1. Dubuque, IA: Kendall/Hunt.

9

Climb Performance and Level Acceleration to Measure Excess Power

The climb performance of an aircraft has a direct bearing on many aspects of flight planning and piloting. In flight planning, a pilot needs to know how long it will take for the aircraft to climb to cruise altitude and the maximum altitude that the aircraft can reach. This has a direct impact on fuel usage, range, and the total time of the flight. Similarly, a pilot needs to know the airspeed to be flown in order to achieve maximum climb performance, whether that is the airspeed for maximum climb gradient or the speed for minimum time to climb. Climb performance may be predicted analytically, as we will do in this chapter, and measured experimentally through flight test.

An aircraft's climb performance is directly dependent on the total excess energy, or power, available to do work (to climb to higher altitude, in this case). This excess energy is directly related to the difference between propulsive power available and the aerodynamic drag acting on the aircraft. Both power available and aerodynamic drag vary across a range of airspeeds, so the excess power and climb performance are a function of airspeed.

We will first develop the theory of climb performance, with a baseline assumption of steady climbs. We will then turn our attention to energy methods, which are useful for rapid climbs by high-performance aircraft. Next, we will discuss flight testing methods for finding the climb performance, which can be a direct measurement of rate of climb across a range of airspeeds. We will also see that the climb performance can be indirectly measured by performing a level, full-throttle acceleration. We will conclude our chapter with some example flight test data, which compares the two approaches.

9.1 Theory

Our analysis of an aircraft in climb begins with the equations of motion for the aircraft along a direction parallel to the aircraft's flight path. Based on the forces acting on the aircraft, illustrated in Figure 9.1, the equation of motion along the flight path is

$$T\cos(\alpha - \varepsilon) - D - W\sin\gamma = m\frac{dV}{dt},$$ (9.1)

where T, D, and W are the thrust, drag, and weight, respectively; m is the vehicle mass; V is the true airspeed; γ is the angle of climb (angle between the flight path and a horizontal ground reference plane); α is the angle between the aircraft's zero lift line and its flight path, and ε is the thrust angle (for vectored thrust) – the angle between the thrust line and the zero lift line. If we neglect the thrust angle (ε) and assume that α is small (in other words, thrust is aligned with the direction of flight) and multiply through by V, we have

$$TV - DV - WV\sin\gamma = mV\frac{dV}{dt}.$$ (9.2)

Introduction to Flight Testing, First Edition. James W. Gregory and Tianshu Liu.
© 2021 John Wiley & Sons Ltd. Published 2021 by John Wiley & Sons Ltd.
Companion website: https://www.wiley.com/go/flighttesting

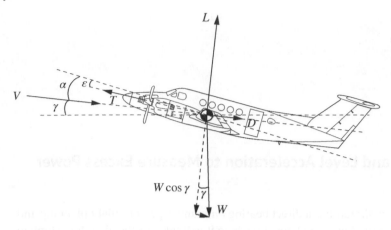

Figure 9.1 Forces acting on an aircraft in climb.

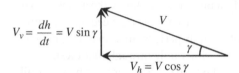

Figure 9.2 Velocity triangle for an aircraft in a steady climb.

Referring to a velocity triangle for the aircraft in this flight condition (Figure 9.2), we see that the rate of climb is $dh/dt = V \sin \gamma$, such that Eq. (9.2) becomes

$$TV - DV - W\frac{dh}{dt} = mV\frac{dV}{dt}. \tag{9.3}$$

Dividing through by aircraft weight ($W = mg$) and moving rate of climb to the right side give

$$\frac{(T-D)V}{W} = \frac{dh}{dt} + \frac{V}{g}\frac{dV}{dt}. \tag{9.4}$$

The term on the left side of Eq. (9.4) is defined as the specific excess power of the aircraft,

$$P_s = \frac{(T-D)V}{W} = \frac{P_A - P_R}{W}, \tag{9.5}$$

where TV is the power available (P_A) from the engine at a particular altitude and DV is the power required (P_R) for flight at that same altitude. The specific excess power simply represents how much "leftover" power is available from the engine to climb to a higher altitude or accelerate, normalized by the aircraft weight. In other words, this is the power available above and beyond that which is required to overcome drag in level flight. Combining Eqs. (9.4) and (9.5), we have

$$P_s = \frac{(T-D)V}{W} = \frac{dh}{dt} + \frac{V}{g}\frac{dV}{dt}. \tag{9.6}$$

This equation represents the two ways that excess power may be applied to increase the energy of the aircraft: the first term on the right side is the rate of climb (time rate of change of potential energy), and the second term is acceleration (time rate of change of kinetic energy).

9.1.1 Steady Climbs

For time being, we will assume that the aircraft is in steady, unaccelerated flight, such that the second term on the right side of Eq. (9.6) vanishes. Thus, for a steady climb, the rate of climb may be expressed as

$$\frac{dh}{dt} = \frac{(T-D)V}{W} = \frac{P_A - P_R}{W} = P_s. \tag{9.7}$$

Since both thrust and drag depend on the altitude (density) of the aircraft, the steady climb performance of an aircraft depends on aircraft weight, altitude, and airspeed.

For the purposes of this analysis, it is convenient to assume that the drag characteristics of the aircraft are the same in a climb as they are in steady, level flight. (In other words, the development of expressions for drag will make a small angle assumption, such that $\cos \gamma \cong 1$ and that no component of the weight vector acts along the longitudinal axis of the aircraft.) Briefly recapitulating our discussion of level flight performance in Chapter 7, the drag acting on the aircraft is

$$D = D_0 + D_i = \frac{1}{2}\rho V^2 S(C_{D_0} + C_{D_i}), \tag{9.8}$$

where the first term is parasitic drag and the second term is induced drag (Figure 9.3). Assuming that the parasitic drag coefficient is constant, that the induced drag coefficient is parabolic ($C_{D_i} = C_L^2/\pi \, AR \, e$), and that lift equals weight ($C_L = 2W/\rho SV^2$), we have

$$D = D_0 + D_i = \frac{1}{2}\rho V^2 S C_{D_0} + \frac{2W^2}{\pi b^2 e \rho V^2}. \tag{9.9}$$

Since the power required for steady, level flight is the product of airspeed and drag, we can multiply Eq. (9.9) by V to obtain expressions for induced power, parasitic power, and total power required,

$$P_R = P_0 + P_i = \frac{1}{2}\rho V^3 S C_{D_0} + \frac{1}{2}\rho V^3 S C_{D_i} = \frac{1}{2}\rho V^3 S C_{D_0} + \frac{2W^2}{\pi b^2 e \rho V}. \tag{9.10}$$

These quantities are plotted in Figure 9.4, showing the relative contributions of the power to overcome parasitic drag (the first term in Eq. (9.10)) and the power to overcome induced drag (the second term) to the overall power required. From a piloting perspective, an important feature of

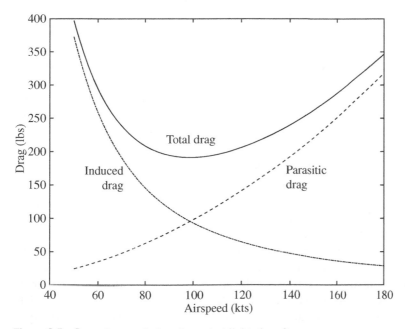

Figure 9.3 Drag characteristics of a typical light aircraft.

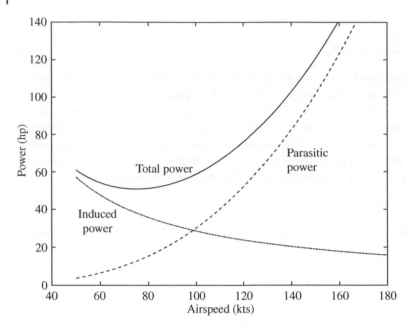

Figure 9.4 Power characteristics of a typical light aircraft. The "back side" of the power curve is clearly seen at airspeeds below 75 kts, where an increase in power is required to fly more slowly.

the power curve is the behavior at low speeds. The power required for steady, level flight actually increases when the aircraft decelerates from an already low speed (below 75 knots, in this case). This is known as the "back side" of the power curve and can have dangerous consequences if a pilot is caught unaware of this aspect of aircraft performance. A pilot can decelerate by pitching up, but if power is not simultaneously increased, the aircraft will start to descend. A novice pilot may then be tempted to increase the aircraft's pitch further, thus exacerbating the situation and potentially leading to stall if the angle of attack is increased by too much. The reason the overall power required increases on the "back side" of the curve is due to the increase in induced drag when the wing is working harder to produce lift to sustain flight at a lower airspeed.

Turning our attention now to climb performance, it is clear that the power required (Eq. (9.10)) is a function of altitude (density), as well as airspeed. The characteristic power required curve is shown in Figure 9.5 for three altitudes routinely encountered in flight for light aircraft. With all other conditions being equal, the power required curve shifts up and to the right as altitude increases. Power required for flight at low airspeed increases with altitude, and at high airspeed, the power decreases with altitude. The optimum (minimum) power condition increases slightly and shifts to higher airspeed as the altitude increases.

Typical power available curves for the same three altitudes are also plotted in Figure 9.5. Power available curves are estimated from engine performance data (see Chapter 7) and assumed propeller efficiency curves. The shaft horsepower of the engine decreases at altitude since less air is available for the combustion process and the engine must be leaned in order to maintain the proper stoichiometric ratio. The reduction in power at lower airspeeds is due to the characteristic efficiency curve for the propeller (see Chapter 7). The key feature of these power curves, in the context of climb performance, is the difference between the power available and the power required. Recall from Eq. (9.7) that the rate of climb is directly related to the specific excess power, which is the difference between power available and power required. For this notional aircraft, the specific excess power is shown in Figure 9.6, expressed as rate of climb. There is a specific airspeed

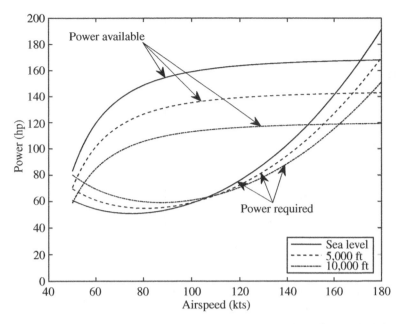

Figure 9.5 Power required and power available characteristics for a typical light aircraft.

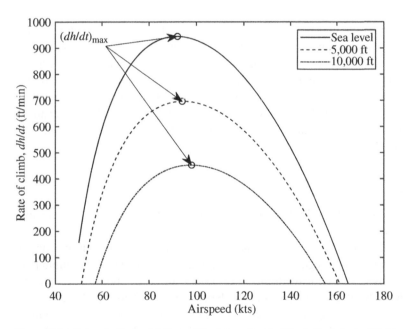

Figure 9.6 Rate of climb variation with airspeed and altitude for a typical light aircraft.

associated with best climb performance (maximum rate of climb), and this airspeed changes with altitude. The best climb speed increases as altitude increases in this case, due to the characteristic shapes of the power available and power required curves.

Assuming that the aircraft is flown at its best climb speed throughout a climb, the maximum rate of climb can be plotted as a function of altitude (Figure 9.7) to estimate the maximum altitude that the aircraft can fly. This maximum altitude can be expressed as either the service ceiling (h_{svc}),

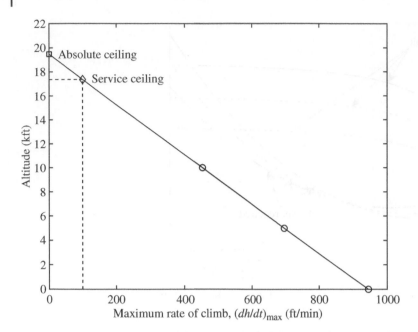

Figure 9.7 Extrapolation of maximum rates of climb to define service and absolute ceilings.

defined as the altitude at which the maximum rate of climb is 100 ft/min, or the absolute ceiling (h_a), where the maximum rate of climb diminishes to 0 ft/min. The service ceiling serves as a practical maximum altitude, since atmospheric conditions often make it difficult to maintain peak climb performance, and the ceiling is a corner of the flight envelope. These ceilings are fundamentally limited by the amount of excess power available from the aircraft to climb to higher altitudes. If the rate of climb data is known (from flight test or analysis) at several altitudes, a linear fit can be applied to the data in order to extrapolate and find the service and absolute ceilings. Presuming a linear fit, the expression is given by

$$h = h_a \left[1 - \frac{(dh/dt)_{max}}{(dh/dt)_{max,SL}} \right], \tag{9.11}$$

where $(dh/dt)_{max}$ is the maximum rate of climb at a given altitude h and $(dh/dt)_{max, SL}$ is a constant value at sea level conditions.

For a given engine and airframe, the climb performance then is heavily dependent on the air density. On a hot day, the density drops significantly, which can dramatically lower the climb performance of the aircraft. This is why density altitude is so important in aviation – high-density altitudes (low density) adversely affect the takeoff and climb performance of aircraft. For example, on a 75 °F standard pressure day, an airport with a field elevation of 6500 ft above sea level would have a density altitude of nearly 9000 ft. It is important to recognize, then, that the ceilings shown in Figure 9.7 are density altitudes, which only correspond to true altitudes on a standard day.

The functional relationship (curve fit) between altitude and maximum rate of climb, used to define the ceilings in Figure 9.7, can also be used to determine the time to climb between two given altitudes. If the inverse of rate of climb is plotted as a function of altitude (Figure 9.8), the curve can simply be integrated between two altitudes to determine the time to climb,

$$\Delta t = \int_1^2 \frac{dh}{(dh/dt)_{max}}. \tag{9.12}$$

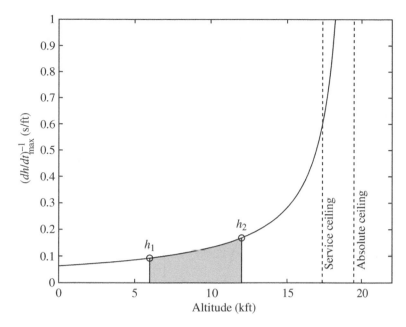

Figure 9.8 Instantaneous time to climb, which asymptotes to infinity at the absolute ceiling. The time required to climb between two altitudes can be found by integrating the function between these limits (the shaded area under the curve).

Maximum rate of climb as a function of altitude can be found by solving Eq. (9.11) for $(dh/dt)_{max}$,

$$\left(\frac{dh}{dt}\right)_{max} = \left(\frac{h_a - h}{h_a}\right)\left(\frac{dh}{dt}\right)_{max,SL}. \tag{9.13}$$

The inverse of this function represents the increment in time required to climb the next small increment of altitude. If we plot $(dh/dt)_{max}^{-1}$ for our notional aircraft over a range of altitudes (Figure 9.8), we see that the incremental time to climb increases significantly as the service ceiling is reached and asymptotes to infinity at the absolute ceiling. Substituting Eq. (9.13) into (9.12) and integrating, we find

$$\Delta t = \frac{h_a}{(dh/dt)_{max,SL}} \ln\left(\frac{h_a - h_1}{h_a - h_2}\right) \tag{9.14}$$

for the time to climb from altitude h_1 to h_2 if the absolute ceiling and maximum rate of climb at sea level are known, and assuming a linear dependence of $(dh/dt)_{max}$ with altitude. In the example illustrated in Figure 9.8, it takes 739 seconds (12.3 minutes) to climb from h_1 (6000 ft) to h_2 (12,000 ft) on a standard day.

It is also important to recognize that maximum rate of climb is not the complete story when it comes to climb performance. Pilots often encounter airfields with surrounding obstacles that must be cleared shortly after takeoff. In this limiting case, it is best to fly the aircraft on a flight path with the highest angle relative to the ground, which maximizes altitude gained for a given *horizontal distance* flown. This condition is distinctly different from the best rate of climb condition, which maximizes altitude gained in a given amount of *time*. Critically, the airspeed for best climb angle is not the same as the airspeed to be flown for best rate of climb. Both airspeeds are clearly shown in the pilot's operating handbook (POH), and the airspeed for best climb angle is always lower than the airspeed for best rate of climb. In piloting nomenclature, the airspeed for best climb angle is expressed as V_x and the airspeed for best rate of climb is expressed as V_y.

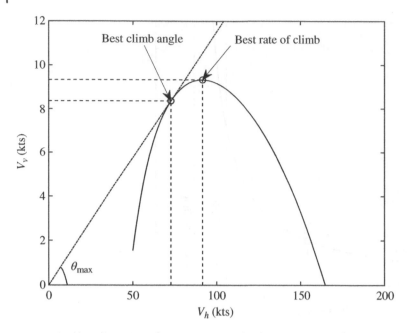

Figure 9.9 Climb hodograph, illustrating the difference between best climb angle and best rate of climb. Note that the *x*-axis is the horizontal component of the airspeed.

The difference between these two airspeeds is best illustrated by a climb hodograph, shown in Figure 9.9 for our notional aircraft at sea level on a standard day. The climb hodograph is constructed by plotting the vertical speed (V_v, which is the same as dh/dt) versus the horizontal speed (V_h) of the aircraft for all airspeeds flown by the aircraft. The vertical and horizontal speeds form two legs of a right triangle, with the true airspeed on the hypotenuse (see Figure 9.2). The best rate of climb condition is defined in a straightforward manner as the maximum V_v (or dh/dt) on the hodograph, in the same way as this condition was defined on Figure 9.6. The condition for best climb angle is defined by plotting a line from the origin to a tangent point on the performance curve such that the slope of this line is maximized. This line is shown in Figure 9.9, whereas the slope of that line is plotted in Figure 9.10 as a function of airspeed. These figures clearly show the difference between V_x and V_y, with the speed for best climb angle being lower than the speed for best rate of climb. Flight testing of climb performance involves directly measuring the rate of climb at various airspeeds on the climb hodograph (Figure 9.9) and repeating those measurements at different altitudes.

9.1.2 Energy Methods

We will now revisit our discussion on specific excess power, this time relaxing our assumption of steady flight. This will allow us to develop an entirely different method for measuring climb performance. Returning our attention to Eq. (9.6), we see that excess power can be used to climb, accelerate, or some combination of the two. Fundamentally, the specific excess power can be applied to increase the overall energy of the aircraft – either by climbing to a higher altitude (increasing potential energy) or by accelerating to a faster airspeed (increasing kinetic energy). This concept forms the basis for an energy approach to climb performance, which is particularly relevant for high-performance aircraft such as fighter jets.

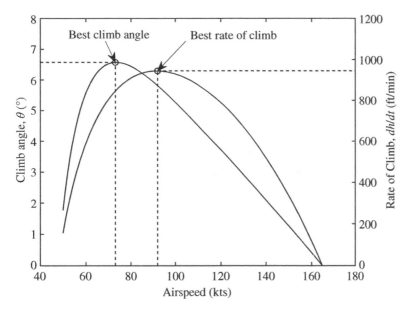

Figure 9.10 Airspeeds for best climb angle and best rate of climb.

Our discussion of energy methods for climb performance begins with a definition of the total energy of the aircraft per unit mass,

$$e_{\text{total}} = \frac{1}{2}V^2 + gh, \tag{9.15}$$

which is the sum of the kinetic and potential energy of the aircraft. This expression can be divided by gravitational acceleration to define the energy height of the aircraft,

$$h_e \equiv \frac{e_{\text{total}}}{g} = \frac{V^2}{2g} + h, \tag{9.16}$$

which has units of distance. Energy height is an important and useful concept for analysis of air-craft performance, since it is a clear representation of the overall energy of an aircraft. Contours of constant energy height are plotted in Figure 9.11 across a range of altitudes and airspeeds. (Note that Figure 9.11 is simply a plot of Eq. (9.16); it is not tied to any particular aircraft.) These contours indicate that an aircraft can trade altitude (potential energy) and airspeed (kinetic energy) without changing the overall energy of the aircraft. This could be accomplished by an aircraft changing flight condition from one point along a contour to another point on the same contour (say, going from point A to point B in Figure 9.11). The total energy of the aircraft remains constant throughout this maneuver (following the $h_e = 3000$ ft energy contour), so no energy above and beyond that which is required to sustain steady, level flight at point A is required to accelerate and descend to point B. However, if a pilot wishes to increase the energy of the aircraft by climbing to higher altitude or accelerating, then power must be added. For example, starting from flight condition B (150 kts at 2000 ft, or $h_e = 3000$ ft), the pilot can climb to an altitude of 3000 ft at the same airspeed of 150 kts (point C on the diagram). This flight condition is now on the $h_e = 4000$ ft energy contour, which is an increase in energy height of 1000 ft from the initial condition of $h_e = 3000$ ft. Or, the pilot could accelerate from 150 to 212 kts at the same altitude of 2000 ft (moving from point B to point D, again increasing the energy height from 3000 to 4000 ft). Finally, a pilot could accelerate and climb from point B to a speed of 184 kts at an altitude of 2500 ft (point E), which is also at an energy height of 4000 ft. In all of these cases, the same amount of energy must be added in order to

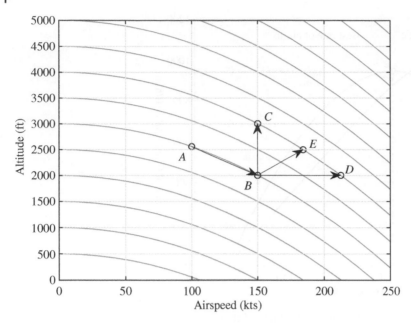

Figure 9.11 Contours of constant energy height.

change the flight condition. (For this discussion, we have been neglecting any dependency of drag on altitude or airspeed, and just considering the change in energy of the flight vehicle.)

We will now return to the aircraft's equation of motion expressed in Eq. (9.3) and continue our derivation in a form suited for energy methods. The right-hand side can be re-cast such that

$$TV - DV - W\frac{dh}{dt} = m\frac{d}{dt}\left(\frac{V^2}{2}\right) \tag{9.17}$$

and the rate of climb term can be moved to the right side,

$$TV - DV = m\frac{d}{dt}\left(\frac{V^2}{2}\right) + mg\frac{dh}{dt}. \tag{9.18}$$

Re-grouping the right side of Eq. (9.18),

$$TV - DV = m\frac{d}{dt}\left(\frac{V^2}{2} + gh\right), \tag{9.19}$$

we recognize that the term within the parentheses is the specific total energy of the aircraft, given by Eq. (9.15). Thus, Eq. (9.19) can also be expressed as

$$TV - DV = m\frac{d}{dt}(e_{\text{total}}). \tag{9.20}$$

Dividing through by the aircraft weight, we establish a relationship between the specific excess power and the time rate of change of the aircraft's energy height:

$$P_s = \frac{TV - DV}{W} = \frac{1}{g}\frac{d}{dt}(e_{\text{total}}) = \frac{dh_e}{dt}. \tag{9.21}$$

Equation (9.21) is another representation of the idea that excess power can be used to climb, accelerate, or some combination of the two. In other words, power is applied to change energy of the aircraft, expressed as energy height (the sum of specific kinetic energy and specific potential energy, divided by gravitational acceleration).

Returning our attention to the plot of energy contours shown in Figure 9.11, energy must be expended in order to move an aircraft to a contour of increased energy height. For example, this could be a constant-airspeed climb (moving from point B to C), a constant-altitude acceleration (moving from point B to D), or an accelerated climb (moving from point B to E). Excess power must be available in order to increase the energy height of the aircraft in any of those maneuvers. The magnitude of an aircraft's specific excess power at a given flight condition describes how quickly the aircraft can climb or change airspeed. It is important to recognize that an aircraft's specific excess power depends on both airspeed and altitude, due to the effects of these two critical parameters on the aircraft's drag and engine performance (see Figure 9.5, for example). In order to quantitatively express an aircraft's ability to climb or accelerate, contours of specific excess power (Eq. (9.21)) may be plotted as a function of altitude and Mach number in the same way that contours of energy height were plotted. This method was first developed by Rutowski (1954) and has been used extensively for quantifying the performance of high-performance aircraft (Small and Prueher 1977). Figure 9.12 illustrates contours of specific excess power for a supersonic fighter jet, based on an analysis of data for a non-afterburning jet at maximum takeoff weight (data from appendix A.3 of Filippone 2006).

Substantial insight on the aircraft's performance can be gleaned from an energy chart such as the one in Figure 9.12. First, it is evident that the aircraft's specific excess power decreases rapidly with altitude. If a pilot wishes to extract maximum performance from the aircraft for climbs and accelerations (in air-to-air engagement with another aircraft, for example), the best condition for maneuvering is at high subsonic Mach numbers near sea level. Also, since the diagram plots specific excess power, a pilot would want to minimize aircraft weight (perhaps by dumping fuel) in order to optimize maneuverability. The performance envelope – the limit for steady flight conditions, or the boundary for ceiling and airspeed – is represented by the contour for zero specific excess power.

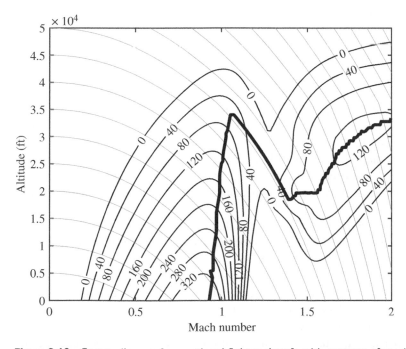

Figure 9.12 Energy diagram for a notional fighter aircraft, with contours of specific excess power (labeled black contours, units of ft/s) superposed on contours of energy height (unlabeled gray contours). The thick line represents the flight profile for minimum time to climb from sea level to high-altitude supersonic flight. Source: Data from Filippone (2006).

Beyond this $P_s = 0$ contour the aircraft (in a given flap configuration) cannot sustain steady, level flight (and clearly cannot climb or accelerate) due to a lack of power available. The "back side" of the power curve is also readily visible in Figure 9.12: the drop-off in the contours at lower subsonic airspeeds (for a constant altitude) indicates that increased power is required to fly more slowly.

Optimum climb profiles of airspeed and altitude may be determined from energy charts such as Figure 9.12 for a notional fighter aircraft. Returning to the relationship between specific excess power and the time rate of change of energy height, Eq. (9.21), we can re-write it as

$$dt = \frac{dh_e}{P_s}. \tag{9.22}$$

Time to climb can be found by integrating between two energy heights,

$$\Delta t = t_2 - t_1 = \int_1^2 \frac{dh_e}{P_s}, \tag{9.23}$$

which is a generalized form of Eq. (9.12). Thus, in order to minimize the time to climb, a pilot should maximize P_s with respect to energy height throughout the climb. This leads to an altitude/airspeed flight profile that follows the locus of the maxima of the P_s contours (with respect to energy height). An important distinction to make is that this maximum condition is not the peak of a P_s contour in Figure 9.12 (which would be the maximum with respect to *altitude*), but is the maximum with respect to *energy height*. This can be determined graphically for a given P_s contour by finding the point on the contour where it is tangent to the highest possible contour of energy height. The thick black line in Figure 9.12 represents a locus of such points – note that the line is slightly to the right of the peak P_s values. Thus, the flight profile of Mach number and altitude to be flown for minimum time to climb is represented by the thick line in Figure 9.12.

It may be surprising to observe that the climb profile is not a continuous increase in altitude once sonic conditions are reached. The primary reason for this is the presence of a transonic drag rise between subsonic and supersonic speeds, which results from compressibility effects at Mach numbers greater than the drag divergence Mach number. The peak drag occurs near a Mach number of 1.2 for this notional aircraft, and then the drag decreases again at higher Mach numbers. This is evident in Figure 9.12 through the dramatic decrease in specific excess power near low supersonic speeds. This transonic drag rise leads to some dramatic implications for aircraft performance. For example, it is not possible for a pilot flying in the aircraft represented by Figure 9.12 to perform a level acceleration from a flight condition of Mach 0.9 and 35,000 ft to a supersonic Mach number of 1.4 at the same altitude. This is because there is a "valley" of specific excess power between these two flight conditions where the $P_s = 0$ contour falls between the two. Thus, a pilot in this aircraft who wishes to accelerate to the new flight condition must first dive to a lower altitude where specific excess power is available to then accelerate to supersonic speeds and climb back to the original altitude of 35,000 ft.

Returning to the idea of minimum time to climb, the flight profile from subsonic low-altitude flight to supersonic high-altitude flight is a very interesting one (the entirety of the dark line in Figure 9.12). As shown in Eq. (9.23), one must continue to minimize specific excess power with respect to energy height throughout the climb. However, some ambiguity arises in the transonic regime due to the local increase in drag and its associated drop in excess power available. There is a flight condition (defined by Mach and altitude) in the initial locally optimized subsonic climb where a new supersonic branch in the profile becomes a global optimum. In other words, the locus of optimal points will have a discontinuity between the subsonic and supersonic branches due to the strong effects of the transonic drag rise. Both the subsonic and supersonic branches are clearly illustrated by the dark line in Figure 9.12. Connecting the two branches is a line following a contour of constant energy height. This portion of the profile represents a constant-energy zoom dive where altitude (potential energy) is traded for airspeed (kinetic energy) to accelerate through the transonic drag rise in a minimum amount of time. Thus, an optimal climb to supersonic speeds (minimum time to climb) for this notional fighter aircraft will actually include a constant-energy descent during a portion of the overall climb profile!

Note that the flight profile indicated by Figure 9.12 is descriptive rather than prescriptive. It would be difficult for a pilot to precisely control altitude and airspeed following this profile throughout the climb, and the sharp transitions between the climb segments and the zoom dive are unrealistic. However, the flight profile does represent the general locus of conditions that should be flown for minimum time to climb. One final point regarding energy diagrams such as Figure 9.12 is worth noting. Each aircraft will have a unique energy diagram based on its drag throughout the flight envelope (both density and airspeed have first-order effects) and its engine performance (strongly dependent on altitude). Different aircraft with different energy diagrams will have different conditions where performance is optimized (generally determined by maximizing specific excess power). When two fighter aircraft are engaged in air-to-air combat, maximum advantage in maneuverability is achieved when there is the greatest difference in flight performance between the two aircraft. To aid in determination of this flight condition, differential specific excess power diagrams may be created for any two aircraft, which would simply be created by subtracting the energy diagram of an opposing aircraft from the diagram for the pilot's aircraft. Regions of maximum positive differential specific excess power are where the pilot wishes to remain for optimal advantage.

This discussion of energy methods is also relevant to relatively low-powered general aviation aircraft and unmanned aerial vehicles (UAVs) since it leads to an alternate method for determining climb performance. Referring back to Eq. (9.6), specific excess power can be quantified in flight test by either directly measuring steady climb performance (first term on the right side) or by measuring the acceleration performance in level flight (second term on the right side). A level acceleration flight test can measure the specific excess power of the aircraft across a range of airspeeds by simply recording the time-resolved velocity and acceleration throughout the level acceleration maneuver. Once specific excess power across a range of airspeeds is known from a level acceleration flight test, then the steady climb performance can be found from Eq. (9.6) (by setting $dV/dt = 0$). Key assumptions involved with the level acceleration flight test method are that unsteady drag effects are neglected, that the drag characteristics (induced and parasitic) are the same for accelerations and climbs, that no unsteady propeller interactions are present, and that the power available is always a maximum for the instantaneous airspeed. In other words, the instantaneous value of measured specific power in a level acceleration is assumed to be equal to the steady value of specific power at that flight speed (transient effects are neglected).

9.2 Flight Testing Procedures

We will now take these two ideas – direct measurement of rate of climb and level acceleration – and discuss how to implement them in a practical flight testing environment for measuring specific excess power. Best piloting practices are described, in order to optimize data quality. We will also consider the various data sources that must be recorded in order to effectively analyze the results and present them in standard form.

9.2.1 Direct Measurement of Rate of Climb

The first method for documenting the climb performance of an aircraft is to directly measure the time to climb. Since the maximum performance of the aircraft is desired for this test, each climb should be conducted at full throttle. For this technique, a series of pressure altitudes should be selected for which the climb performance will be measured. (Note that the altimeter should be referenced to 29.92 inHg in order to measure pressure altitude.) For each test altitude, the test pilot should start and stop the climb at altitudes $\pm \Delta h/2$ relative to the target altitude, ensuring that a steady climb has been established before the initial altitude ($h_{\text{target}} - \Delta h/2$) has been reached, and maintained at least until the final altitude ($h_{\text{target}} + \Delta h/2$) is reached. The beginning and ending altitudes for each climb should be selected such that sufficient time elapses for the averaging of data, but should not be so large that the test program is unnecessarily lengthy. For example, for a target altitude of 3000 ft MSL and $\Delta h = 400$ ft, a steady climb should be established by 2800 ft and maintained through at least 3200 ft. The rate of climb ($dh/dt = \Delta h/\Delta t$) is most easily measured by recording the time at the beginning and end of the steady portion of the climb, using a stopwatch. (Note that the reading on the vertical speed indicator on the aircraft will be subject to phase delays and fluctuations in the reading, making it less suitable for this measurement.) For convenience, the climb testing is often done in conjunction with glide flight tests (Chapter 10), where a sawtooth pattern of alternating climbs and descents is flown in order to save time. In order to correct for non-standard conditions, the takeoff weight must be recorded, with fuel burn and outside air temperature (OAT) measured throughout the flight. If comparison with drag polar data is desired, engine parameters such as engine speed (revolutions per minute (rpm)), manifold pressure, and percent power should also be recorded during each climb.

For each test altitude, a range of airspeeds should be selected in order to quantify the aircraft performance throughout the excess power curve (*e.g.*, see Figure 9.6). The airspeed test points should cover the range from just above stall to near the maximum airspeed, such that the shape of the excess power curve is adequately resolved. Repeat runs for the same flight condition may be desired in order to improve the statistics of the data. It is particularly important to conduct climb testing on a calm day; updrafts, downdrafts, atmospheric turbulence, and vertical wind shear can all have a substantial impact on the quality of the recorded data. It is best if the flight path is flown over terrain where thermal updrafts or downdrafts are unlikely to be found; avoid parking lots, bodies of water, or other terrain features that have different heat capacities from the surrounding terrain that could result in significant updrafts or downdrafts depending on the time of day. The flight path should be oriented in the crosswind direction, such that wind shear effects are minimized. For optimal results, data should also be acquired on a flight path flown in the opposite direction, in order to cancel wind shear effects.

9.2.2 Measurement of Level Acceleration

In some scenarios, it may be more feasible to measure the level acceleration performance of an aircraft instead of a direct measurement of time to climb. For example, flight operations of UAVs could be limited to low altitudes due to regulatory and safety concerns, which would limit the range of altitudes available to establish a steady climb. Or, in other scenarios, a level acceleration can be used to measure specific excess power much more quickly than by a series of climbs at different airspeeds, thus saving on test time and cost.

Referring to Eq. (9.6), we will take $dh/dt = 0$ and make direct measurements of $V(t)$ and dV/dt. Traditional methods of data acquisition (hand-written recordings) may be too slow to provide the necessary sample rate (on the order of 1 sample per second). Thus, digital DAQ techniques are ideal for recording the time-resolved velocity and acceleration data needed for the level acceleration method. Time-resolved velocity is best recorded by a Global Positioning System (GPS) receiver, which typically has an effective sample rate falling within the range of 1–10 Hz.

However, note that GPS measures ground speed, so the local wind speed must be accounted for in order to convert to true airspeed (see Chapter 8 for details on how to fly a cloverleaf maneuver in order to measure wind speed). The best approach for finding the acceleration is to numerically differentiate the time-resolved velocity measurement. An accelerometer could possibly be used, but this is a challenging measurement due to the need for coordinate transformations between the accelerometer triad and the aircraft body axes, and from the body axes to the direction of flight. (Note that the aircraft attitude is changing throughout the maneuver in order to control angle of attack and maintain fixed altitude.) Also, accelerometers typically have a higher level of noise than what would be obtained by differentiating the GPS velocity time history. In addition, measurements of takeoff weight and fuel burned are needed in order to determine the aircraft weight at the time of the maneuver. Pressure altitude and OAT should also be recorded.

A level acceleration is performed by first setting up the aircraft in slow flight at an airspeed as close to stall as reasonable. Then, maximum power is impulsively applied in a smooth motion, and altitude is held constant while the aircraft accelerates. Active management of the aircraft attitude will be required throughout the maneuver, with significant nose-down stick pressure required at higher airspeeds in order to avoid a climb. If an autopilot is available on the aircraft, the pilot may wish to use the altitude hold feature while performing this maneuver. Depending on the type of aircraft being tested, a single level acceleration maneuver may require about 90 seconds to complete. Wind speed can be estimated by performing a constant-speed cloverleaf maneuver immediately before or after the level acceleration.

9.3 Data Analysis

Once specific excess power data are recorded (by either method), it must be corrected for non-standard atmospheric conditions and non-standard weight. The basis for these corrections is discussed as follows. There are several different correction approaches proposed in the literature (Herrington et al. 1966; Smith 1981; Kimberlin 2003; Ward et al. 2006). Here, we will focus only on a correction of the altimeter reading for non-standard temperature (a small effect), and correction for non-standard weight (a large effect), largely following Smith's (1981) suggested procedure. Note that the analysis of flight data should be performed with respect to true airspeed, rather than indicated airspeed or ground speed. (See Chapter 8 for details on the various airspeed definitions and how to convert between them.)

A correction for non-standard temperature is required due to the assumption of standard temperature in the calibration of the altimeter. When referenced to 29.92 inHg, the altimeter records pressure altitude as long as standard temperature conditions are present. If not, then the altimeter reading must be corrected in order to obtain the actual height above sea level (Herrington et al. 1966; Smith 1981).

Building off the material presented in Chapter 2, we will start with a form of the hydrostatic equation written for a change in altitude measured in climb performance testing,

$$\Delta p = -\rho g \Delta h = -\frac{p}{RT} g \Delta h. \tag{9.24}$$

For an altimeter calibrated to standard temperature, $T = T_{std}$, the change in pressure corresponding to a measured change in altitude (Δh_{meas}) is

$$\Delta p = -\frac{p}{RT_{std}} g \Delta h_{meas}. \tag{9.25}$$

The actual change in altitude (Δh_{act}) that would have been indicated by the altimeter in response to that same change in pressure, if the altimeter had been calibrated to the actual temperature, is

$$\Delta p = -\frac{p}{RT_{act}}g\Delta h_{act}. \tag{9.26}$$

Since both (9.25) and (9.26) are expressions for the same change in pressure that the aircraft experiences, the two may be equated in order to obtain our correction factor,

$$\Delta h_{act} = \Delta h_{meas}\frac{T_{act}}{T_{std}}, \tag{9.27}$$

where temperature is in absolute units (K or °R). This expression can be equivalently expressed as

$$\left(\frac{dh}{dt}\right)_{act} = \left(\frac{dh}{dt}\right)_{meas}\frac{T_{act}}{T_{std}}. \tag{9.28}$$

Correction for non-standard weight is needed since the aircraft weight has a non-trivial impact on the climb performance (a more heavily loaded airplane will climb more slowly). Thus, the rate of climb values need to be corrected to a standard weight, which is the maximum gross takeoff weight of the aircraft. Kimberlin (2003) provides an approach to weight correction, termed the power independent of weight (PIW)-climb independent of weight (CIW) method (similar to the PIW-velocity independent of weight (VIW) method discussed in Chapter 7). We will discuss here an alternative approach, detailed by Smith (1981) and Herrington et al. (1966) that is based on differentiating specific excess power with respect to weight in order to find the weight correction terms.

Returning to our definition of rate of climb as a function of specific excess power (Eq. (9.7)),

$$\frac{dh}{dt} = P_s = \frac{TV - DV}{W}, \tag{9.29}$$

we need to differentiate P_s with respect to aircraft weight in order to determine the sensitivity of rate of climb to variations in weight. Since drag is a function of aircraft weight, this gives

$$\frac{dP_s}{dW} = -\frac{TV}{W^2} + \frac{DV}{W^2} - \frac{V}{W}\frac{dD}{dW}. \tag{9.30}$$

Noting that drag is composed of parasitic and induced components, we can substitute Eq. (9.9) into Eq. (9.30) and differentiate drag with respect to weight. Assuming small changes in weight such that small finite differences equal the differentials gives the correction factor

$$\Delta P_s = -\Delta W\left(\frac{P_{s,meas}}{W_{test}} + \frac{4}{\rho V b^2 \pi e}\right), \tag{9.31}$$

where $\Delta P_s = P_{s,std} - P_{s,meas}$, $\Delta W = W_{std} - W_{test}$, W_{std} is the aircraft's maximum gross takeoff weight, W_{test} is the actual aircraft weight at the time of the measurement, and the Oswald efficiency factor (e) may be estimated if unknown. The first term in Eq. (9.31) is a weight correction factor resulting from inertial effects (more power required to change the energy height of a heavier aircraft). The second term in Eq. (9.31) is related to an increase in the induced drag of a heavier aircraft. This correction factor can be applied to both rate of climb data and level acceleration data by working with the standardized specific excess power, $P_{s,std}$.

9.4 Flight Test Example: Cirrus SR20

We will conclude with an example of flight testing of specific excess power for a Cirrus SR20 (sixth generation), with both rate of climb and level acceleration flight tests. This data set is for the same aircraft tested for the level flight drag polar (Chapter 7), where the aircraft had a parasite drag coefficient of $C_{D_0} = 0.0287$, an Oswald efficiency factor of $e = 0.7016$, and a wing span of $b = 38.3$ ft.

Table 9.1 Flight data from a Cirrus SR20 in climb at a pressure altitude of 3000 ft.

Fuel burned (gal)	IAS (kts)	OAT (°C)	h_1 (ft)	h_2 (ft)	Δt (s)
3.5	75	19	2900	3240	25.95
3.9	85	18	2900	3100	13.40
4.2	95	18	2880	3100	20.84
4.5	105	18	2820	3000	21.26
4.9	115	18	2700	2900	23.91
5.4	125	18	2700	2800	21.48

Table 9.2 Flight data from a Cirrus SR20 in climb at a pressure altitude of 6000 ft.

Fuel burned (gal)	IAS (kts)	OAT (°C)	h_1 (ft)	h_2 (ft)	Δt (s)
10.4	75	13	5800	6000	24.32
10.6	85	13	5620	5900	29.99
11.0	95	13	5760	5900	18.49
11.3	105	13	5540	5800	31.37
11.8	115	13	5760	5900	25.52
12.2	125	13	5570	5800	82.35

The pre-flight weight of the aircraft for this test was 2901 lb, and the rated maximum takeoff weight is 3150 lb. The aircraft was flown at pressure altitudes of 3000 ft (Table 9.1) and 6000 ft (Table 9.2). In the data analysis, the measured rate of climb was computed by $dh/dt = (h_2 - h_1)/\Delta t$. Pressure was found from the measured pressure altitude referenced to the standard atmosphere, density was determined from $\rho = p/RT$, and density altitude was determined from the standard atmosphere. True airspeed was calculated from $V = V_i\sqrt{\sigma}$, with the assumption that indicated, calibrated, and equivalent airspeeds were all equal (see Chapter 8 for a more refined approach). Aircraft weight for each test point was calculated from $W_{test} = W_{std} - f_{burned}\rho_{fuel}$, where f_{burned} is the fuel burned and $\rho_{fuel} = 6$ lb/gal. The corrected rate of climb values were found by applying the correction factors in Eqs. (9.28) and (9.31), with the understanding that $P_s = dh/dt$ in these steady climbs.

Reduced data for the two pressure altitudes are shown in Tables 9.3 and 9.4 and plotted in Figure 9.13 along with theoretical data. The theoretical prediction was based on estimates of power available and power required (based on the drag polar). Power available was estimated from engine and propeller data (Chapter 7), with shaft power of 165 hp at a density altitude of 4000 ft and 150 hp at a density altitude of 7000 ft. With an assumed engine speed of 2700 rpm and a propeller diameter of 74 in., the propeller efficiency was calculated for each airspeed (from calculations of J and C_p). The effect of density altitude on the measured and predicted climb performance is clear, with diminishing rate of climb at higher airspeed. There is reasonably good agreement between theory and measurement, particularly at the higher airspeeds. Note that there is a discontinuity in the measured excess power data at both altitudes, near airspeeds of 110–120 knots. This is likely due to the lower angle of attack required at these higher airspeeds, and the wing entering the range of angles of attack where there is a greater extent of laminar flow (entering the "drag bucket").

Table 9.3 Derived quantities from flight test data at a pressure altitude of 3000 ft.

TAS (kts)	Measured dh/dt (=$P_{s,meas}$) (ft/min)	Density altitude (ft)	W_T (lb)	ΔW (lb)	T/T_{std}	ΔP_s (ft/min)	Corrected dh/dt (ft/min)
79.8	786	4145	2880	270	1.035	−147.1	667
90.2	896	4030	2878	272	1.032	−150.4	773
100.9	633	4030	2876	274	1.032	−119.0	534
111.5	508	4030	2874	276	1.032	−102.0	422
122.1	502	4030	2872	278	1.032	−97.7	420
132.7	279	4030	2869	281	1.032	−72.5	216

Table 9.4 Derived quantities from flight test data at a pressure altitude of 6000 ft.

TAS (kts)	Measured dh/dt (=$P_{s,meas}$) (ft/min)	Density altitude (ft)	W_T (lb)	ΔW (lb)	T/T_{std}	ΔP_s (ft/min)	Corrected dh/dt (ft/min)
83.5	493	7138	2839	311	1.036	−141.6	369
94.6	560	7138	2837	313	1.036	−139.7	441
105.7	454	7138	2835	315	1.036	−120.6	350
116.9	497	7138	2833	317	1.036	−119.7	395
128.0	329	7138	2830	320	1.036	−95.8	245
139.1	168	7138	2828	322	1.036	−72.9	101

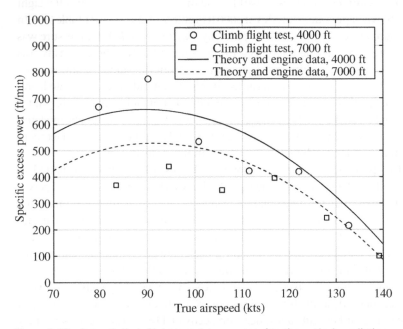

Figure 9.13 Rate of climb flight test data compared to theoretical prediction.

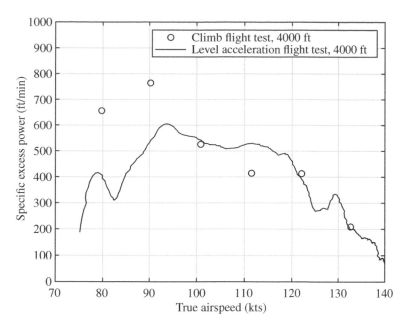

Figure 9.14 Rate of climb flight test data compared with level acceleration data.

A level acceleration flight test was also performed immediately following the last data point shown in Table 9.1, at a pressure altitude of 3000 ft, and with fuel burned of 7.2 gal. The complete maneuver involved a level acceleration from 75 KTAS (knots true airspeed) up to 142 KTAS, which took place over a period of 92.5 seconds. Ground speed was measured with a GPS receiver, and true airspeed was assumed to be equal to the ground speed for this test (this assumption is only valid under calm winds, or when the maneuver is conducted in the crosswind direction). The GPS-measured velocity data were low-pass filtered with a second order Butterworth filter with a cut-off frequency of 0.15 Hz, in order to remove noise from the signal. Acceleration as a function of time was calculated from the first derivative of the filtered velocity signal with respect to time. Based on these input data, the specific excess power was calculated from Eq. (9.6) with $dh/dt = 0$ and a correction factor ΔP_s from Eq. (9.31) was applied to each data point in the velocity time history. (The correction factor ranged from about 140 ft/min at low speeds, to about 60 ft/min at high speeds, with an average value near 100 ft/min.) Figure 9.14 shows the results of the level acceleration flight test, overlaid on the rate of climb data from the same pressure altitude of 3000 ft. There is reasonable agreement between the two approaches, particularly at higher airspeeds. Fluctuations in the level acceleration data set may be due to the difficulty in maintaining constant altitude throughout the maneuver.

Finally, the maximum rate of climb data for the direct measurement of climb at pressure altitudes of 3000 and 6000 ft and the maximum excess power from the level acceleration test at a pressure altitude of 3000 ft are collected into a single comparison with aircraft manufacturer data in Figure 9.15. The manufacturer (POH) data are the maximum rate of climb provided for two weights across a range of altitudes. For this representation, the data are plotted as a function of density altitude and all results are presented for a standardized weight of maximum takeoff weight. The rate of climb data for the lower altitude do not agree with the POH data very well, but the level acceleration data and the higher-altitude data both show good agreement. This may be due to the sparse data set selected for testing – the quality of these results may be improved by testing at more airspeeds throughout the drag polar. This would enable finer resolution of the potential laminar drag bucket effect and facilitate selection of the maximum value of excess power.

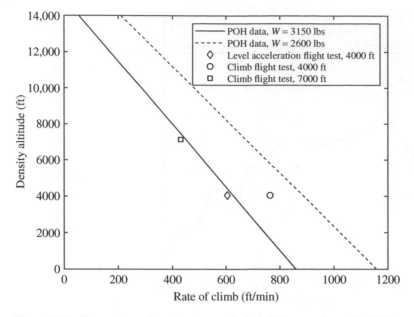

Figure 9.15 Comparison of maximum rate of climb flight test data with POH data.

Nomenclature

AR	aspect ratio
b	wing span
C_{D_i}	induced drag coefficient
C_{D_0}	parasite drag coefficient
C_L	lift coefficient
C_p	power coefficient
D	drag
D_i	induced drag
D_0	parasite drag
e	Oswald efficiency factor
e_{total}	specific total energy of the aircraft
f_{burned}	fuel burned
g	gravitational acceleration
h	altitude
h_a	absolute ceiling
h_e	energy height
h_{svc}	service ceiling
h_{target}	target altitude for flight testing
J	propeller advance ratio, V/ND
m	aircraft mass
p	pressure
P_A	power available from the engine, TV
P_i	induced power

P_R	power required for level flight, DV
P_s	specific excess power
P_0	parasite power
R	gas constant for air
S	wing area
t	time
T	thrust
T	temperature
T_{std}	standard temperature
V	velocity (true airspeed)
V_h	horizontal component of velocity
V_i	indicated airspeed
V_v	vertical component of velocity, dh/dt
V_x	airspeed for best climb angle
V_y	airspeed for best rate of climb
W	aircraft weight
W_{std}	standard weight (typically maximum takeoff weight)
α	angle of attack
ε	thrust angle
γ	climb angle
ρ	air density
ρ_{fuel}	density of fuel
σ	density ratio, ρ/ρ_{SL}

Subscripts

act	actual
max	maximum
meas	measured
SL	sea level
std	standard
test	test condition

Acronyms and Abbreviations

CIW	climb independent of weight (standardized rate of climb)
DAQ	data acquisition
GPS	Global Positioning System
KTAS	knots true airspeed
OAT	outside air temperature
PIW	power independent of weight (standardized weight)
POH	Pilot's operating handbook
rpm	revolutions per minute
UAV	unmanned aerial vehicle
VIW	velocity independent of weight (standardized velocity)

References

Filippone, A. (2006). *Flight Performance of Fixed and Rotary Wing Aircraft*. Reston, VA: American Institute of Aeronautics and Astronautics.

Herrington, R.M., Shoemacher, P.E., Bartlett, E.P., and Dunlap, E.W. (1966). *Flight Test Engineering Handbook*, USAF Technical Report 6273. Edwards AFB, CA: US Air Force Flight Test Center.

Kimberlin, R.D. (2003). *Flight Testing of Fixed-Wing Aircraft*. Reston, VA: American Institute of Aeronautics and Astronautics.

Rutowski, E.S. (1954). Energy approach to the general aircraft performance problem. *Journal of the Aeronautical Sciences* 21 (3): 187–195.

Small, S.M. and Prueher, J.W. (1977). *Fixed Wing Performance. Theory and Flight Test Techniques*. U.S. Naval Test Pilot School, Patuxent River, MD, DTIC Accession Number ADA061239.

Smith, H.C. (1981). *Introduction to Aircraft Flight Test Engineering*. Basin, WY: Aviation Maintenance Publishers.

Ward, D., Strganac, T., and Niewoehner, R. (2006). *Introduction to Flight Test Engineering*, 3e, vol. 1. Dubuque, IA: Kendall/Hunt.

10

Glide Speed and Distance

In this chapter we'll study the glide performance of aircraft and discuss flight test methods for measuring this performance. The glide ratio (horizontal distance flown relative to a loss of altitude) is directly related to the aerodynamic efficiency of the aircraft, expressed as the lift-to-drag ratio. This determination of glide ratio is critical for informing the pilot on how far the aircraft can glide if engine power is lost at a particular altitude.

A poignant example of the critical nature of glide performance is the ditching of US Airways flight 1549 in the Hudson River on January 15, 2009 (Figure 10.1). Shortly after takeoff from New York's LaGuardia airport, the Airbus A320 encountered a flock of Canada geese at an altitude of about 2800 ft. Both engines ingested birds, resulting in a near-complete loss of power. From a maximum altitude of about 3000 ft, Capt. Chesley Sullenberger had very few options available for safely landing the aircraft in a heavily populated urban area. He immediately notified air traffic control of the emergency situation and explored the possibility of returning to LaGuardia (KLGA) or landing at nearby Teterboro, NJ (KTEB). Ultimately, Capt. Sullenberger determined that these landing options were too far away relative to the aircraft's altitude and glide performance. He piloted the aircraft along the Hudson river and ditched in the river adjacent to 48th Street in Manhattan. Despite the impact of the fuselage landing directly on water, and the frigid waters of the Hudson, there were no fatalities in the incident, leading many to refer to the successful ditching as the "Miracle on the Hudson."

One of the most critical things for the success of this ditching is that Capt. Sullenberger immediately flew the aircraft at the airspeed for best glide while his co-pilot, Jeffrey Skiles, worked through the emergency checklists (Brazy 2009; Smith 2009). Pitching the aircraft for best glide speed is critical for extending the range and maximizing the number of landing options in an engine-out condition. The flight manual for the aircraft (as well as the onboard flight management system) specifies the best glide speed that must be flown in these conditions. Flight testing in the certification phase of the aircraft directly informed the glide performance specifications in the flight manual.

This chapter focuses on the theory and methods for flight testing of aircraft glide performance. The first section will introduce the theory of glide, beginning with a review of the drag forces acting on an aircraft and a discussion of why an optimum lift-to-drag ratio exists. Following this, we'll study the forces acting on an aircraft in glide and show how the lift-to-drag ratio is directly related to the glide path angle or glide ratio. We'll then turn our attention to two methods for analytically determining the glide performance of an aircraft. The first method involves plotting a graph of horizontal and vertical speeds in a glide and finding optimum conditions on this chart, while the second method involves direct manipulation of the equations to find an optimum analytical solution. Both of these methods will show that there exist specific airspeeds to be flown in order to optimize glide performance (smallest glide angle, or maximum time aloft). Following this

Introduction to Flight Testing, First Edition. James W. Gregory and Tianshu Liu.
© 2021 John Wiley & Sons Ltd. Published 2021 by John Wiley & Sons Ltd.
Companion website: https://www.wiley.com/go/flighttesting

Figure 10.1 US Airways flight 1549, moments after gliding to a water landing on the Hudson River on January 15, 2009. Source: Greg Lam Pak Ng.

theoretical foundation, we'll discuss flight test methods for measuring the glide path angle in flight, data acquisition methods, and analysis techniques to reduce the data into standardized form.

10.1 Theory

Determination of the glide performance for an aircraft is critical for determining the distance that an aircraft can fly in the event of engine failure. Glide testing can also be a straightforward way to measure the lift-to-drag ratio of the aircraft, if the drag on a windmilling propeller is neglected. Since the lift-to-drag ratio L/D is a measure of aerodynamic efficiency, it has a substantial impact on many aspects of aircraft performance, including aircraft range and endurance, as well as glide performance.

We'll now consider the underlying theory, in order to understand the key parameters that affect glide performance. Key principles to keep in mind are that the glide path angle (angle between the aircraft flight path and the horizon) is the key parameter to optimize. We'll see that the glide path angle is proportional to the lift-to-drag ratio, so best glide will occur at $(L/D)_{max}$. Finally, we'll see that L/D depends strongly on airspeed and that there is an optimal airspeed that should be flown for achieving $(L/D)_{max}$. We'll work through the theory of glide performance using both analytical and graphical approaches.

10.1.1 Drag Polar

We first introduced the drag polar in Chapter 7 (Level Flight), where the contributions of induced drag and parasitic drag to the overall aircraft drag were discussed. We also considered various methods for estimating the drag polar of an aircraft, including engineering estimates based on historical data as well as modern computational tools. In order to establish the aerodynamic principles of gliding flight, we'll recap the discussion of the drag polar here, and provide a little more depth into the physics of aircraft drag. The following discussion assumes steady, level flight conditions, and for the time being we'll simplify our discussion by assuming that lift equals weight. (Note that, strictly speaking, lift is not equal to weight in a glide, since a component of the weight vector offsets the drag).

Lift and drag are given by

$$L = \frac{1}{2}\rho V^2 S C_L \cong W,$$ (10.1)

and

$$D = \frac{1}{2}\rho V^2 S C_D,$$ (10.2)

where V is the freestream velocity (true airspeed). The drag polar for an aircraft is an expression for the drag coefficient, comprised of the parasitic drag coefficient (C_{D_0}) and the induced drag coefficient (C_{D_i}),

$$C_D = C_{D_0} + C_{D_i} = C_{D_0} + \frac{C_L^2}{\pi\, \text{AR}\, e},$$ (10.3)

where AR is the aspect ratio of the wing and e is the Oswald efficiency factor. The parasitic drag coefficient can be assumed constant with respect to airspeed (neglecting Reynolds and Mach effects), but the induced drag coefficient varies with flight velocity since the required lift coefficient in steady flight depends on airspeed (for a given aircraft weight),

$$C_L = \frac{2W}{\rho V^2 S}.$$ (10.4)

When these relations are substituted into Eq. (10.2), the influence of parasitic and induced drag becomes clear:

$$D = \frac{1}{2}\rho V^2 S C_{D_0} + \frac{2W^2}{\pi b^2 e \rho V^2}.$$ (10.5)

The parasitic drag (the first term on the right side) is due to viscous drag and pressure drag resulting from (hopefully small) regions of separated flow. This term is invariant with lift, but varies with the square of airspeed.

The induced drag – the second term on the right side of Eq. (10.5) – is a strong function of the lift produced by the wing (thus, it's sometimes referred to as "drag due to lift"). The induced drag is related to the generation of wing tip vortices (see Figure 10.2), which represent wasted kinetic energy. With a wing in a lifting configuration, the pressure on the bottom side of the wing is higher

Figure 10.2 Smoke visualization of wing tip vortices on a Boeing 727. Source: NASA 1973.

than the upper-side pressure. This pressure gradient leads to airflow around the wing tips from bottom to top, which generates the wing tip vortices. The wasted kinetic energy for the tip vortices directly translates into increased drag. Flow velocity that was in the freestream direction ahead of the aircraft is redirected into swirling flow in the tip vortices, which is rotational energy that does not contribute to lift generation (thus, it is wasted energy). Since the tip vortices are a byproduct of lift production, they are the primary source of drag induced by lift. Equation (10.5) illustrates that the induced drag increases as the airspeed decreases (varying inversely with the square of velocity). This is because the lift coefficient (set by the angle of attack) must be higher at lower airspeeds, such that lift can be maintained equal to weight. Also note that a heavier aircraft will have more induced drag at a given airspeed, since the lift coefficient has to be higher to sustain that flight condition. Induced drag is also a function of the wing design, with a large wing span (high AR) producing less induced drag.

A plot of the total aircraft drag as a function of airspeed (Eq. (10.5)) is shown in Figure 10.3. This illustrates how the drag dramatically changes with airspeed. At low speeds in steady, level flight, the aircraft must be pitched to high angle of attack in order to generate high lift coefficient and offset the lower available dynamic pressure (see Eq. (10.1)), while maintaining $L = W$. The wings are doing much more work on the flow at low flight speed, creating stronger downwash and greater induced drag (see Eq. (10.3)). The parasitic drag contribution at low speed, on the other hand, is relatively low due to the low dynamic pressure. Thus, at lower speeds, the induced drag dominates the overall drag production. Since parasitic drag is a parabolic function of airspeed and induced drag decreases with the square of velocity (Eq. (10.5)), the parasitic drag term dominates the overall drag at higher speeds. This is because the aircraft is pitched to relatively low angle of attack, since sufficient lift can be produced with high dynamic pressure and low lift coefficient. In essence, the lift Eq. (10.1) must be satisfied, no matter how small or great the dynamic pressure is for a given flight condition. Thus, at higher speeds, the parasitic drag dominates, and the induced drag diminishes greatly. At a particular airspeed within the flight envelope, the contributions of the induced and parasitic drag components are equal, which coincides with the velocity for minimum

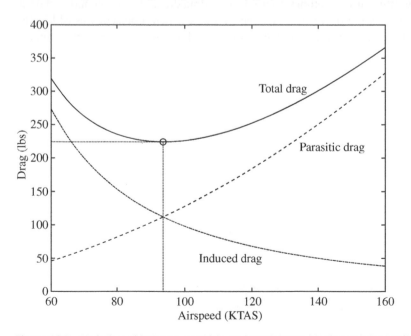

Figure 10.3 Variation of induced, parasitic, and total drag with airspeed for a typical light aircraft.

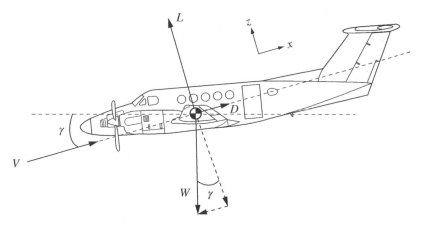

Figure 10.4 Forces acting on an aircraft in glide, assuming no thrust or drag from the propeller.

drag (see Figure 10.3). Since lift is constant throughout the flight envelope (for steady, level flight), the minimum drag condition corresponds to $(L/D)_{\text{max}}$.

10.1.2 Gliding Flight

We'll now extend this discussion of lift-to-drag ratio to a gliding flight condition. The link between glide distance and lift-to-drag ratio is best illustrated by Figure 10.4. Here the lift acts perpendicular to the freestream, drag acts parallel to the freestream, and weight acts in the downward direction. (Note that the lift is not equal to the full aircraft weight in this case!) Here, we have neglected any residual thrust produced by the propeller. The glide path angle, γ, is between horizontal and the direction of flight (the freestream velocity vector). Note that glide path angle is not the same as angle of attack (α), which is the angle between the wing chord and the freestream velocity vector.

Based on the force diagram in Figure 10.4, and assuming steady flight, we can write equations of motion for the perpendicular (z) and parallel (x) directions (with respect to the freestream),

$$\sum F_z = 0 = W \cos \gamma - L, \tag{10.6}$$

and

$$\sum F_x = 0 = W \sin \gamma - D. \tag{10.7}$$

Solving for lift and drag, we have

$$L = W \cos \gamma, \tag{10.8}$$

and

$$D = W \sin \gamma. \tag{10.9}$$

When we form a ratio of Eqs. (10.8) and (10.9), we find that

$$\frac{L}{D} = \frac{1}{\tan \gamma}. \tag{10.10}$$

Solving Eq. (10.10) for glide path angle, and referring to the vector diagrams in Figure 10.5, we have

$$\gamma = \tan^{-1} \left(\frac{1}{L/D} \right) = \tan^{-1} \left(\frac{V_v}{V_h} \right) = \tan^{-1} \left(\frac{h}{R} \right) = \sin^{-1} \left(\frac{V_v}{V} \right) = \sin^{-1} \left(\frac{h}{s} \right). \tag{10.11}$$

Here, V_v and V_h are the vertical and horizontal components of aircraft velocity in the glide, R is the horizontal distance flown as the aircraft descends through an altitude change h, and s is the distance

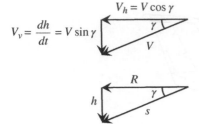

$$V_v = \frac{dh}{dt} = V \sin \gamma$$

Figure 10.5 Velocity and distance triangles for aircraft motion in glide.

traveled along the flight path. Equation (10.11) illustrates that higher L/D, or higher aerodynamic efficiency, will directly translate into a smaller glide path angle (γ) and longer glide distance (R) for a given altitude change (h). The glide path angle and freestream velocity are related to the vertical and horizontal components of velocity, as shown in Figure 10.5, and the velocity triangle is similar to the distance triangle in the lower half of the figure. For our flight testing work, the most direct representation of glide distance is to write the force balance for drag (Eq. (10.7)) as

$$D = W\frac{V_v}{V}, \tag{10.12}$$

where we have substituted in a value of the glide path angle from Eq. (10.11). Also, from the velocity triangles in Figure 10.5, the key expressions we'll use for flight testing are

$$V_v = V \sin \gamma, \tag{10.13}$$

and

$$V_h = V \cos \gamma. \tag{10.14}$$

10.1.3 Glide Hodograph

A useful way to illustrate the relationship between glide path angle and L/D is the glide hodograph (Figure 10.6). The glide hodograph is a presentation of vertical velocity (V_v) versus horizontal velocity (V_h) across a range of airspeeds. Each point on the glide hodograph represents the necessary sink rate for a given horizontal speed. The glide angle can be directly read from the diagram if the x- and y-scales are plotted with equal spacing. The angle of a line extending from the origin to each point on the diagram represents the glide path angle (see Figure 10.6).

A glide hodograph based on analytical predictions can be formed from the drag polar (Eq. (10.3)). First, a vector of lift coefficients is defined, ranging from small values associated with high-speed cruise up to high values near stall ($C_{L,\,\mathrm{max}}$). At each lift coefficient, the associated drag coefficient is calculated from the drag polar (C_{D_0} and C_{D_i}, Eq. (10.3)). The lift-to-drag ratio, C_L/C_D, is then calculated for each point in the initially defined vector of C_L values. Recognizing that $L/D = C_L/C_D$, we can rewrite Eq. (10.11) and find the glide path angle for each point from

$$\gamma = \tan^{-1}\left(\frac{1}{C_L/C_D}\right). \tag{10.15}$$

The true airspeed for each point may be determined by solving

$$L = W \cos \gamma = \frac{1}{2}\rho V^2 S C_L, \tag{10.16}$$

for airspeed,

$$V = \sqrt{\frac{2W \cos \gamma}{\rho S C_L}}, \tag{10.17}$$

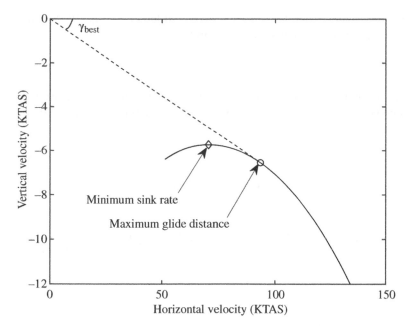

Figure 10.6 Sample glide hodograph for a typical light aircraft.

and substituting in appropriate values for each variable. This result completely defines the velocity triangle (Figure 10.5) for each of the values in the presumed vector of lift coefficients. Thus, the vertical and horizontal components of velocity can be found from Eqs. (10.13) and (10.14), and plotted on the hodograph.

The glide hodograph in Figure 10.6 illustrates that there are separate optimum airspeeds for glide performance: one for maximizing time aloft (corresponding to the airspeed for minimum sink rate, V_v), and a separate point for maximizing glide distance (corresponding to the airspeed for minimum glide path angle). The minimum sink rate condition can be useful for ensuring maximum time aloft without engine power, but this condition is rarely encountered in practice. Typically, an aircraft that has lost power needs to extend the glide for the longest feasible distance, so the condition for minimum glide angle should be flown, which corresponds to the flight condition for $(L/D)_{max}$, as noted by Eq. (10.15).

10.1.4 Best Glide Condition

The glide hodograph illustrates that glide performance is a strong function of airspeed. In order to optimize glide performance (maximizing glide distance, and minimizing the glide path angle), the pilot must fly at the airspeed for the best glide condition. As seen in Eq. (10.15), this best glide condition is the same airspeed at which $(C_L/C_D)_{max}$ is achieved. We'll now derive an expression for the airspeed for best glide.

Starting with the expression for airspeed shown in Eq. (10.17), we need to substitute in values for the glide path angle and the lift coefficient. We'll select values associated with $(C_L/C_D)_{max}$, since we now know that this is the best glide condition. First, we'll focus our attention on identifying the lift and drag coefficients for the best glide condition. Taking the drag polar, we can write

$$\frac{C_L}{C_D} = C_L \left[C_{D_0} + \frac{C_L^2}{\pi \, \text{AR} \, e} \right]^{-1}. \tag{10.18}$$

Since C_L/C_D must be maximized relative to lift coefficient, we can differentiate Eq. (10.18) with respect to C_L and set the resulting expression equal to zero to find

$$C_{D_0} = \frac{C_L^2}{\pi \, AR \, e}, \tag{10.19}$$

which must be satisfied at the condition for best glide. In other words, when lift-to-drag ratio is maximized for optimal glide performance, the parasitic drag must be equal to the induced drag. Solving Eq. (10.19) for lift coefficient,

$$C_L = \sqrt{C_{D_0} \pi \, AR \, e} \tag{10.20}$$

at the condition for best glide. Also, substituting Eq. (10.19) into the drag polar, we find that

$$C_D = 2C_{D_0} = 2C_{D_i} \tag{10.21}$$

for best glide. Recognizing that Eqs. (10.20) and (10.21) are true at the best glide condition, we can form a ratio of the two,

$$\left(\frac{C_L}{C_D}\right)_{\max} = \sqrt{\frac{\pi \, AR \, e}{4C_{D_0}}}. \tag{10.22}$$

By substituting Eq. (10.22) into Eq. (10.15), we have

$$\gamma_{\min} = \tan^{-1}\left[\left(\frac{4C_{D_0}}{\pi \, AR \, e}\right)^{\frac{1}{2}}\right]. \tag{10.23}$$

We're now ready to substitute these results into Eq. (10.17) to identify the speed for best glide. A typical value for the minimum (best) glide angle provided by Eq. (10.23) for a general aviation aircraft is approximately 4° to 6°. Making a small angle assumption such that $\cos \gamma_{\min} \approx 1$, and substituting Eq. (10.20) into (10.17) we have

$$V_{\text{glide}} = \left(\frac{4}{C_{D_0} \pi \, AR \, eS^2}\right)^{\frac{1}{4}} \left(\frac{W}{\rho}\right)^{\frac{1}{2}}, \tag{10.24}$$

which illustrates how best glide speed depends on the drag polar of the aircraft (C_{D_0} and e) and wing properties (b and S), which are all constant for a given aircraft.

Aircraft weight and density both have an impact on the airspeed for the best glide condition – we can see this in a plot of L/D as a function of true airspeed (see Figure 10.7(a)), where decreased weight shifts the curve to the left and the optimum glide speed decreases. However, from Eq. (10.22) we see that aircraft weight has no impact on the actual value of $(L/D)_{\max}$, which is a function of only the drag polar and the geometry of the wing. Pilots of gliders work with the weight-dependency of best glide speed by changing the amount of ballast that they carry on board in order to maintain optimal glide at the desired airspeed (a heavier glider has faster optimal speed).

Similarly, higher altitude (lower density) shifts the curve to the right, as shown in Figure 10.7(b). Keep in mind, however, that the airspeed used in Eq. (10.24) and plotted here is true airspeed. Since indicated airspeed is based on the assumption of standard sea level density (see Chapter 8, $V_i \cong V_e = V\sqrt{\sigma}$) then the indicated airspeed for best glide does not change with altitude. And, as above, the actual value of $(L/D)_{\max}$, and thus the glide path angle, does not change with altitude as long as the pilot maintains the indicated airspeed for best glide.

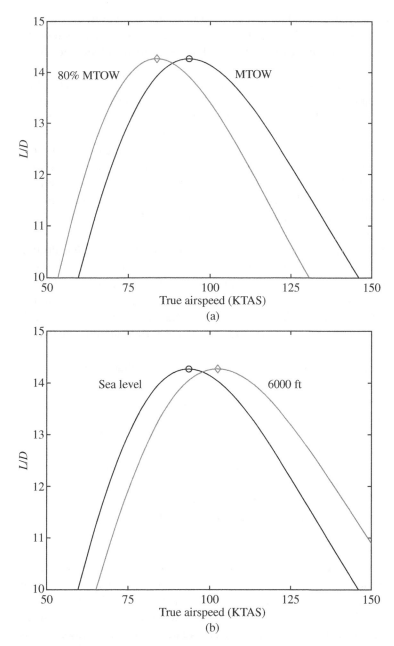

Figure 10.7 Dependency of the lift-to-drag characteristics on (a) aircraft weight, and (b) altitude. Note that these are plotted as a function of true airspeed; if presented as a function of indicated airspeed, the altitude curves will collapse onto the sea level curve.

10.2 Flight Testing Procedures

The flight test procedure for glide flight testing is very similar to climb flight tests (see Chapter 9). The main objective is to measure the rate of descent (vertical speed) at each of a series of indicated airspeeds. Since each airspeed set point should be conducted at the same mean altitude, the pilot must climb between each glide test point. This requires a series of climbs and glides, where an efficient approach to flight testing will combine the climb and glide tests into one flight, interlacing the climbs and glides.

The number and range of airspeed test points to be flown should capture the curvature of the drag polar (see Figure 10.3), such that the best glide speed can be identified. (Airspeed test points that are faster and slower than the anticipated best glide speed should be flown.) The altitude interval to be flown for each glide should enable sufficient distance (time) for the glide to be stabilized before beginning data acquisition. The altitude range of the glide (Δh) should be centered on the desired mean test altitude (h_{target}), and be of sufficient range to enable averaging of the data. For example, if h_{target} is 4000 ft, and Δh is 400 ft, then a steady glide should be established during a descent by the time 4200 ft is reached and maintained through 3800 ft. This provides 400 ft of steady gliding flight for data acquisition. Multiple target altitudes can be flown in order to verify the collapse of data onto a single drag polar curve.

Data to be recorded include altitude, time, airspeed, temperature, and aircraft weight. For altitude measurement, the altimeter should be set to measure pressure altitude, with a reference pressure of 29.92 inHg. Starting and ending altitudes of the stabilized glide are recorded, and a stopwatch used to record the start and end times to determine the time interval (Δt) of the glide. Throughout the stabilized glide the indicated airspeed must be held constant at the desired set value and recorded. At each glide test point, the outside air temperature (OAT) and total fuel burned should be recorded, and the takeoff weight must be known.

For each glide set point, the throttle is pulled to idle at the beginning of the glide. This approach is preferable to a complete shutoff of the engine for safety of flight, since there is a small risk that the engine may not restart from a shutoff condition. However, the engine-idle approach will result in residual thrust being produced, or possibly a small amount of drag, when the engine is operating at idle with a rotating propeller. If the aircraft has a constant-speed propeller, the prop control should be adjusted to feather the propeller and minimize drag. One approach that some researchers have employed to reduce the impact of residual thrust or drag is to install a load sensor between the engine and the firewall (Norris and Bauer 1993). The pilot then adjusts the engine and propeller controls in the glide such that the residual thrust can be zeroed out. Certain effects such as the drag from the slipstream on the fuselage, cooling drag, etc. cannot be readily controlled; however, the flight test engineer should be cognizant of these physical effects.

The pilot should select flight conditions that are optimized for quality flight test data. The presence of wind gusts, updrafts, or downdrafts can be particularly problematic for glide testing. In general, avoid performing this flight test on a gusty day or in unstable atmospheric conditions with convective currents and turbulence (aircraft with low values of wing loading will be particularly susceptible to gusts). Vertical gradients of wind speed or direction (wind shear) can be accounted for in data analysis (see Herrington et al. 1966 for corrections for velocity gradients). However, it is best to avoid the effects of wind shear by conducting the glide tests in the crosswind direction, or by testing on a day with uniform winds. Also, be aware of updrafts and downdrafts occurring over bodies of water or large parking lots due to the different thermal capacities of water, asphalt, and soil. For each glide in the flight test, attention should be paid to atmospheric conditions: if wind gusts or a wind shear is encountered, the test point should be repeated or a note of the anomaly should be made.

The preceding discussion is focused on the methods needed to assess the drag polar and find the best glide speed. Results from this approach to glide flight testing will likely have more uncertainty in the determination of the drag polar, relative to the level flight test methods discussed in Chapter 7. However, the two methods can be compared, with the advantages and disadvantages of each being different. In the glide test, a major source of uncertainty is the residual thrust produced by the engine and propeller with the engine at idle. For the level flight test, the major limitation is the requirement of engine performance data, which must be obtained by ground testing (typically from engine manufacturer data from a pressure-controlled test cell with a dynamometer). Propeller performance data (efficiency) is also required for that approach, with that data also coming from the manufacturer.

Finally, we'll consider the glide flight test guidelines provided by the FAA in Advisory Circular 23-8C (FAA 2011, pp. 56–57). A series of glides at different airspeeds are recommended at an airspeed range that is 10–15% higher than the best rate of climb speed. Since this specific test is to evaluate the engine-off glide performance, the FAA guidance suggests that the test be done with the engine inoperative, with no effort made to stop the propeller from turning if it is freely windmilling. A constant-speed (variable-pitch) propeller should be adjusted to the minimum-drag (feathered) configuration. If engine-off glide testing is performed, there is a small risk of the engine not restarting, so the FAA guidance recommends performing these tests within safe gliding distance of an airfield.

10.3   Data Analysis

The primary objective of the data analysis is to develop a glide hodograph based on experimental data. Each data point in the hodograph is comprised of the vertical and horizontal components of velocity appearing in the velocity triangle of Figure 10.5. The measured vertical speed (V_v) and freestream velocity (V) form the base and hypotenuse of the triangle, respectively. The glide path angle can be found from Eq. (10.13) and the horizontal speed can be calculated from Eq. (10.14). Before doing these calculations, however, we must apply a series of corrections to the data, to standardize for weight, altitude, and temperature.

First, the measured vertical rate of descent must be corrected for non-standard temperature by

$$\left(\frac{dh}{dt} \right)_{corr} = \left(\frac{dh}{dt} \right)_{meas} \left(\frac{T_{test}}{T_{std}} \right), \tag{10.25}$$

where T_{test} is the measured outside air temperature at the test altitude, and T_{std} is the standard day temperature at that altitude (in absolute units of °R or K). Details on this correction factor were provided in Chapter 9, where the correction factor was formed from the ratio of the hydrostatic equation applied at measured and corrected conditions.

Next, it's important for both V_v and V to be the same type of airspeed. The temperature-corrected dh/dt provided by Eq. (10.25) is a true airspeed, but the freestream velocity measured in flight is an indicated airspeed. The measured rate of descent will change with altitude when expressed as a true airspeed, so should be standardized as an equivalent airspeed in order to make it independent of altitude. This is done by

$$V_v = \sqrt{\sigma} \left(\frac{dh}{dt} \right)_{corr}, \tag{10.26}$$

where $\sigma = \rho/\rho_{SL}$. For most general aviation applications with minimal error on the airspeed indicator and flight at low speed and altitude, the measured indicated airspeed can be assumed equal to equivalent airspeed (see Chapter 8 for details on airspeed types). Thus, the glide hodograph is best constructed as a function of equivalent airspeed in order to remove altitude effects.

Both the vertical speed and the freestream velocity need to be corrected for non-standard weight. This correction factor is the same as the one developed for velocity independent of weight (VIW) in Chapter 7, which we'll summarize here. Taking the lift equation, combining with Eq. (10.6) and solving for velocity, we have

$$V = \left(\frac{2W \cos \gamma}{\rho S C_L} \right)^{\frac{1}{2}}. \tag{10.27}$$

Forming a ratio of Eq. (10.27) at standard and test conditions and cancelling like terms leads to

$$V_{std} = V_{test} \left(\frac{W_{std}}{W_{test}} \right)^{\frac{1}{2}}, \tag{10.28}$$

where we have assumed that $\cos \gamma_{std} \cong \cos \gamma_{test}$. Equation (10.28) should be applied to both the vertical speed V_v, as well as airspeed V.

Once corrections for non-standard temperature, altitude, and weight have been applied, the vertical and horizontal speed data can be plotted on a glide hodograph (such as Figure 10.6). From this presentation of the data, the best glide speed at maximum takeoff weight (MTOW) can be identified from the condition for the minimum glide path angle (minimum V_v/V_h). The best glide speed can be found for other weight conditions by solving Eq. (10.28) for V_{test}, and V_{std} is the best glide speed at MTOW. Since the aircraft weight in flight will always be less than maximum takeoff weight, the actual best glide speed will be less than the MTOW value. For example, for an aircraft with an actual weight of 2600 lb and a MTOW of 3150 lb, the best glide speed will be about 9% lower than the MTOW value.

If a pilot wishes to maximize the actual (tapeline) horizontal distance over the ground that can be covered in a glide, it is important to maintain the indicated airspeed for best glide throughout the glide. If this is done, then the minimum glide path angle will be achieved throughout, and this is independent of the altitude interval because the optimal glide path angle is not a function of density. Maximum horizontal glide distance can be found from the velocity triangle in Figure 10.5 and solving Eq. (10.11) for R,

$$R = \frac{h_{initial} - h_{final}}{\tan \gamma_{min}}. \tag{10.29}$$

(Note that the velocity triangles in Figure 10.5 are defined with positive change in altitude in the downward direction.) Equation (10.29) provides the value of maximum horizontal glide distance as a function of altitude that is reported in the Pilot's Operating Handbook (POH).

10.4 Flight Test Example: Cirrus SR20

We'll conclude with an example of glide flight testing on a Cirrus SR20 G6 (this data is from the same flight where climb performance was evaluated in Chapter 9). The drag polar for this aircraft, as measured in the flight test reported in Chapter 7, is $C_{D_0} = 0.0287$ and $e = 0.7016$. The specified maximum takeoff weight for the aircraft is 3150 lb, and the actual takeoff weight for this flight test was 2901 lb. The aircraft has a wing span of 38.3 ft and a wing area of 144.9 ft². The glide flight test data presented here was conducted at a pressure altitude of 3000 ft. A series of glides were conducted at the indicated airspeeds noted in Table 10.1. Data for the correction factors is shown in Table 10.2, along with the final values of vertical and horizontal velocity, with the glide hodograph provided in Figure 10.8.

Based on the flight test data, the best glide angle for this aircraft is estimated to be 4.1°, achieved at an equivalent airspeed of 90 knots equivalent airspeed (KEAS) (assumed equal to indicated airspeed in this case) when the aircraft is at MTOW. If the pilot were to lose power at an altitude

Table 10.1 Flight data from a Cirrus SR20 in glide at a pressure altitude of 3000 ft.

Fuel burned (gal)	IAS (kts)	OAT (°C)	h_1 (ft)	h_2 (ft)	Δt (s)
3.8	80	18	3020	2900	14.81
4.1	85	18	3100	2900	14.14
4.3	95	18	2960	2800	12.85
4.7	105	18	2780	2600	15.60
5.2	115	18	2620	2450	10.58
6.2	125	18	3500	3000	17.51

Table 10.2 Derived quantities from flight test data at a pressure altitude of 3000 ft.

IAS (kts)	Measured dh/dt (ft/min)	W_{test}/W_{std}	T/T_{std}	ρ/ρ_{SL}	Corrected V (KEAS)	Corrected V_v (KEAS)	Corrected V_h (KEAS)	γ (°)
80	−486	0.914	1.032	0.887	83.7	−4.9	83.5	3.3
85	−849	0.913	1.032	0.887	89.0	−8.5	88.5	5.5
95	−747	0.913	1.032	0.887	99.4	−7.5	99.2	4.3
105	−692	0.912	1.032	0.887	109.9	−7.0	109.7	3.6
115	−964	0.911	1.032	0.887	120.5	−9.7	120.1	4.6
125	−1713	0.909	1.032	0.887	131.1	−17.2	130.0	7.6

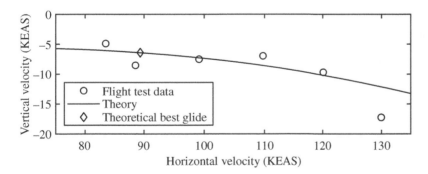

Figure 10.8 Glide hodograph of flight test data and theoretical data based on a drag polar with $C_{D0} = 0.0287$ and $e = 0.7016$.

of 8000 ft above ground level and maintain this best glide condition throughout the descent, the aircraft could travel an estimated 18 nautical miles in the glide. This compares favorably with the POH values, which specify a maximum glide ratio of 9.11 and a best glide speed of 100 knots indicated airspeed (KIAS). This POH glide ratio corresponds to a best glide angle of 6.3°, and a horizontal glide distance of 12 nautical miles if the best glide condition is maintained. The reason for the difference between the flight test estimate and the POH value is likely a difference between how the tests were conducted. With the data presented here, the engine was operated at idle with a spinning propeller, whereas the flight testing for the POH values was likely done with the engine off. The difference between these cases is the residual thrust produced by the engine during the engine-idle glide tests. A residual power of 37 hp at a glide speed of 100 KIAS (or, 120 lb of residual thrust) is sufficient to resolve the difference between the two cases.

Nomenclature

AR	aspect ratio
b	wing span
C_D	drag coefficient
C_{D_i}	induced drag coefficient
C_{D_0}	parasite drag coefficient
C_L	lift coefficient
D	drag
e	Oswald efficiency factor
F	force
h	altitude
L	lift
R	horizontal glide distance (range)
s	glide distance along the flight path
S	wing area
t	time
T	temperature
V	true airspeed
V_e	equivalent airspeed
V_h	horizontal component of true airspeed vector
V_i	indicated airspeed
V_v	vertical component of true airspeed vector
W	aircraft weight
x	direction parallel to the flight path
z	direction perpendicular to the flight path
α	angle of attack
γ	glide path angle
ρ	density
σ	density ratio, ρ/ρ_{SL}

Subscripts

best	best
corr	corrected
final	final
glide	glide
initial	initial
max	maximum
min	minimum
std	standard
target	target
test	test condition

Acronyms and Abbreviations

FAA	Federal Aviation Administration
IAS	indicated airspeed
KEAS	knots equivalent airspeed
KIAS	knots indicated airspeed
MTOW	maximum takeoff weight
OAT	outside air temperature
POH	pilot's operating handbook
SL	sea level
VIW	velocity independent of weight (standardized velocity)

References

Brazy, D.P. (2009). *Cockpit Voice Recorder 12 – Factual Report of Group Chairman*. Washington, D.C.: National Transportation Safety Board, NTSB Accident ID DCA09MA026.

Federal Aviation Administration. (2011). Flight Test Guide for Certification of Part 23 Airplanes. Advisory Circular 23-8C, U.S. Department of Transportation, Federal Aviation Administration.

Herrington, R.M., Shoemacher, P.E., Bartlett, E.P., and Dunlap, E.W. (1966). *Flight Test Engineering Handbook*, AFTR 6273, AD 636392, 5-11–5-12.

NASA Armstrong Research Center. (1973). NASA Photo ECN-3831. http://www.nasa.gov/centers/dryden/multimedia/imagegallery/B-727/ECN-3831.html (accessed 4 February 2016).

Norris, J. and Bauer, A.B. (1993). Zero-thrust glide testing for drag and propulsive efficiency of propeller aircraft. *Journal of Aircraft* 30 (4): 505–511. https://doi.org/10.2514/3.46372.

Smith, R.G. (2009). *Flight Data Recorder 10A – Factual Report of Group Chairman*. Washington, DC: National Transportation Safety Board, NTSB Accident ID DCA09MA026.

11

Takeoff and Landing

Takeoff and landing (TO&L) are the beginning and ending phases of a flight, respectively. Takeoff is defined as the process by which an aircraft is brought from standstill to a safe flight condition (Dekker and Lean 1959). Similarly, landing is the process by which an aircraft is brought from a safe flight condition to standstill. From a technical perspective, TO&L is a more complicated maneuver that involves acceleration and deceleration of an aircraft on the ground and in air. It is affected by a number of factors such as wind, runway slope, aircraft weight, air density, air temperature, pilot technique, and runway surface conditions. Flight testing of TO&L requires measurements of the aircraft's time-resolved position in these dynamic conditions. The primary objective of TO&L flight testing is to establish the required runway length for a given aircraft.

First, we will discuss the equations of motion that provide mathematical models for simulations to predict TO&L performance. The different phases of TO&L including the ground roll, rotation, transition, and climb/descent are described, and the total TO&L distances are the sums of the component distances in these phases. We will conclude with a discussion of flight testing procedures and data acquisition methods, with some examples of TO&L performance of a Cessna R182 provided.

11.1   Theory

TO&L performance data are difficult to quantify in a systematic way since they strongly depend on pilot technique, which is highly variable. Nevertheless, the basic physical laws are still applicable to TO&L. As shown in Figure 11.1, the takeoff can be generally decomposed into two phases: the ground phase and the air phase. In the ground phase, an aircraft is accelerated by its propulsion system from a standstill to the takeoff speed. In a maneuver termed rotation, the pilot pitches the nose up to initiate liftoff during the takeoff run. In the air phase of takeoff (immediately following rotation and liftoff), the aircraft is transitioned from horizontal motion into a steady climb until an obstacle height is cleared.

The landing maneuver can be thought of as a reciprocal process to takeoff, with similar segments. The air phase of landing involves a steady descent of the aircraft on final approach, a landing flare to bleed off airspeed, and a slow descent to touchdown. In the ground phase of landing the aircraft is decelerated from the touchdown location to a standstill. From a kinematic standpoint, TO&L are reciprocal processes, resulting in the equations of motion for TO&L having the same mathematical form. In the following discussions, we will develop theoretical relations to estimate the ground

Introduction to Flight Testing, First Edition. James W. Gregory and Tianshu Liu.
© 2021 John Wiley & Sons Ltd. Published 2021 by John Wiley & Sons Ltd.
Companion website: https://www.wiley.com/go/flighttesting

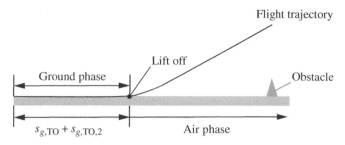

Figure 11.1 Takeoff diagram.

distance covered in each portion of the TO&L phase of flight. We will conclude this section by providing some simplified estimates of the TO&L distances.[1]

11.1.1 Takeoff Ground Roll

The dynamical equation for an aircraft accelerating during the ground roll in takeoff is

$$\left(\frac{W}{g}\right)\frac{dV}{dt} = T - R - W\varphi, \tag{11.1}$$

where V is the true airspeed of the aircraft in the ground roll, T is the thrust, R is the total resistance to the takeoff roll, g is the gravitational constant, W is the aircraft weight, and φ is the slope angle of the moving direction relative to the horizon ($\varphi > 0$ for uphill) which accounts for a sloped runway. Figure 11.2 shows the force diagram in the ground roll phase. In general, the thrust is a function of velocity. For a reciprocating engine/propeller combination, the thrust is inversely proportional to airspeed,

$$T = \frac{k_0}{V}, \tag{11.2}$$

where k_0 is a constant. For propeller-driven aircraft, this constant can be taken as the power transmitted to the flow by the propeller, $k_0 = P = \eta_{\text{prop}}P_{\text{shaft}}$, since power is equal to the product of thrust and velocity. (See Chapter 7 for details on estimating propeller efficiency, η_{prop}, and engine

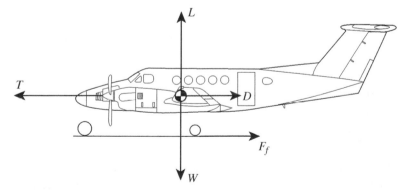

Figure 11.2 Force diagram for takeoff ground roll.

---

1 Another approach to estimating takeoff and landing distance is through scaling laws based on historical aircraft performance data. A brief description of this approach is provided in an online supplement, "Scaling Relations for Takeoff Performance Estimation."

shaft power, P_{shaft}.) For a turbojet engine, T is reasonably constant with V throughout the takeoff ground roll. For a turbofan engine, T decreases slightly with V during the ground roll; a model provided by Anderson (1999) is

$$T = k_1 - k_2 V + k_3 V^2, \tag{11.3}$$

where k_1, k_2, and k_3 are empirical coefficients.

The total resistance R is the combination of aerodynamic drag and ground friction force $R = D + F_f$, where $F_f = \mu_r(W - L)$ is the rolling friction force and μ_r is the rolling friction coefficient for the tires on the runway. Total resistance can be expressed in terms of the aerodynamic parameters as

$$R = qS\left(C_{D_0} + \frac{C_L^2}{\pi \, \text{AR} \, e'}\right) + \mu_r(W - qSC_L), \tag{11.4}$$

where $q = 1/2\rho V^2$ is the dynamic pressure. The empirical factor, e', is the Oswald efficiency factor that is modified by ground effect. With the wing closer to the ground (such as for an aircraft with a wing mounted underneath the fuselage), the tip vortices interact with the ground in a way that reduces the induced drag. This adjusted efficiency factor (e') can be related to the traditional Oswald efficiency factor (e) by

$$e' = e\frac{1 + (16h_{\text{wing}}/b)^2}{(16h_{\text{wing}}/b)^2}, \tag{11.5}$$

where h_{wing} is the height of the wing above the ground and b is the wing span. This result comes from an approximation based on lifting line theory, with image vortices provided by the ground plane (see McCormick 1995 for details).

By differentiating Eq. (11.4) with respect to C_L and setting the change equal to zero, we have the optimum lift coefficient $C_{L,\text{opt}} = 0.5\pi \, \text{AR} \, e' \, \mu_r$ for achieving the minimum resistance. This is an interesting theoretical result, which balances the reduction of induced drag at lower angle of attack (or, C_L) with the reduction of rolling resistance at higher angle of attack. However, how to practically operate an aircraft to approach this optimal state strongly depends on the pilot skills and aircraft limits such as ground clearance for the tail.

The kinematical equation for an aircraft accelerating during the ground roll in takeoff is

$$\frac{ds_{g,\text{TO}}}{dt} = V \pm V_{\text{wind}}, \tag{11.6}$$

where $s_{g,\text{TO}}$ is the takeoff ground roll distance, V is the true airspeed, V_{wind} is the wind speed, and the signs "+" and "−" denote a tail wind and a head wind, respectively. Combination of Eqs. (11.1) and (11.6) yields

$$ds_{g,\text{TO}} = \frac{W(V \pm V_{\text{wind}})dV}{g(T - R - W\varphi)}. \tag{11.7}$$

Integration of Eq. (11.7) gives the takeoff ground roll,

$$s_{g,\text{TO}} = \int_0^{V_{\text{TO}}} \frac{W(V \pm V_{\text{wind}})dV}{g(T - R - W\varphi)}, \tag{11.8}$$

where V_{TO} is the takeoff velocity or lifting-off velocity. For our purposes, the takeoff velocity can be taken to be 20% higher than the stall speed of the aircraft in the clean configuration (flaps retracted), such that $V_{\text{TO}} = 1.2 \, V_{s1}$. In terms of velocity, the takeoff ground roll is expressed as

$$s_{g,\text{TO}} = \int_0^{t_{g,\text{TO}}} V(t)dt \tag{11.9}$$

where $t_{g,\text{TO}}$ is the time of the takeoff ground roll.

Table 11.1 Values of runway surface friction coefficient.

	Range of μ_r	
Types of surface	Brakes off	Brakes on
Dry concrete/asphalt	0.03–0.05	0.3–0.5
Wet concrete/asphalt	0.05	0.15–0.3
Icy concrete/asphalt	0.02	0.06–0.10
Hard turf	0.05	0.4
Firm dirt	0.04	0.3
Soft turf	0.07	0.2
Wet grass	0.08	0.2

Source: Anderson (1999) and Raymer (2006).

The relevant parameters to the takeoff roll are the aircraft weight (W), thrust available (T), air density (ρ), wind direction and velocity (V_{wind}), runway slope (φ), and rolling friction coefficient (μ_r). The relevant aerodynamic parameters are (C_L, C_{D_0}, e), and the parameters in the thrust model are k_0 for Eq. (11.2) or k_1, k_2, and k_3 for Eq. (11.3). For a given aircraft, the aerodynamic parameters in cruise flight are known, but they could be considerably changed in takeoff due to the ground effect and the use of high-lift devices. The friction parameter, μ_r depends on the surface materials and roughness. The values of μ_r for several runway surfaces are listed in Table 11.1 (Anderson 1999; Raymer 2006). When a set of the parameters is given, Eq. (11.8) can be numerically integrated to predict the time history of V in the ground roll. It is possible to compare the predicted and measured time histories of V to estimate a set or subset of the parameters as an optimization problem. Formally, it is written as

$$\|V_{\text{pre}}(t) - V_{\text{meas}}(t)\| \to \min, \tag{11.10}$$

where V_{pre} and V_{meas} are the predicted and measured ground-roll velocities, respectively, and $\|\bullet\|$ is a norm for the distance between two functions such as the L2 norm.

11.1.2 Landing Ground Roll

Equation (11.1) is also valid for an aircraft decelerating during the landing ground roll. The thrust T in landing could be positive or negative, depending on whether the thrust-reversing devices are implemented. Since aerodynamic decelerating devices are usually employed in landing, the parasite drag coefficient (C_{D_0}) is significantly increased. The slope angle $\varphi > 0$ is for uphill landing. The main difference between the landing and takeoff ground rolls is that the brakes are applied during landing. The braking friction coefficient that is used for μ_r in Eq. (11.4) is a strong function of runway surface conditions (wet or icy surfaces), whether or not tires are slipping, and how hard the pilot is applying the brakes (see Table 11.1). It is useful to compare the time histories of the predicted and measured V in landing, and a set or subset of the parameters can be determined as an optimization problem. Similar to Eq. (11.8), the landing ground roll is expressed as

$$S_{g,L} = \int_{V_{\text{TD}}}^0 \frac{W(V \pm V_{\text{wind}})dV}{g(T - R - W\varphi)} \tag{11.11}$$

where V_{TD} is the touchdown velocity.

11.1.3 Rotation Distance

During the takeoff acceleration, once the aircraft reaches a safe velocity, the pilot will pitch the nose up with the elevator control. The wheels of the main landing gear (for a tricycle configuration) remain on the ground through the rotation maneuver, which is performed to increase the angle of attack and increase the lift in a short period of time. The conclusion of the rotation is when the main wheels liftoff from the ground and free flight begins. The rotation distance is the length of runway traveled from the point when the pilot starts the pitch-up maneuver to the time when the main wheels leave the ground. For landing, rotation consists of the duration from initial touch of the main wheels on the ground to the point when the nose is lowered to the runway and brakes are applied. This distance is denoted by either $s_{g, \text{TO}, 2}$ for takeoff or $s_{g, L, 2}$ for landing. Reasonable estimates for these distances are

$$s_{g,\text{TO},2} = V_{\text{TO}} t_r \tag{11.12}$$

and

$$s_{g,L,2} = V_{\text{TD}} t_r, \tag{11.13}$$

where t_r is the average time used to carry out the takeoff or landing rotation maneuver, respectively. For most pilots this time interval is on the order of 1–2 seconds for takeoff and 3–4 seconds for landing.

11.1.4 Transition Distance

During takeoff, the aircraft must transition from level acceleration along the runway surface to a steady climb. During landing, an aircraft transitions from a steady descent on final approach, through a flare maneuver to the point of touchdown. The horizontal distance that the aircraft travels while making these transitions is estimated as the first part of the air distance in a takeoff, or the last air segment in landing. In a simplified case, the flight trajectory in this transition is assumed to be circular, as shown in Figure 11.3. To maintain the circular motion along the circular arc with a radius of R_c, a centrifugal force is provided by the additional lift generated by the lifting surfaces, $i.e.$,

$$\Delta n = \frac{V^2}{gR_c} = \frac{\Delta L}{W}, \tag{11.14}$$

where $\Delta L = L_{\text{tran}} - W$ is an increment of the lift in the transition, L_{tran} is the lift in the transition phase, and V can be either V_{TO} in takeoff or V_{TD} in landing. Therefore, the radius R_c is given by

$$R_c = \frac{V^2}{g\Delta n} = \frac{V^2}{g(L_{\text{tran}}/W - 1)}. \tag{11.15}$$

Furthermore, we express $L_{\text{tran}} = 0.5\rho V^2 S C_{L, \text{tran}}$ and $W = 0.5\rho V_s^2 S C_{L,\text{max}}$ in a limiting case, and then we have

$$R_c = \frac{V^2}{g\left(\frac{V^2 C_{L,\text{tran}}}{V_s^2 C_{L,\text{max}}} - 1\right)}, \tag{11.16}$$

where $C_{L, \text{tran}}$ is the lift coefficient in the transition, $C_{L, \text{max}}$ is the maximum lift coefficient, and V_s is the stall velocity. The velocity used in Eq. (11.16) can be the average velocity throughout the maneuver, which may be approximated as the takeoff velocity. When R_c is known, as shown in Figure 11.3, the transition distance is given by the projection of the circular arc on the ground, $i.e.$,

$$s_{\text{tran}} = R_c \left| \sin \theta_{\text{CL}} \right|, \tag{11.17}$$

where θ_{CL} is the climb angle of the aircraft at the end of the circular arc.

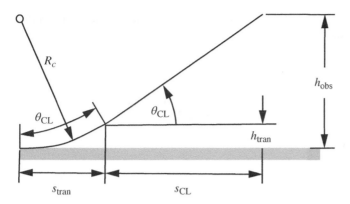

Figure 11.3 Transition distance geometry.

11.1.5 Climb Distance

The second part of the air distance is the distance covered during the initial climb before the standard obstacle distance is cleared in takeoff or during the final approach after the obstacle is cleared but before the landing flare begins. If the altitude at the end of the takeoff transition is greater than the obstacle clearance height or if the landing flare begins above the obstacle clearance height, this distance has no meaning and is not included in the air distance. As shown in Figure 11.3, the altitude at the end of the takeoff transition or the height at the beginning of the landing flare is estimated by

$$h_{\text{tran}} = R_c(1 - |\cos\theta_{\text{CL}}|). \tag{11.18}$$

If h_{tran} is greater than the obstacle clearance height, the steady climb distance will not be included in calculation of the air distance. If h_{tran} is smaller than the obstacle clearance height, as indicated in Figure 11.3, the air distance covered during the initial climb for the obstacle clearance height h_{obs} is

$$s_{\text{CL}} = \frac{h_{\text{obs}} - h_{\text{tran}}}{|\tan\theta_{CL}|}. \tag{11.19}$$

For aircraft certified under 14 CFR part 23, the required obstacle clearance height is $h_{\text{obs}} = 50$ ft.

11.1.6 Total Takeoff and Landing Distances

The total TO&L distances are expressed as the sum of the distances of the different stages, *i.e.*,

$$s_{\text{TO}} = s_{g,\text{TO}} + s_{g,\text{TO},2} + s_{\text{tran}} + s_{\text{CL}} \tag{11.20}$$

and

$$s_L = s_{g,L} + s_{g,L,2} + s_{\text{tran}} + s_{\text{DC}}, \tag{11.21}$$

where s_{DC} is the distance for steady descent in landing that corresponds to s_{CL} in takeoff.

11.1.7 Simple Estimations

Besides the above theoretical method, an approximate estimate for the ground roll can be obtained based on a balance between the work done in the takeoff and the gain in the kinetic energy,

$$s_{g,\text{TO}} = \frac{W}{g(T-R)_{\text{mean}}}\left(\frac{V_{\text{TO}}^2}{2}\right), \tag{11.22}$$

where an average value for force is used in the estimate for work. For the air distance phase of the takeoff, a similar estimate for the takeoff air distance is

$$s_{a,\text{TO}} = \frac{W}{(T-R)_{\text{mean}}} \left(h_{\text{obs}} + \frac{V_{\text{obs}}^2 - V_{\text{TO}}^2}{2g} \right), \tag{11.23}$$

where the gain in potential energy is also included, and V_{obs} is the velocity at h_{obs}.

Similarly, the landing ground roll and air distance are, respectively,

$$s_{g,L} = -\frac{W}{g(T-R)_{\text{mean}}} \left(\frac{V_{\text{TD}}^2}{2} \right) \tag{11.24}$$

and

$$s_{a,L} = -\frac{W}{(T-R)_{\text{mean}}} \left(h_{\text{obs}} + \frac{V_{\text{obs}}^2 - V_{\text{TD}}^2}{2g} \right). \tag{11.25}$$

The total distances in TO&L are, respectively,

$$s_{\text{TO}} = s_{g,\text{TO}} + s_{a,\text{TO}} \tag{11.26}$$

and

$$s_L = s_{g,L} + s_{a,L}. \tag{11.27}$$

11.2 Measurement Methods

There are several traditional methods for measurements of aircraft's position and velocity in flight testing in TO&L (Kimberlin 2003). The sighting bar method that is the simplest method, which uses several sighting bars with known heights located at known positions along a runway to provide landmarks for tracking a test aircraft in TO&L. Using stopwatches, observers can record times at which the aircraft starts to run, lifts off, or touches down. Since the horizontal and vertical information of sighting bars are known, the TO&L distances of the aircraft can be estimated along with the averaged speed. Furthermore, accelerometers on a test aircraft can be used as a simple instrument to record the continuous acceleration history in TO&L. Then, the acceleration history can be integrated to obtain the velocity and position data.

Optical techniques can be used, such as the strip camera method, movie theodolite (cine-theodolite), the onboard theodolite method, or videogrammetry[2]. A series of strip pictures or film pictures of a test aircraft in TO&L can be recorded by using a strip camera or a movie camera. Then, the TO&L distances can be obtained by processing the pictures. In contrast to the strip camera and movie camera methods, the onboard theodolite method involves a camera mounted on a test aircraft to record the images of runway lights and other objects on the ground. The aircraft's position is then determined relative to the designated ground coordinate system. The optical methods are applicable in visible conditions in a relatively small range. For a large range in all weather conditions, a ground-based radar tracking system is more suitable.

The most useful measurement technique for TO&L performance measurement is the Global Positioning System (GPS) (described in detail in Chapter 3). For quality data of TO&L performance, it is best to use a correction approach such as real-time kinematic (RTK) GPS or differential global positioning system (DGPS). In the flight testing environment, DGPS position errors are less than 3 m in latitude and longitude and less than 4 m in altitude, while RTK GPS can offer cm-scale

---

2 Videogrammetry methods are discussed in an online supplement.

accuracy. Since the flight range for TO&L is limited to several kilometers, a local coordinate system is more convenient than the geophysical coordinate system, and they can be transformed. Germann (1997) evaluated the accuracy of DGPS in measuring the TO&L ground rolls and air distances, finding an accuracy that was within the 10 ft error bounds of the traditional video theodolite measurements. However, the accuracy assessment of DGPS measurements of the air phase and landing was inconclusive since the liftoff and touchdown points could not be exactly determined by DGPS. These levels of effort for DGPS are reasonable for TO&L, but can be improved upon through RTK GPS, which is an increasingly common and low-cost approach, especially for unmanned aerial vehicle (UAV) flight testing.

11.3 Flight Testing Procedures

11.3.1 Standard Flight Procedures

Before discussing flight test procedures, we will provide an overview of basic aircraft operations in the airport environment. A series of TO&L are typically done by remaining in the airport environment and flying the aircraft in a rectangular circuit with one leg based on the runway – this is referred to as the airport traffic pattern, which is illustrated in Figure 11.4. Most traffic patterns involve left-hand turns, but some runways may use a right-hand traffic pattern in order to minimize community noise or to deconflict with traffic on adjacent runways or other airports. The portions of the standard traffic pattern are the upwind leg, crosswind leg, downwind leg, base leg, and final approach. We will work our way through the traffic pattern by following the flight of an aircraft from takeoff through landing. In order to minimize required runway length, the aircraft should always be flown into the wind (*i.e.*, the runway should be selected such that a maximum headwind component is present, which effectively provides the aircraft with a "head start" of airspeed when the ground speed is zero). When an aircraft remains in the traffic pattern, the maximum altitude

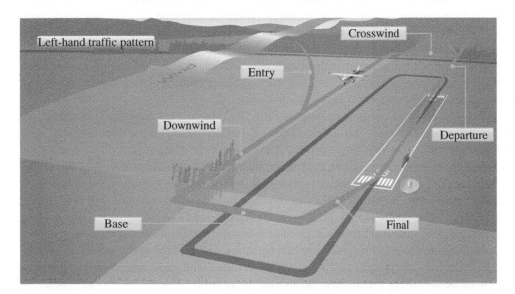

Figure 11.4 Standard airport traffic pattern for takeoff and landing. Source: Federal Aviation Administration 2016.

that is flown for much of the pattern is referred to as the traffic pattern altitude, which is usually 1000 ft above ground level (AGL).

The upwind leg, or departure segment, is the air distance from the takeoff point up to the point of turning crosswind. On the upwind leg, once a steady positive rate of climb has been established and a safe altitude has been reached the flaps and landing gear are retracted. A clean configuration is maintained for much of the traffic pattern. After reaching an altitude of around 700 ft AGL on the upwind leg, the pilot will turn to the crosswind leg and continue to climb up to pattern altitude. After flying crosswind for a short distance, the pilot turns downwind to follow a path parallel to the runway and offset from the runway centerline by about half a mile. The pilot stabilizes flight on the downwind leg at constant altitude (pattern altitude) and begins to set up the aircraft for the landing phase. For most general aviation aircraft, an appropriate flight speed on the downwind leg will be between 90 and 110 knots indicated airspeed (KIAS).

Once the aircraft is at a downwind location that is in line with the intended touchdown point ("abeam the numbers," referring to the runway number painted at the end of the runway), the pilot pulls back on power, configures one notch of flaps, and establishes the aircraft in a steady descent. When the runway numbers are in a location approximately 135° from the nose (45° behind the wing), the pilot turns the aircraft to enter the base leg of the traffic pattern. A steady descent is maintained on the base leg at a lower flight speed than the downwind leg, and the landing flap configuration is set (increased flap deflection). Finally, as the runway centerline is approached, the pilot initiates a turn to final approach such that the aircraft is lined up with the runway at the completion of the base-to-final turn. The aircraft is now on a final approach segment. The ideal configuration is for the aircraft to be in a steady descent on a glide path of approximately 3° below horizontal. The pilot will maintain pitch and power to achieve a stabilized descent at the proper approach velocity.

The key to a successful landing is proper energy management (both kinetic and potential energy) and maintaining stabilized flight conditions throughout. If these criteria are not satisfied, then the pilot should abort the landing (termed a "go around," which involves throttle-up, flap retraction, and maintaining positive control) and continue in the air in the upwind direction to initiate a new lap of the traffic pattern to attempt a new landing. Hazardous conditions for landing would be a low-altitude, low-velocity approach (too little energy), or a high-altitude, high-velocity approach. Energy can be traded between potential and kinetic (*i.e.*, trading altitude for airspeed) by throttle and control stick inputs, but large corrections should be avoided. The ideal final approach segment involves an aircraft set up in landing configuration well before landing, with a stabilized descent (constant airspeed and descent rate) all the way to flare.

The pilot can infer approach angle by assessing the "sight picture" of the runway position and orientation in the wind screen of the aircraft. This can vary depending on the dimensions (length and width) of the runway, with a larger runway making it appear that the aircraft is closer to the ground than typical. Most runways also provide supplemental visual guidance to the pilot in the form of light bars to the side of the runway, which indicate whether the aircraft is above, below, or on the desired glide slope. For example, the precision approach position indicator (PAPI) is a series of four horizontal lights on the side of the runway. If two white and two red lights are visible, then the aircraft is on glide slope. Conversely, if three or four red lights are visible in the row, then the aircraft is below glideslope; if three or four white lights are visible in the row, then the aircraft is above glide slope. If the aircraft glide slope cannot be corrected with small control inputs, then the landing should be aborted by executing a "go around" procedure.

Once the aircraft crosses the runway threshold, the pilot pulls the power back to idle and maintains the descent to a low altitude of about one wing span above the runway. At this point, the pilot begins the landing flare, where the nose is gradually pitched up to bleed off airspeed and gradually decrease altitude to a gentle touch of the main landing gear on the runway. (A successful flare to a gentle touchdown is perhaps the most difficult portion of the landing – easier said than done!). After main gear touchdown, the pilot gradually releases the control stick to allow the nose wheel to touch down, and appropriate braking is applied until a safe velocity is reached where the aircraft can be turned to exit the runway.

11.3.2 Flight Test Procedures

Takeoff performance can be evaluated with various aircraft configurations and piloting approaches. The typical flight procedure recommended by the POH is to start at the end of the runway while holding brakes, increase the throttle to full power and wait for a moment for full power to develop, and initiate the takeoff roll with brake release. The beginning of the maneuver corresponds with brake release and continues through climb-out to an altitude of 50 ft AGL. During the takeoff roll, the pilot tracks the centerline of the runway using light differential braking, nose wheel steering, or rudder deflection (depending on the aircraft design configuration and instantaneous flight speed). Once the aircraft velocity reaches rotation speed, the pilot pulls back on the control stick to pitch the nose up. Since this rapidly increases the angle of attack, lift rapidly builds to a point where $L \geq W$ and the aircraft lifts off the ground. Immediately following liftoff, the pilot pitches the nose of the aircraft to maintain an airspeed for best climb angle, V_x (see Chapter 9 for a discussion of climb performance). This flight condition is maintained at least until an altitude of 50 ft AGL. At this point the data collection is complete, and the pilot can continue flight in the standard airport traffic pattern to return for a landing on the same runway.

Takeoff performance can be evaluated under a range of operating conditions. For example, a soft field takeoff length could be evaluated, simulating takeoff from a wet or muddy turf runway. This maneuver involves minimizing the weight on the nose gear such that it does not get stuck in the soft turf. To execute a soft field takeoff, the pilot maintains forward motion of the aircraft at all times, transitioning seamlessly from taxi to takeoff. Full power is applied to initiate the takeoff roll, and the control stick is held full aft in order to pitch the nose up as much as possible and minimize weight on the nose gear. As dynamic pressure builds during the takeoff roll, the nose will pitch up, such that the pilot must gradually release pressure on the control stick in order to avoid premature liftoff and stall. At a certain low airspeed, the aircraft will liftoff the ground but will not be able to fly safely out of ground effect without stalling. Thus, the pilot will maintain flight at a very low altitude in order to build up airspeed with the aircraft in ground effect. Once at a safe airspeed, the pilot will then pitch the nose for the speed best climb angle (V_x), and climb at this speed until passing an altitude of 50 ft AGL.

Takeoff flight testing can also be done with various flap settings. The standard approach is to use the takeoff flap setting, which is typically a small angle such as 10°. Takeoff performance can also be assessed with the aircraft in a clean configuration (no flap deflection), which will require longer runway length, but may offer better performance on initial climb-out. For no-flap takeoff flight tests, it is important to use a long runway with plenty of safety margin.

In the landing test, the pilot will maneuver the aircraft through the standard traffic pattern. Once on final approach, the desired flap setting should be set with sufficient time to stabilize flight, with the airspeed and descent rate stabilized on the desired conditions with minimal adjustments to pitch or power required for stabilized flight. The pilot will fly the final approach through an altitude of 50-ft AGL, where the data record begins. Shortly after the aircraft touches down, the pilot lowers the nose wheel to the ground and initiates standard braking to bring the aircraft to a complete stop on the runway. All instrumentation, including the GPS system or/and the videogrammetric system, must be operating during TO&L.

11.3.3 Data Acquisition

The aircraft weight should be recorded at the beginning of the flight test, and fuel burn tracked throughout the flight in order to determine aircraft weight at the time of each TO&L. Local surface winds should be directly measured by a portable anemometer. Alternatively, the pilot can contact the air traffic controller at a towered airport to request a report of the wind direction and magnitude at that moment. [Note that winds reported by the meteorological aerodrome report (METAR) or automated weather observing system (AWOS) can be up to one hour old.] Measurements of pressure altitude (by setting an altimeter to a reference pressure of 29.92 inHg) and outside air temperature should be recorded.

Takeoff ground roll and air distance can be measured by GPS to record position and velocity as a function of time. An alternate means of recording velocity and altitude as a function of time is to video record the airspeed indicator and altimeter throughout the takeoff or landing. If advanced data acquisition methods are available (*e.g.*, with a smartphone or other data acquisition device), data from the gyroscopes, accelerometers, magnetometer, and barometer (if available) are useful data sets for identifying the aircraft conditions throughout TO&L. Also desirable are engine settings such as percent power, engine speed (RPM, revolutions per minute), manifold pressure, and fuel flow rate.

If GPS measurement techniques are not available, the point of liftoff from the runway can be estimated to within about ±50 ft by stationing observers along the length of the runway and identifying the liftoff point relative to features along the length of the runway (*e.g.* runway edge lights are typically spaced constant intervals of no greater than 200 ft, with a distance that can be readily measured by online mapping tools).

11.3.4 Data Analysis

The ultimate objective of this flight test is to determine the required field length for TO&L under various aircraft configurations. Thus, the position and velocity as a function of time should be extracted from the GPS data set, with the total takeoff or landing length (Eq. (11.26) or (11.27)) determined from the ground and air distances. The point of liftoff from the runway can be determined by careful analysis of gyroscope, acceleration, and altimeter data.

The data must be reduced to standard conditions, which is maximum takeoff weight at sea level on a standard day with zero winds. In order to make these adjustments, we introduce the following correction factors. However, since TO&L performance are highly variable and subject to pilot skill and technique, the correction factors we introduce here will be based on historical empirical results, rather than mathematical analysis (see Lush 1953 or Herrington et al. 1966).

The most significant correction to the takeoff distance is for the effects of wind. Taking V_{wind} to be the *component* of the wind velocity vector aligned with the runway direction (headwind is positive), we have the empirical relation (Herrington et al. 1966)

$$s_{g,\text{corr}} = s_g \left(\frac{V_{\text{TO}} + V_{\text{wind}}}{V_{\text{TO}}} \right)^{1.85}, \tag{11.28}$$

where s_g is the measured takeoff ground roll, $s_{g,\text{corr}}$ is the wind-corrected ground roll, and V_{TO} is the takeoff velocity. The correction for air distance from takeoff to clearance of a 50-ft obstacle is given by (Kimberlin 2003)

$$s_{a,\text{corr}} = s_a + V_{\text{wind}} t, \tag{11.29}$$

where s_a is the measured air distance, $s_{a,\text{corr}}$ is the corrected air distance, and t is the time from takeoff to the moment when a 50-ft obstacle is cleared.

Corrections for nonstandard weight (W), density ratio ($\sigma = \rho/\rho_{SL}$), engine speed (N), and engine power (P) for an aircraft with a piston engine and constant-speed propeller are provided by (Herrington et al. 1966, Eq. 6.335)

$$S_{g,std} = S_{g,test}\left(\frac{W_{std}}{W_{test}}\right)^{2.6}\left(\frac{\sigma_{std}}{\sigma_{test}}\right)^{-1.7}\left(\frac{N_{std}}{N_{test}}\right)^{-0.7}\left(\frac{P_{std}}{P_{test}}\right)^{-0.9}. \tag{11.30}$$

The air distance measurement is corrected by

$$S_{a,std} = S_{a,test}\left(\frac{W_{std}}{W_{test}}\right)^{2.3}\left(\frac{\sigma_{std}}{\sigma_{test}}\right)^{-1.2}\left(\frac{N_{std}}{N_{test}}\right)^{-0.8}\left(\frac{P_{std}}{P_{test}}\right)^{-1.1} \tag{11.31}$$

for light aircraft.

11.4 Flight Test Example: Cessna R182

TO&L tests of the Cessna R182 were conducted to measure the TO&L distance and speed of the aircraft using an onboard DGPS system (Liu and Schulte 2007). Table 11.2 lists the basic parameters of the Cessna R182. Figure 11.5 shows the altitude, speed and distance of the Cessna R182 aircraft as a function of time during takeoff. The liftoff time was about 25 seconds; the estimated takeoff distance was about 1200 ft, which was larger than the value given by the manufacturer. At the liftoff time, the indicated airspeed was about 80 KIAS. The speed and distance of the aircraft in takeoff calculated by Eq. (11.1) are also plotted in Figure 11.5 for comparison. In calculations, the coefficients in the thrust relation (Eq. (11.3)) are $k_1 = 1557$ N, $k_2 = 1.8$ Ns/m, and $k_3 = 0$. Other

Table 11.2 Basic parameters of the Cessna R182.

Aircraft parameter	Value
Engine	LYC O-540-J3C5D
Horsepower	235
Std fuel	61 gal
Max fuel	80 gal
Cruise speed	156 kts
Stall speed	50 kts
Range	520 nm
Service ceiling	14,300 ft
Takeoff	820 ft
Landing	600 ft
Wingspan	35 ft
Length	28.33 ft
Height	8.75 ft
Empty weight	1782 lb
Gross weight	3100 lb
Aspect ratio	7.5
Wing area	173.6 ft^2

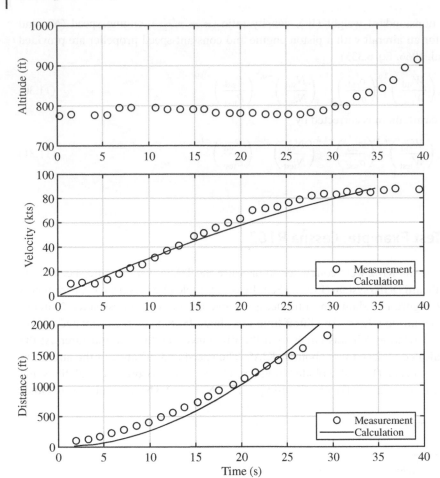

Figure 11.5 The altitude, speed, and distance of the Cessna R182 aircraft during takeoff.

relevant parameters in calculations are $C_{D_0} = 0.020$, $C_L = 0.7$, $e = 0.7$, and $\mu_f = 0.03$. Reasonable agreement is observed between the analytical prediction and the flight test results, particularly considering the wide variability in environmental conditions and piloting technique.

Nomenclature

AR	aspect ratio
b	wing span
C_{D_0}	parasite drag coefficient
C_L	lift coefficient
D	drag
e	Oswald efficiency factor out of ground effect
e'	Oswald efficiency factor in ground effect
F_f	friction force
g	gravitational acceleration
h	height above ground level
k_i	thrust model constants ($i = 0, 1, 2,$ or 3)

n	load factor
N	engine speed (revolutions per unit time)
P	power
q	dynamic pressure
R	resistance force
R_c	radius of circular arc for transition to climb
s	distance
s_a	air distance
s_g	ground distance
S	wing area
t	time
T	thrust
V	airspeed
V_s	stall speed
V_{s1}	stall speed, clean configuration
V_{wind}	wind speed
V_x	speed for best climb angle
W	aircraft weight
φ	angle of runway with respect to level (positive is an uphill slope)
η	propeller efficiency
μ_r	rolling coefficient of friction for tires
ρ	density
σ	density ratio, ρ/ρ_{SL}
θ_{CL}	climb angle

Subscripts

CL	climb
corr	corrected
DC	descent
L	landing
max	maximum
mean	mean
min	minimum
meas	measured
obs	obstacle
opt	optimum
pre	predicted
prop	propeller
r	rotation
shaft	shaft
std	standard
SL	sea level
test	test condition
TD	touchdown
TO	takeoff
tran	transition between ground and air segments
wing	wing

Acronyms and Abbreviations

AGL	above ground level
AWOS	automated weather observing system
DGPS	differential GPS
GPS	Global Positioning System
KIAS	knots indicated airspeed
METAR	meteorological aerodrome report
PAPI	precision approach position indicator
RPM	revolutions per minute
RTK	real-time kinematic
TO&L	takeoff and landing
UAV	unmanned aerial vehicle

References

Anderson, J.D. (1999). *Aircraft Performance and Design*. Boston: McGraw-Hill.

Dekker, F.E.D. and Lean, D. (1959). "Takeoff and landing performance," Chapter 8. In: *AGARD Flight Test Manual*, vol. 1. New York: Pergamon Press.

Federal Aviation Administration. (2016). *Airplane Flying Handbook, FAA-H-8083-3B*. United States Department of Transportation, Federal Aviation Administration. https://www.faa.gov/regulations_policies/handbooks_manuals/aviation/airplane_handbook/

Germann, K. P. (1997). Flight test evaluation of a differential global positioning system sensor in runway performance testing. MS thesis, Department of Aerospace Engineering, Mississippi State University.

Herrington, R. M., Shoemacher, P. E., Bartlett, E. P., and Dunlap, E. W. (1966). *Flight Test Engineering Handbook*. USAF Technical Report 6273, Edwards AFB, CA: US Air Force Flight Test Center. Defense Technical Information Center Accession Number AD0636392, https://apps.dtic.mil/docs/citations/AD0636392.

Kimberlin, R.D. (2003). *Flight Testing of Fixed-Wing Aircraft*. Reston, VA: American Institute of Aeronautics and Astronautics.

Liu, T. and Schulte, M. (2007). Flight Testing Education at Western Michigan University. AIAA Paper 2007-0700, Reno, NV: Aerospace Sciences Meeting.

Lush, K. J. (1953). Standardization of Take-Off Performance Measurements for Airplanes. Technical Note R-12, Edwards AFB, CA: U.S. Air Force Air Research and Development Command, U.S. Air Force Flight Test Center, Defense Technical Information Center Accession Number AD0024338, https://apps.dtic.mil/docs/citations/AD0024338.

McCormick, B.W. (1995). *Aerodynamics, Aeronautics, and Flight Mechanics*, 2e. New York: Wiley.

Raymer, D.P. (2006). *Aircraft Design: A Conceptual Approach*, 4e. Reston, VA: American Institute of Aeronautics and Astronautics.

12

Stall Speed

The stall speed of an aircraft is one of the most critical pieces of information that a pilot will want to know before taking off in any aircraft. Stall occurs when the flow separates from the wing, leading to a substantial loss of lift. This is a potentially dangerous condition if the pilot is unprepared for stall and if it occurs close to the ground. Thankfully, recovery from a stalled condition is a straightforward and benign process when done at altitude: the pilot lowers the nose to reduce the angle of attack, adds power, allows lift to reestablish, and then resumes normal flight. Stall maneuvers are routinely flown in flight training and currency training, with the maneuver performed at sufficiently high altitude to allow a safe recovery. Safety-minded pilots will practice the maneuver often in order to maintain proficiency and will maintain flight speeds well above the stall speed. Thus, safe airmanship requires intimate familiarity with the stall speed – the lowest airspeed that can be flown in steady, level flight.

Since stall is a nonlinear, three-dimensional aerodynamic phenomenon, it is difficult to predict precisely the full aircraft's stall characteristics based on computational fluid dynamics (CFD) or wind tunnel testing. Thus, flight testing is the only definitive way to establish the stall characteristics of the aircraft. We will focus in this chapter on flight testing of level stall, and in Chapter 13, we will study stall conditions in turning flight.

This chapter will start with a discussion of the aerodynamic theory for stall, which begins with viscous effects. We will review the boundary layer, effects of pressure gradient on boundary layer development, and the conditions that define boundary layer separation (the fundamental mechanism leading to stall). Readers with a strong background in aerodynamic fundamentals may wish to skim through or skip this section, but it is included here for those who would benefit from a concise review. Following this, we will discuss three-dimensional aspects of the wing geometry as they relate to stall, and the techniques that aircraft designers employ to lower the stall speed or improve aircraft controllability in stall. We will conclude the theory section with a brief summary of some first-order techniques for predicting the stall speed for an aircraft.

We will next turn our attention to flight testing methods for stall, defining appropriate piloting technique for identifying the stall speed. Several correction factors for nonstandard deceleration rate, center of gravity (CG) location, and weight will be introduced. We will then conclude the chapter with an example of digital data acquisition (DAQ) techniques applied to stall speed measurement on a Cirrus SR20.

Introduction to Flight Testing, First Edition. James W. Gregory and Tianshu Liu.
© 2021 John Wiley & Sons Ltd. Published 2021 by John Wiley & Sons Ltd.
Companion website: https://www.wiley.com/go/flighttesting

12.1 Theory

For steady level flight conditions, there is a direct link between lift production (L) and airspeed (V),

$$L = W = \frac{1}{2}\rho V^2 S C_L, \tag{12.1}$$

where W is the aircraft weight, ρ is the density, S is the wing area, and C_L is the lift coefficient. Assuming that the minimum speed for low-speed flight is not power limited (due to the increase in induced drag at low speeds), the minimum speed that the aircraft can fly is governed by stall,

$$V_S = \sqrt{\frac{2W/S}{\rho C_{L,\text{max}}}}, \tag{12.2}$$

where the maximum lift coefficient is a fixed value for the wing. The higher the maximum lift coefficient, the more slowly the aircraft can fly before reaching stall. Beyond setting the minimum flight speed for the aircraft, stall characteristics also dictate the approach and landing speeds of an aircraft since the pilot will want to fly a slow and safe airspeed above stall speed (typically 20–30% higher). The landing speed also dictates the required runway length, since the kinetic energy of the aircraft (proportional to the square of velocity) must be dissipated by braking.

Aerodynamic stall is a result of flow separation, where the streamlines no longer follow the curvature of the airfoil on the wing, leading to a loss of lift (see Figure 12.1). At low angle of attack, the streamlines follow the curvature of a well-designed airfoil (see the $\alpha = 6°$ case in Figure 12.1), leading to normal lift production. But, as the angle of attack increases, there comes a critical angle at which flow separation occurs – this defines the stall condition ($\alpha = 15°$ in Figure 12.1, which illustrates trailing edge stall). At angles of attack higher than the critical angle, the extent of flow

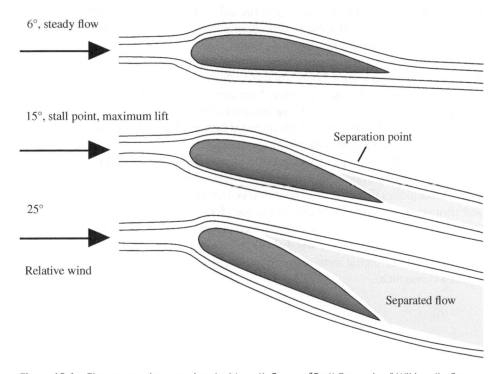

Figure 12.1 Flow separation associated with stall. Source: "Stall Formation," Wikimedia Commons.

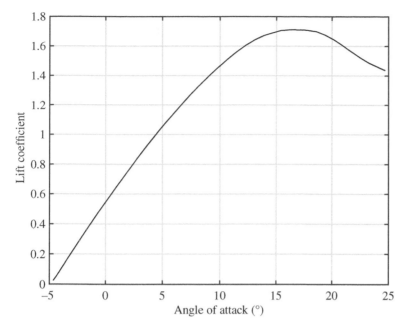

Figure 12.2 Lift coefficient data for a typical airfoil (SM701, calculated in XFOIL). Source: Wikimedia Commons user Botag, "Lift Curve," Wikimedia Commons.

separation increases until the flow is fully separated. This trend in separation behavior results in a maximum lift coefficient occurring at the stall angle of attack and then a decrease in C_L at angles beyond the critical angle of attack, as shown in Figure 12.2.

Stall characteristics and flow separation are inherently viscous phenomena, with intricate ties to the viscous boundary layer. We will provide a brief overview of two-dimensional boundary layer characteristics and how the boundary layer behavior leads to separation and stall on an airfoil. We will then shift to a view of the entire wing to consider how three-dimensional stall develops.

12.1.1 Viscous Boundary Layers

A boundary layer is the thin region of fluid close to the surface of the wing that is affected by viscosity. In the near-wall region the local velocity goes to zero such that the no-slip condition is satisfied at the wall. At the leading edge of the wing the boundary layer starts in the laminar state, where the streamlines within the boundary layer are steady and parallel (*i.e.*, zero wall-normal velocity). Natural instabilities start to develop and grow some distance downstream of the leading edge, ultimately leading to transition of the boundary layer from laminar to turbulent. The turbulent boundary layer is much thicker, with unsteady fluctuations within the viscous region leading to enhanced momentum transport in the wall-normal direction, which draws higher momentum fluid closer to the wall. Typical time-averaged profiles of velocity across the boundary layer are illustrated in Figure 12.3, which are normalized by the boundary layer thickness (δ) and the freestream velocity (U_∞), such that the shapes can be compared. The laminar boundary layer velocity profile is from a numerical solution of the Blasius equation, and the turbulent boundary layer velocity profile is from the one-seventh power law (an engineering approximation), $u/U_\infty = (y/\delta)^{1/7}$, where u is local velocity and y is the wall-normal direction. The profiles shown in Figure 12.3 illustrate how the turbulent boundary layer has a much fuller profile, where

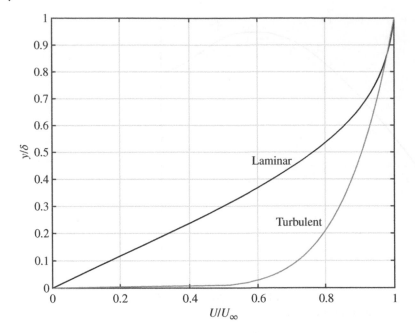

Figure 12.3 Examples of laminar and turbulent boundary layer velocity profiles.

high-momentum fluid is closer to the wall and the velocity gradient is much higher at the wall (compared to the laminar profile).

We will now consider how boundary layer development depends on the distribution of pressure around the airfoil (imposed by the inviscid outer flow). Near the leading edge of the airfoil, the upper surface pressure decreases rapidly until reaching the suction peak, and then increases again more gradually with chordwise distance to the trailing edge. Decreasing pressure in the downstream direction ($\partial p/\partial s < 0$) is referred to as a favorable pressure gradient, since this condition leads to delayed transition, thinning of the boundary layer (fuller profile), and a boundary layer state that is more resistant to separation. Increasing pressure ($\partial p/\partial s > 0$), on the other hand, is referred to as an adverse pressure gradient since it promotes transition, thickens the boundary layer, and makes the boundary layer more susceptible to separation.

12.1.2 Flow Separation

Separation occurs when the local flow velocity (within the boundary layer) decreases to a point where the flow reverses direction (see Figure 12.4), and the wall-normal velocity gradient goes to zero $(\partial u/\partial y)_{y=0} = 0$. These effects occur since the streamwise pressure gradient affects the shape of the boundary layer profile, retarding the local flow velocity in an adverse pressure gradient. This is aptly illustrated by the Navier–Stokes equation applied to the boundary layer (assuming steady, 1D, incompressible flow with no body forces),

$$\rho u \frac{\partial u}{\partial s} = -\frac{\partial p}{\partial s} + \mu \frac{\partial^2 u}{\partial y^2}, \tag{12.3}$$

where an adverse pressure gradient on the right side leads to decreasing velocity on the left side. In essence, the momentum of the flow close to the wall (within the boundary layer) cannot overcome the increasing pressure of adverse pressure gradient, so the local flow within the boundary layer decelerates to zero in the near-wall region.

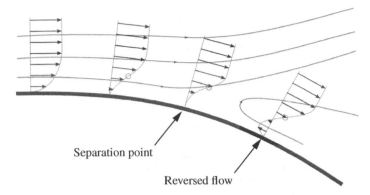

Separation point

Reversed flow

Figure 12.4 Boundary layer profiles illustrating flow separation. Source: Image by Olivier Cleynen, "Boundary layer separation." Licensed under CC BY 3.0.

As the flow progresses downstream, particularly in an adverse pressure gradient, the local velocity decelerates and the wall-normal velocity gradient at the wall goes to zero (the point of separation). Separation characteristics also depend on the nature of the boundary layer profile. A fuller velocity profile – one that has higher momentum fluid closer to the wall – will be more robust toward boundary layer separation. Thus, a turbulent boundary layer of the same thickness as a laminar boundary layer will be much more resistant to separation (see Figure 12.3).

12.1.3 Two-Dimensional Stall Characteristics

Depending on the nature of the 2D airfoil section, the state of the boundary layer, and other factors, there can be several different types of stall behavior. Trailing edge stall occurs when the boundary layer separates near the trailing edge, leaving a recirculating region of reversed flow. For this aft portion of the airfoil, the streamlines no longer follow the airfoil curvature and a mild loss of lift is experienced. As angle of attack increases, the separation point continues to move upstream, leaving a larger region of recirculated flow. This continues until the separated region completely engulfs the airfoil and it is fully stalled. By nature, trailing edge stall is a gradual phenomenon, as shown by the lift curve for the NACA 0021 in Figure 12.5. Leading edge stall, on the other hand, occurs when the boundary layer separates near the leading edge. This can occur when there is a strong suction peak with a strong adverse pressure gradient immediately following the suction peak. Downstream of the separation location the flow is massively separated, with a large recirculating region of reversed flow. Leading edge stall is characterized by a relatively sharp drop-off in lift as the angle of attack increases. Thin airfoil stall is yet another type, which results from a sharp leading edge and a separation bubble that forms near the leading edge (since the flow cannot navigate the strong pressure gradients associated with streamline curvature over the sharp leading edge). At some distance downstream of the leading edge separation bubble, however, the flow reattaches and lift is still nearly as high as the fully attached case. As angle of attack increases, the leading edge separation bubble enlarges progressively, resulting in a progressive loss of lift. Trailing edge stall (see Figure 12.5) is a very gradual process relative to thin airfoil and especially leading edge stall, and the maximum lift is typically not as high. For airfoils commonly used on general aviation (GA) aircraft, leading or trailing edge stall are the dominant phenomena.

Reynolds number can also have a substantial impact on the stall characteristics. Defined as $Re = \rho U_\infty c/\mu$, where c is airfoil chord and μ is dynamic viscosity, the Reynolds number can be thought of as the ratio of inertial to viscous forces. High Reynolds number flows are dominated by inertial effects, while low - Re flows have strong viscous effects. High Reynolds number implies

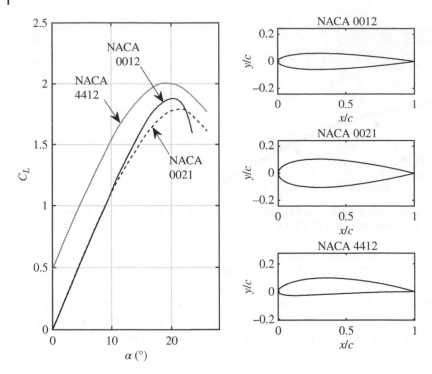

Figure 12.5 Stall characteristics of several canonical airfoil sections, illustrating leading edge stall (NACA 0012), trailing edge stall (NACA 0021), and the effects of camber (NACA 4412) at $Re = 10 \times 10^6$.

thinner boundary layers with earlier transition. Both of these effects lead to higher momentum flow near the wall, making the boundary layer more resistant to separation. These effects are illustrated by the lift curves for a NACA 0012 airfoil shown in Figure 12.6. The angle of attack for stall increases with Reynolds number, but the pre-stall lift curve slope remains close to what is predicted by thin airfoil theory (0.11/°). Typical general aviation aircraft have a chord Reynolds number on

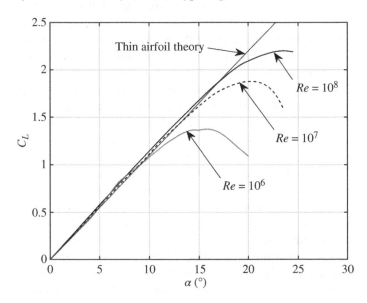

Figure 12.6 Effects of Reynolds number on airfoil stall characteristics.

the order of 1–10 million, depending on airspeed and altitude. (For example, an aircraft with a 5-ft chord wing cruising at standard seal level conditions at 100 KTAS will have a chord-based Reynolds number of 5.4×10^6.)

12.1.4 Three-Dimensional Stall Characteristics

Three-dimensional stall over an aircraft wing is much more complex than two-dimensional stall over an airfoil, since there are a number of complicating factors. Vortical flows are present at the wing/body junction and the wing tip, and these have an impact on boundary layer development. Furthermore, there can be spanwise flows (particularly on swept wings) that impact the boundary layer stability characteristics and can lead to earlier separation. Aircraft designers typically try to design a wing such that the flow will first stall at the wing root rather than the wing tip.

Root stall is greatly preferred over tip stall for several important reasons. First, the aircraft ailerons are located at the wing tips and stalled flow at that location would lead to a loss of roll control authority. Second, any asymmetry in lift production at the wing tips (if one wing tip stalls before the other) will lead to a substantial rolling moment due to the larger moment arm. These two factors combined can easily cause the aircraft to enter a spin during stall. Finally, root stall generates unsteady flow that interacts with the fuselage and tail to induce buffeting, which serves as a warning indicator to the pilot of imminent deep stall. Different wing planforms will have different naturally occurring spanwise progressions of stall (see Figure 12.7), based on the effects of local chord Reynolds number, crossflows within the boundary layer, and other effects. These naturally occurring stall patterns can be disadvantageous (*e.g.*, tip stall on a highly tapered or highly swept wing), as discussed in Sweberg and Dingeldein (1945) or FAA (2016). Thus, pilots and designers work to avoid tip stall conditions. Some of the strategies for controlling the three-dimensional stall progression are discussed next.

12.1.5 Stall Control

A number of strategies are available for stall control. The most commonly implemented tool for changing $C_{L,\max}$ and reducing the stall speed is the deployment of flaps. Trailing edge flaps effectively change the camber of the wing, allowing it to pitch to a much higher angle of attack before reaching stall. With multielement flaps, there are a number of advantageous effects that are employed: each element starts with a fresh boundary layer that has not yet thickened, the gaps between airfoil elements provide a venturi effect that provides higher momentum flow to energize boundary layers, and the multielement nature of the setup changes the loading (and pressure distribution) on each element. It is common practice on nearly every flight of an aircraft to deploy the flaps, which lowers the stall and approach speeds. Thus, in flight testing it is important to characterize the stall speed at various flap deployment settings.

Another common strategy for controlling stall is to install vortex generators (VGs), which can be introduced in the design phase or added to some aircraft as an aftermarket product. VGs are simple triangular tabs that protrude upward from the wing surface at an angle to the local flow (see Figure 12.8). Physically, one can think of a VG as a very small wing protruding upward from the main wing – it will produce a small amount of force in the spanwise direction (the direction and magnitude of this force depend on the angle of the VG with respect to the local flow, just like angle of attack). Since the VG is producing a net force, with associated pressure differences, a tip vortex forms at the top of the VG. This small vortex, oriented in the streamwise direction and starting at the VG, serves to entrain higher momentum fluid from the freestream and draw it down close to the wing surface. This energizes the boundary layer, making it more robust to separation.

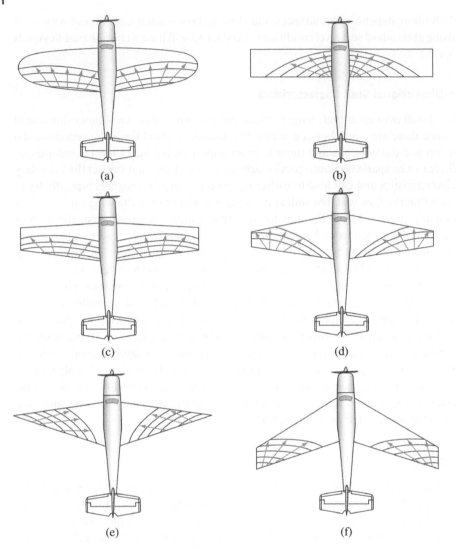

Figure 12.7 Stall progression for various wing planforms. (a) Elliptical wing, (b) regular wing, (c) moderate taper wing, (d) high taper wing, (e) pointed tip wing, and (f) swept wing. Source: Modified from Federal Aviation Administration (2016).

Another strategy available to aircraft designers is wing twist – either geometric or aerodynamic. Geometric twist simply involves changing the local geometric angle of the wing as a function of span. Since root stall is preferable, the geometric pitch angle at the root is usually the highest, with a gradual reduction in the outboard spanwise direction (this design characteristic is referred to as washout). Geometric twist usually amounts to no more than a few degrees, but is often used to force root stall. Aerodynamic twist can also be implemented, which involves tailoring the local airfoil section as a function of span to control the value of local $C_{L,\max}$ that can be achieved. This spanwise tailoring of the airfoil section or camber can help control the wing's three-dimensional stall characteristics (*e.g.*, see the different stall characteristics of airfoil sections in Figure 12.5).

Stall strips are another design feature for controlling stall. These strips are triangular-shaped wedges placed on the leading edge of the wing (see Figure 12.8), which are set to promote stall at a desired spanwise location on the wing. These essentially force the airfoil to have a lower value of

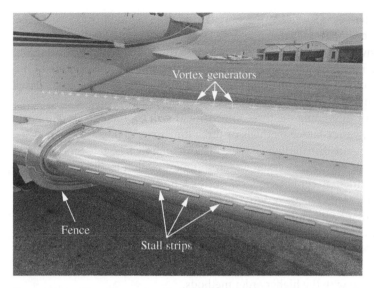

Figure 12.8 Three examples of stall control devices on the wing of a Cessna Citation Excel: a fence, stall strips, and vortex generators.

$C_{L,\max}$ at the locations where they are positioned. When the strips are carefully placed, the designer can ensure that stall occurs first at the desired spanwise location, in an effort to avoid tip stall.

The local thickness-to-chord ratio can also be tailored by controlling wing taper. A larger-chord airfoil may tend to separate earlier since the boundary layer has had more time to develop and thicken to the point where it is less robust to separation. This is one of the reasons for a tapered wing, with a root chord greater than the tip chord.

The spanwise chord distribution may also be tailored to introduce a notch, or snag, on the leading edge. This discontinuity in the spanwise chord distribution leads to a discontinuity in the spanwise lift distribution, which produces a streamwise-oriented vortex (similar to the reason why a tip vortex is produced). The streamwise-oriented vortex typically arrests the outward spanwise propagation of stalled flow in order to protect the tip region and preserve roll control even in deep stall. Another way of producing this streamwise vortex at a particular mid-span location is through a fence, which in some cases wraps around the leading edge and interrupts spanwise flows (see Figure 12.8).

12.1.6 Stall Prediction

Approaches are available for predicting $C_{L,\max}$; however, there can be substantial uncertainty associated with these methods. For two-dimensional airfoils, there is a host of experimental data from wind tunnel testing available. Common data sources include Abbott and von Doenhoff (1959), which was compiled from Jacobs et al. (1933) and Abbott et al. (1945). Also, the University of Illinois Urbana-Champaign has compiled and maintains a comprehensive database of airfoil coordinates and data (Selig 2020). Two-dimensional airfoil stall characteristics can also be approximated through analytical tools such as panel codes with associated boundary layer formulations to facilitate coupled viscous–inviscid interactions (*e.g.*, Drela and Giles 1987; Drela 1989). However, it is important to recognize that panel codes cannot simulate separated flows, so any stall predictions are at best an approximation.

If the spanwise variation in local airfoil section, the chord distribution, and geometric twist are known, then a first-order approach such as Shrenk's method may be used to estimate the three-dimensional stall characteristics (Shrenk 1940). Shrenk's method essentially estimates the spanwise lift distribution for the wing by breaking it up into basic and additional lift components. The basic lift distribution is invariant with angle of attack, while the additional lift distribution varies with pitch angle. Once these two functional distributions are known, the additional lift profile is scaled up with angle of attack until the point when the C_L for any airfoil section along the span first reaches $C_{L,\,\mathrm{max}}$ (in other words, stall is identified in a quasi-2D sense).

Computational approaches for estimating the three-dimensional stall characteristics of a wing are also available. For example, vortex-lattice methods or three-dimensional panel codes may be used to estimate the pre-stall pressure distribution on the wing as it approaches stall. Since both of these techniques are inherently inviscid methods; however, some estimation techniques must be used to determine the stall condition. Recent examples of vortex-lattice methods include the AVL code by Drela (2017), Tornado by Melin (2010), or VSPAERO by NASA (2020). Higher-fidelity stall prediction must employ computational simulation of the full unsteady Navier–Stokes equations with appropriate turbulence models that must be validated. These full CFD simulations can be quite expensive if the three-dimensional boundary layer flow is appropriately modeled. In all of these analytical approaches, however, there remains uncertainty in the stall estimation since approximations are involved, even with the higher order methods.

12.2 Flight Testing Procedures

12.2.1 Flight Characteristics

The fundamental parameter to be measured in stall flight testing is the indicated (or equivalent) airspeed at which stall occurs, as well as any undesirable stall characteristics that affect stall recovery or aircraft controllability. Aircraft configuration will have a significant impact on the aerodynamic characteristics of the wing, so stall characteristics should be evaluated for each configuration. The most common stall conditions are the landing and clean configurations. In the landing configuration, the landing gear is extended, and the flaps are deployed with maximum deflection – the stall speed associated with the landing configuration is termed V_{S0}. The clean configuration is done with landing gear and flaps fully retracted, with a stall speed denoted V_{S1}. It may also be desirable to test the stall speed of the aircraft for the takeoff configuration, with flaps in the takeoff position and landing gear retracted. Depending on the aircraft configuration, the power may be set to idle (landing and clean configurations) or full throttle (clean and takeoff configuration). Besides aircraft configuration and power setting, several other factors may have an impact on the stall characteristics, including wind gusts, aircraft weight and balance (CG), and the deceleration rate leading up to stall. The significance of each of these characteristics will be discussed as follows, starting with the basic piloting procedure for stall flight test.

The flight test is started with the aircraft at a trimmed flight condition for an airspeed 50% greater than the anticipated stall speed (noted as $1.5V_{S0}$ or $1.5V_{S1}$ for flap deployed or clean configuration, respectively). Aircraft stall is induced simply by decelerating the aircraft (pitching up) in a quasi-steady manner until the stall condition is reached, while holding altitude constant. Stall is identified when one of three conditions is reached (FAA 2011): when an uncontrollable nose drop occurs, when downward pitch motion results from activation of a safety device (*e.g.*, a stick pusher), or when the pitch control reaches a stop (limit of actuation range) – this would be the minimum speed obtained with the control at the stop (2 seconds minimum time with the control at the stop). As soon as stall is encountered, the proper recovery procedure is to lower the nose and add full power to recover airspeed and then gradually pitch up until stabilized, level flight is reestablished. The indicated airspeed is recorded at the point of stall, with the aircraft handling and controllability characteristics carefully noted.

The stall must be entered in a quasi-steady manner, since dynamic effects can dramatically change the stall characteristics – a dynamic stall vortex can form in a rapid-pitch up maneuver, and as this vortex convects over the wing surface the lift can be momentarily sustained in a way that is impossible in a static stall. Thus, the deceleration rate leading to stall must be low – FAA guidance stipulates that the deceleration rate must be less than 1 knot/s. Figure 12.9 depicts how the deceleration rate is defined, with a two-point linearization of the velocity time history between $1.1V_S$ and V_S – the slope of this line should be less than 1 knots/s.

Stall speed varies under a number of different aircraft conditions, so the testing program should be planned to configure the aircraft in each of these configurations. Flap settings are of first order importance, since they have a significant impact on the pressure distribution on the wing and the value of $C_{L,\,max}$ that can be achieved. In addition to this, the aircraft should be stalled in both engine idle and full power settings. Power-on stalls are associated with takeoff, when maximum takeoff power is used. The stall event is influenced by the engine in two primary ways: when the aircraft is pitched to high angle of attack, a component of the thrust vector acts against the weight, causing less loading on the wing for a given airspeed. Thus, the stall speed for a power-on stall is typically lower than that for a power-off stall. Furthermore, the spiraling flow associated with the prop wash will have an asymmetric effect on the local angle of attack experienced by each wing. For a single-engine aircraft with a propeller rotating clockwise (as viewed by the pilot), the right wing will experience more downwash and the left wing will experience upwash. This upwash on the left wing leads to higher local angle of attack and the potential for the left wing to stall first. Power off stalls remove these two effects, isolating the wing aerodynamics as the primary factor dictating stall behavior. Power-off stalls are most often experienced in flight on final approach, or in the traffic pattern when the engine setting is close to idle. Beyond the effects of flap setting and engine throttle setting, the

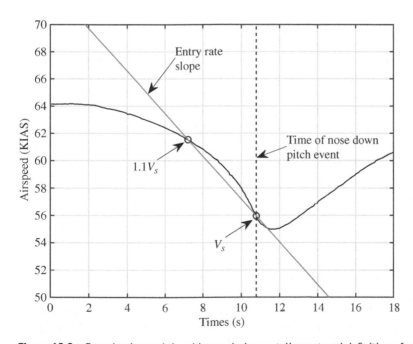

Figure 12.9 Sample airspeed time history during a stall event and definition of entry rate slope. Source: Based on Federal Aviation Administration (2011).

configuration of the landing gear on a retractable gear aircraft can have an impact on stall. This is due to the deployment of the landing gear slightly changing the lift distribution on the wing, and is particularly relevant for low-wing aircraft.

The aircraft's CG also has an impact on the stall characteristics: when the CG is in the forward-most location, the aircraft is most susceptible to stall (the stall speed will be the highest). Why is this the case? It is related to the stability of the aircraft and the amount of downforce produced by the tail to maintain trimmed flight. In the most forward CG condition, the distance between the CG and the center of pressure (*i.e.*, the distance between the lines of action of the weight and lift vectors) is a maximum. In this case, a greater nose-down pitching moment (about the CG) is produced due to the larger moment arm between the CG and the center of pressure. Thus, a larger tail force is required to balance the higher nose-down moment. When higher tail downforce is produced, the lift produced by the wing itself must increase in order to balance the forces and maintain level flight. Thus, in a trimmed condition with the CG at the most forward location, more lift must be produced by the wing at that airspeed (the wing is working harder!). This leads to a trim condition with higher angle of attack and lower stall margin at a given airspeed, making the forward CG location most susceptible to stall.

The stall event is typically identified as the point at which the aircraft cannot sustain level flight – stall may be accompanied by a loss of altitude, a change in the pitching moment, and/or a brief sensation of falling. Once the initial stall event is identified, the pilot immediately recovers from the stall by lowering the nose, and if in a power-off stall, applying full throttle. In some cases with power-on stalls, it is not possible to force the aircraft into a condition where altitude is lost. Before reaching this point, the control deflection is maximized such that the aircraft has reached a nose-high limit. If this condition is reached first under steady flight, then this nose-high airspeed defines the stall speed.

For all stall maneuvers, flight safety is paramount. Before conducting each stall, the pilot should perform two 90° clearing turns, one in each direction, in order to look for other aircraft and to make the test aircraft more readily visible to any aircraft in the surrounding area. The stalls should be conducted at a safe height that allows for plenty of altitude to recover from the stall before coming close to the ground – recovery altitudes of at least 2000 ft above ground level are typically sufficient. Coordinated flight must be maintained with a sufficient amount of right rudder, to correct for the left-turning tendencies encountered in a nose-high attitude close to stall (torque reaction, the spiraling slipstream, and asymmetrical loading on the propeller). If coordinated flight is not maintained, the aircraft can stall asymmetrically and rapidly enter a spin.

As with all flight tests, it is best if the stalls are conducted in still air such that wind gusts do not adversely impact the measurements. Stalls into a steady wind are acceptable, since the stall event is an aerodynamic phenomenon that directly depends on indicated airspeed.

12.2.2 Data Acquisition

Data to be acquired in flight include the indicated airspeed associated with the stall event, the rate of decay of airspeed leading up to stall, fuel burn since takeoff (to determine aircraft weight), CG location, outside air temperature, pressure altitude, local altimeter setting, and aircraft configuration for the stall (flaps, landing gear, power on/off). It is also very instructive to tape tufts of yarn on the wing and record their motion on video camera. This unequivocally establishes the existence of stall conditions and reveals the type and extent of the stall phenomenon. Tufts can be cut from colored yarn or string (select a color that contrasts with the paint color of the wing, in order to enhance visibility). Optimal length may be 4–6 in. long such that the tufts are easily visible and will follow the local flow characteristics. The tufts can be taped to the wing using common clear adhesive tape or masking tape, and the tufts should be arranged in a regular grid pattern with chordwise and spanwise spacings of about $0.25c$ or less.

If modern DAQ techniques are used, on manned or unmanned aircraft, then some digitized form of airspeed must be recorded. This could be digital acquisition of the dynamic pressure from a test pitot probe in undisturbed flow, which would be the most robust method of airspeed measurement. Alternatively, if only Global Positioning System (GPS) data is available, the recorded ground speed may be converted to calibrated airspeed (correct for winds aloft, compressibility, and altitude) by the same methods discussed in Chapter 8. Accelerometers may be used to identify the stall event and the associated sudden loss of lift, which would manifest as a momentary drop in measured acceleration in the z-direction.

12.3 Data Analysis

As we have done for many previous flight tests, we must correct for various nonstandard conditions. The key corrections relevant to stall speed are for altitude (often neglected), deceleration rates greater than 1 knot/s leading to stall, CG located at a position other than the forward limit and weight other than maximum takeoff weight. All of the corrections should be done using equivalent airspeed and should be performed in the order as they are presented here.

Altitude effects on the stall speed are primarily a Reynolds effect – as the altitude increases, the Reynolds number decreases and the boundary layer thickens. The increased significance of viscous effects at higher altitude may lead to stall at slightly lower airspeeds (compared to sea level conditions). Gallagher et al. (1992) suggest that the variations in stall speed due to viscous effects (altitude) are on the order of 2 knots per 5000 ft. The specific variation of stall speed with altitude may be found by performing the flight test at several altitudes and extrapolating the results to sea level conditions. These variations tend to be small, so in many cases, the stall speed is assumed to be invariant with altitude.

Note that an explicit correction for density changes is not needed, since stall speed is recorded and used as an indicated (or calibrated) airspeed. If true airspeed were used for reporting the stall speed, then the variation of stall speed would have to be tracked with altitude (see Eq. (12.2)) by the pilot. But, when stall speed is reported as an indicated airspeed, it will not change with altitude. We can see this by combining Eq. (12.2) with the relationship between true airspeed and indicated airspeed (assuming that indicated airspeed, calibrated airspeed, and equivalent airspeed are all equal), to find

$$V_S = V_{S,\text{true}} \sqrt{\sigma} = \sqrt{\frac{2W}{\rho_{\text{SL}} S C_{L,\text{max}}}}, \tag{12.4}$$

where V_S is an indicated airspeed which does not vary with altitude. A more detailed discussion on the different types of airspeeds is in Chapter 8.

Correction to the stall speed may be made if the deceleration rate is greater than 1 knot/s (the FAA-stipulated maximum deceleration rate). Unsteady aerodynamics become important when stall is measured on a decelerating or accelerating aircraft, since changes of vorticity are created through the unsteady motion and there are time scales (lags) associated with this vorticity generation. Correction for nonnegligible deceleration may be accomplished by

$$V_{S,\text{std}} = V_{S,\text{test}} \sqrt{\frac{R + 2m}{R + m}}, \tag{12.5}$$

where $m = 1\,\text{s/ft}$, and R is a deceleration parameter that can be approximated by

$$R = \frac{V_{S,\text{test}}}{0.5c(dV/dt)},$$ (12.6)

where c is wing chord. This correction scheme and the expression for R are empirically derived correlations given by Small and Prueher (1977) without citation. An alternative correction scheme is to perform multiple flight tests with varying deceleration rate and extrapolate to find the stall speed at the desired deceleration rate. This should be done by converting the measured stall speeds to lift coefficients, plotting lift coefficient versus deceleration rate, determining a linear fit to this relationship to find the desired $C_{L,\max}$ at a given deceleration rate, and then converting that lift coefficient back to a stall speed. (The extrapolation is done in terms of $C_{L,\max}$ since the variation with deceleration rate is linear.)

As discussed earlier, FAA certification requirements (2011) stipulate that the stall speed be measured with the CG positioned at the most forward limit, as this will be the condition associated with the highest stall speed. However, the CG often cannot be conveniently positioned at the forward limit for flight testing. In this case, the CG should be placed at the most feasible forward position, and correction to the stall speed made to account for this. Some published correction schemes (*e.g.*, Gallagher et al. 1992) involve multiple flight tests with varying CG locations and an extrapolation of a linear fit to find the stall speed. Others, such as Kimberlin (2003), offer a correction equation (Eq. 4.5 of Kimberlin) of the same form as Gallagher et al. (1992), but with no derivation or justification provided. Thus, an alternative correction scheme based on first principles will be derived here.

Considering the forces applied to the wing-tail system (see Figure 12.10), recall that the downforce produced by the tail is set by the elevator trim such that there are no net moments about the aircraft's CG. A consequence of this tail downforce, however, is that the magnitude of lift produced by the wing must increase such that the lift, weight, and tail downforce are in equilibrium. These two concepts can be expressed in equation form as

$$\sum M_{\text{CG}} = 0 = -L_{\text{wb}}\left(\frac{\bar{c}}{4} - x_{\text{CG}}\bar{c}\right) + L_t\left(l_t + \frac{\bar{c}}{4} - x_{\text{CG}}\bar{c}\right)$$ (12.7)

and

$$\sum F_y = 0 = L_{\text{wb}} - W - L_t,$$ (12.8)

where $\bar{c}$ is the mean aerodynamic chord, L_{wb} is the lift produced by the wing-body combination acting at the aerodynamic center of the wing ($\bar{c}/4$), l_t is the distance between the wing's aerodynamic center and the quarter chord of the horizontal tail, and x_{CG} is the location of the CG expressed as a fraction of mean aerodynamic chord measured from the leading edge. The downforce produced

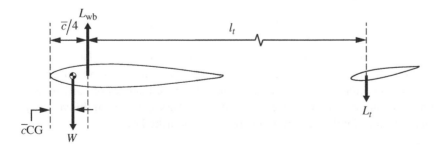

Figure 12.10 Diagram of forces acting in equilibrium on the wing and tail in trimmed flight.

by the horizontal tail, $L_t = L_{wb} - W$, can be substituted into the moment balance and several terms canceled to obtain

$$L_{wb} = \frac{W}{l_t}\left(l_t + \frac{\bar{c}}{4} - x_{CG}\bar{c}\right). \tag{12.9}$$

The lift on the wing-body combination at the incipient stall condition is given by

$$L_{wb} = \frac{1}{2}\rho V_S^2 S C_{L,max}, \tag{12.10}$$

which is the critical condition that defines stall. (Note that the maximum lift coefficient must be defined in terms of the lift on the wing and body, rather than the overall lift on the aircraft, since the tail does not have a direct impact on separation over the wing.) Equating (12.9) and (12.10), we obtain

$$\frac{1}{2}\rho V_S^2 S C_{L,max} = \frac{W}{l_t}\left(l_t + \frac{\bar{c}}{4} - x_{CG}\bar{c}\right). \tag{12.11}$$

This equation can be written for the most forward CG limit (the standard reference condition, denoted with a "std" subscript) and for some arbitrary CG location used in flight test (noted by a "test" subscript). If we form a ratio of these two equations, recognizing that $C_{L,\,max}$ is an inherent property of the wing that does not change based on CG movement, we obtain the correction factor for stall speed:

$$V_{S,std} = V_{S,test}\left[\frac{\left(l_t + \frac{\bar{c}}{4} - x_{CG,std}\bar{c}\right)}{\left(l_t + \frac{\bar{c}}{4} - x_{CG,test}\bar{c}\right)}\right]^{0.5}. \tag{12.12}$$

Note that stall speed may be retained as indicated airspeed, if one desires, and the units of tail moment arm and mean aerodynamic chord must be consistent.

Weight correction is done to account for the substantial impact of aircraft weight on the stall speed. A heavier aircraft must generate more lift at a given airspeed and will reach stall conditions at a higher speed than a lighter aircraft. Thus, the stall speed measured for the aircraft in flight test must be converted to stall speed at some reference weight. In this case, the reference weight is maximum takeoff weight, and the test weight can be measured by subtracting the weight of the fuel burned from the takeoff weight of the aircraft at the beginning of flight test. The weight correction factor may be obtained by writing an equation for the stall speed in terms of aircraft weight (Eq. (12.2)) for both the standard and test conditions, and taking the ratio of the two. This gives

$$V_{S,std} = V_{S,test}\sqrt{\frac{W_{std}}{W_{test}}}, \tag{12.13}$$

where the reference weight W_{std} is the maximum takeoff weight, $V_{S,\,test}$ is the measured stall speed in flight test, and W_{test} is the aircraft weight at the time of stall in flight test.

12.4 Flight Test Example: Cirrus SR20

Sample data from a stall flight test with digital DAQ are shown in Figure 12.11, where both ground speed and acceleration are plotted against time. For these tests the DAQ unit was mounted on the front dashboard of a Cirrus SR20, with the three-axis accelerometer approximately aligned with the axes of the aircraft (precise alignment is not necessary). The stall event is readily identified in the

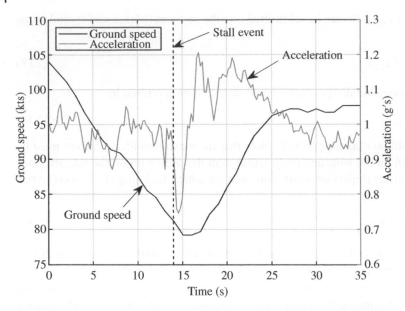

Figure 12.11 Identification of the stall event by cross-plotting acceleration and ground speed; the dashed line indicates stall (81 knots ground speed). Note that the deceleration rate is approximately 1.6 knots/s ground speed.

z-axis accelerometer (approximately perpendicular to the ground plane), where the acceleration rapidly drops from 1.0 to less than 0.8. This sudden drop in g's is an indicator of the bottom momentarily dropping out and the brief sensation of falling that is experienced during a stall. (This is similar to the momentary lightweight feeling in a high-speed elevator when it begins its descent.) The vertical dashed line in Figure 12.11 identifies the point of stall based on the acceleration data and can be used to identify the ground speed associated with stall (81 knots in this case). This value is high relative to the pilot's operating handbook (POH)-defined value of 70 KIAS, but wind information was not available to convert between ground speed and indicated airspeed. It is important to recognize that the velocity measured by GPS is a ground speed and must be converted to indicated airspeed. (Chapter 8 details the relationships between the various airspeeds.) In particular, the cloverleaf patterns described in Chapter 8 can be flown in steady wind conditions in order to acquire enough information to determine true airspeed (and wind speed/direction as a byproduct). One particularly useful feature of the GPS data is that it provides a time record of the ground speed. Under steady wind conditions, the deceleration rate determined from GPS data should provide the same result as if airspeed were recorded directly via a pitot-static system. Thus, GPS ground speed data can be used to verify (and correct for) nonstandard deceleration rates. In the data set shown in Figure 12.11, the deceleration rate is approximately 1.6 knots/s between 0 and 14 seconds, which is above the specified 1 knot/s requirement and the stall speed would require correction.

An example of three-dimensional stall characteristics, as well as methods to control stall development, are shown in the photos in Figure 12.12. Here tufts have been applied to the wing of a Cirrus SR20 for flight test. Local stall conditions are revealed by unsteady motion of the tufts (indicating unsteady, separated flow) or by tufts oriented in the upstream direction (indicating locally reversed flow). Spanwise flows are also visible as the direction of the tufts changes. Figure 12.12(a) shows the wing in the pre-stall condition, with completely attached flow throughout (the tufts are all oriented in the streamwise direction). At the higher angle of attack shown in Figure 12.12(b) (note the angle of the horizon), trailing edge stall is evident from

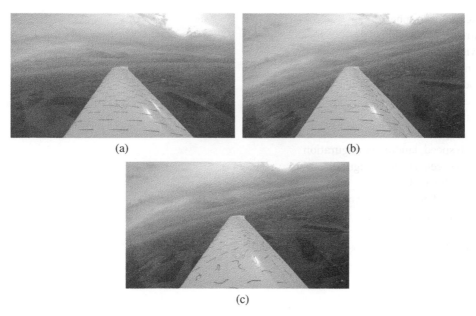

(a)   (b)

(c)

Figure 12.12 Tufts on a Cirrus SR20 showing (a) attached flow, (b) trailing edge stall on the inboard portion of the wing, and (c) deep stall.

the inboard-oriented flow starting at mid-span, and minor unsteadiness evidenced by separated strands of yarn in the tufts. In Figure 12.12(c), the wing is in deep stall, with separated flow extending over the majority of the wing surface. This is indicated by tufts oriented in many different directions, showing the chaotic nature in this snapshot of the unsteady flow. There are two primary regions of separated flow – one initiating at the wing root, where a stall strip is present and propagating outboard to about 30% span. The other significant portion of separated flow is at mid-span, where a notch is present (a discontinuity in the leading-edge geometry), with a stall strip present just inboard of the notch. The strip serves to initiate stall at that location, and the notch produces a strong streamwise-oriented vortex that minimizes outboard propagation of that unsteady stalled flow. This has the benefit of maintaining a region of attached flow near the wing tip throughout the stall, such that roll control authority is preserved.

Nomenclature

c	chord
$\bar{c}$	mean aerodynamic chord
C_L	lift coefficient
F	force
l_t	tail moment arm
L	lift
L_t	lift produced by tail (downforce)
L_{wb}	lift produced by wing-body combination
m	constant
M	moment
p	pressure
R	deceleration parameter

Re	Reynolds number
s	streamwise distance
S	wing area
t	time
u	local velocity within boundary layer
U_∞	freestream velocity
V	airspeed
V_S	stall speed
V_{S0}	stall speed, landing configuration
V_{S1}	stall speed, clean configuration
W	aircraft weight
x_{CG}	center of gravity location (fraction of mean aerodynamic chord)
y	wall-normal distance
α	angle of attack
δ	boundary layer thickness
μ	dynamic viscosity
ρ	density
σ	density ratio, ρ/ρ_{SL}

Subscripts

max	maximum
std	standard
SL	sea level
test	test condition
true	true airspeed
wb	wing-body

Acronyms and Abbreviations

CFD	computational fluid dynamics
CG	center of gravity
FAA	Federal Aviation Administration
GA	general aviation
KIAS	knots indicated airspeed
KTAS	knots true airspeed
NACA	National Advisory Committee for Aeronautics
GPS	Global Positioning System
VG	vortex generator
2D	two-dimensional

References

Abbott, I.H. and von Doenhoff, A.E. (1959). *Theory of Wing Sections: Including a Summary of Airfoil Data*. Mineola, NY: Dover Publications.

Abbott, I.H., von Doenhoff, A.E., and Stivers, L.S. Jr., (1945). Summary of Airfoil Data. *NACA-TR-824*.

Drela, M. (1989). XFOIL: an analysis and design system for low Reynolds number airfoils. In: *Low Reynolds Number Aerodynamics*, Lecture Notes in Engineering, vol. 54 (ed. T.J. Mueller), 1–12. Berlin: Springer-Verlag https://doi.org/10.1007/978-3-642-84010-4_1.

Drela, M. (2017). AVL. http://web.mit.edu/drela/Public/web/avl (accessed 23 August 2020).

Drela, M. and Giles, M.B. (1987). Viscous-inviscid analysis of transonic and low Reynolds number airfoils. *AIAA Journal* 25 (10): 1347–1355. https://doi.org/10.2514/3.9789.

Federal Aviation Administration (2011). Flight test guide for certification of part 23 airplanes. Advisory Circular 23-8C, U. S. Department of Transportation, Federal Aviation Administration (16 November 2011).

Federal Aviation Administration (2016). Pilot's handbook of aeronautical knowledge. FAA-H-8083-25B, United States Department of Transportation, Federal Aviation Administration, Airman Testing Standards Branch, AFS-630, P.O. Box 25082, Oklahoma City, OK 73125. https://www.faa.gov/regulations_policies/handbooks_manuals/aviation/phak/media/pilot_handbook.pdf (accessed 29 December 2020).

Gallagher, G.L., Higgins, L.B., Khinoo, L.A., and Pierce, P.A. (1992). Fixed wing performance. U.S. Naval Test Pilot School Flight Test Manual, USNTPS-FTM-NO. 108 (preliminary), 30 September 1992. http://www.usntpsalumni.com/Resources/Documents/USNTPS_FTM_108.pdf (accessed 29 December 2020).

Jacobs, E.N., Ward, K.E., and Pinkerton, R.M. (1933). The Characteristics of 78 Related Airfoil Sections From Tests in the Variable-Density Wind Tunnel. *NACA-TR-460*.

Kimberlin, R.D. (2003). *Flight Testing of Fixed-Wing Aircraft*. Reston, VA: American Institute of Aeronautics and Astronautics.

Melin, T. (2010). Tornado: a vortex lattice method implemented in MATLAB. http://tornado.redhammer.se (accessed 23 August 2020).

NASA (2020). OpenVSP. http://openvsp.org, v. 3.21.2. (accessed 23 August 2020).

Selig, M. (2020). UIUC airfoil data site. https://m-selig.ae.illinois.edu/ads.html (cited 19 May 2020).

Shrenk, O. (1940). A Simple Approximation Method for Obtaining the Spanwise Lift Distribution. *NACA TM-948*.

Small, S.M. and Prueher, J.W. (1977). Fixed wing performance: theory and flight test techniques. U.S. Naval Test Pilot School Flight Test Manual, USNTPS-FTM-No. 104, July 1977, DTIC accession number ADA061239.

Sweberg, H.H. and Dingeldein, R.C. (1945). Summary of Measurements in Langley Full-Scale Tunnel of Maximum Lift Coefficients and Stalling Characteristics of Airplanes. *NACA-TR-829*. https://ntrs.nasa.gov/api/citations/19930091906/downloads/19930091906.pdf (accessed 29 December 2020).

13

Turning Flight

The turning performance of an aircraft is of special interest for military aircraft, but it also provides helpful perspective on the performance limits of general aviation aircraft. Turn performance can be described by the turn radius, turn rate (or time to turn), bank angle, and the g-forces experienced by the pilot and aircraft, which are all related to one another as well as to the airspeed. The key limiting physics that dictate the turning performance are power available, the drag polar, the maximum lift coefficient, and the load-carrying capability of the aircraft structure. These limits form the boundaries on the operating envelope of the aircraft, expressed as the $V-n$ diagram. The lift required by the wing for a sustained level turn increases as the bank angle increases, leading to an aerodynamic stall limit on the turning performance. This leads to stall in a turn happening at an airspeed higher than normal, which is a hazardous condition for pilots operating near the ground. Higher bank angles lead to higher lift production and, in turn, higher stall speed. This can pose signification limitations on the maneuverability of an aircraft, particularly in high-altitude, mountainous terrain. A pilot flying under these conditions must be careful to avoid flying into a canyon that would be impossible to exit by climbing or turning – one can easily be boxed in by limits on turn radius, stall, and climb performance. Thus, flight test evaluation of turn performance is relevant for not only highly-maneuverable military aircraft, but also for light general aviation aircraft.

In this chapter, we'll discuss the theory of turn performance as it relates to the $V-n$ diagram and the achievable turn rate and turn radius. We'll develop relations involving the related parameters of load factor, turn rate, turn radius, airspeed, bank angle, and maximum lift coefficient. Next, we'll discuss flight testing methods for assessing the turn performance of general aviation aircraft, where the minimum turn radius is most often limited by aerodynamic stall.

13.1 Theory

We will begin our discussion with short derivations of the equations of motion governing aircraft performance in turning flight. Figure 13.1 shows an aircraft in a steady, constant altitude turn of radius R and bank angle ϕ, where both rudder and ailerons are used to turn the aircraft in coordinated flight. The turn is enacted by banking the aircraft with ailerons such that the lift vector tilts and pulls the aircraft through the turn, while rudder is used to maintain proper orientation of the aircraft's longitudinal axis with the flight path. The forces acting on the aircraft during the turn are weight (W), which always acts in the downward direction; and the lift vector (L), which acts perpendicular to the freestream and along the vertical axis of the aircraft's coordinate system. In an inertial frame of reference, the lift vector is tilted away from the vertical in order to pull the aircraft through a turn. The lift vector can be broken down into its vertical and horizontal components.

Introduction to Flight Testing, First Edition. James W. Gregory and Tianshu Liu.
© 2021 John Wiley & Sons Ltd. Published 2021 by John Wiley & Sons Ltd.
Companion website: https://www.wiley.com/go/flighttesting

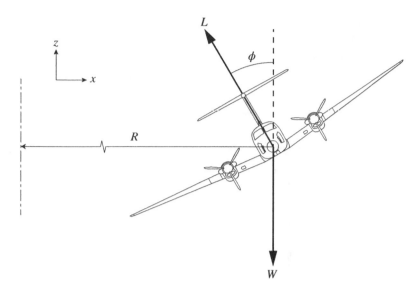

Figure 13.1 Forces acting on an aircraft in a steady turn (inertial frame of reference).

A summation of the forces acting in the vertical direction yields

$$L \cos \phi = W. \tag{13.1}$$

The horizontal component of the lift vector can be related to the centripetal force (F_c), since the aircraft is following a curved flight path. Based on kinematics,

$$L \sin \phi = F_c = \frac{W}{g} \frac{V^2}{R}, \tag{13.2}$$

where W/g is the mass of the aircraft, and V^2/R is the centripetal acceleration based on the tangential velocity, $V = \omega R$ (which is the true airspeed in this case). Note that the turn rate, ω, is the time rate of change of aircraft heading: $\omega = d\theta/dt$. Higher turn rate leads to shorter time required to turn through a given heading change.

Equation (13.1) indicates that the steeper the bank angle (ϕ) is, the greater the magnitude of lift must be in order to sustain level flight. For example, approximately 15% more lift is required for constant-altitude flight in a 30° bank. In contrast, a steep bank of 60° requires 50% greater lift at constant altitude. Equation (13.2) illustrates that the magnitude of the centripetal force pulling the aircraft through the turn also increases as the bank angle goes up.

At this point, it's helpful to define the load factor,

$$n \equiv \frac{L}{W}, \tag{13.3}$$

which describes how much more lift is being generated relative to the weight of the aircraft. Load factor is a nondimensional parameter, but is often reported as g's. Load factor may be related to the bank angle of the aircraft by combining Eq. (13.1) with (13.3) to write

$$n = \frac{1}{\cos \phi}. \tag{13.4}$$

Clearly, in steady, level flight the load factor is unity, and higher bank angles lead to larger load factors (pulling more g's).

Until now, our discussion has been limited to an inertial frame of reference. Let's momentarily shift our frame of reference to that of the aircraft body axes (see Figure 13.2), and remember that

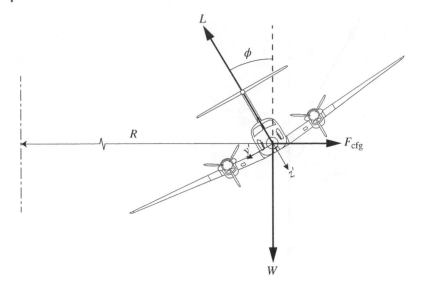

Figure 13.2 Forces acting on an aircraft in a steady turn (aircraft frame of reference).

the aircraft is in a coordinated turn. In this rotating (non-inertial) frame of reference, we see that the centripetal force – the component of lift pulling the aircraft through the turn – is balanced by an apparent centrifugal force. (Recall from physics that the apparent centrifugal force is only valid in the frame of reference of the body moving along a curved path.) The centrifugal force (F_{cfg}) is one that is felt by the occupants of an aircraft: this is the feeling of being pressed down in your seat and weighing more than normal during a turn. The magnitude of the centrifugal force is related to the number of g's pulled in the turn, which is a measure of the load factor (Eq. (13.4)). From Figure 13.2, we can see that

$$\sum F_{y'} = F_{cfg} \cos\phi - W \sin\phi = 0$$
$$\sum F_{z'} = L - W \cos\phi - F_{cfg} \sin\phi = 0. \tag{13.5}$$

The resultant force produced by the sum of the projections of W and F_{cfg} onto the z' axis is equal to the lift vector in the z' direction. This resultant force,

$$L = W \cos\phi + F_{cfg} \sin\phi = nW, \tag{13.6}$$

leads occupants of the aircraft to feel heavier in a turn. From Eq. (13.4), the greater the bank angle, the heaver the occupants will feel. This has implications for flight test – the load factor (n) can be measured in flight and used to determine bank angle.

Beyond the increased load in the z' direction, there is no net force in the lateral (y') direction, since the sine component of the weight is balanced by the component of centrifugal force in the y' direction. This leads to the nonintuitive result that no side force is felt by the occupants of the aircraft (even in a steeply banked turn, as long as coordinated flight is maintained with rudder and aileron control). This can be a disconcerting feeling since it is at odds with the visual picture presented by a tilted horizon and is unlike what is experienced in ground vehicles in aggressive cornering. One implication of this is that a cup of water, filled to the brim and held level with the aircraft floor, won't spill a drop in a steep turn (assuming a coordinated turn in calm air). This also means that, if our eyes are closed or if we are flying in a diffuse cloud, we would not be able to tell the direction of a turn by feel alone. Furthermore, if the angle of the turn is shallow enough, we won't be able to perceive the increased load factor (*e.g.,* a 3.5% increase in weight for a 15° bank, from Eq. (13.4)). These characteristics of turning flight led to numerous accidents in the early days of aviation. Airmail pilots in those days were flying by eyesight and "the seat of their pants." With the pressure they felt to deliver their cargo, the pilots would often press on in poor visibility. Under these conditions, a slow turn could easily develop and the pilot would neither see it, nor feel it (the sensations of the inner ear can be deceptive and unreliable in a sustained turn). If back pressure on the stick is not applied during the turn, the upward component of lift will be less than the weight, and the aircraft will descend. The combination of these factors leads to a slow, turning descent that is known as the "graveyard spiral." These phenomena highlight the need for flight instruments such as the attitude indicator, which allow the pilot of an aircraft to fly without outside visual references.

We'll now derive some relationships between the load factor (n), turn radius (R), rate of turn (ω), and airspeed (V), based on the representation of the aircraft in the inertial frame (Figure 13.1). Since the vertical component of lift is equal to weight (Eq. (13.1)) and the horizontal component of lift is equal to the centripetal force (Eq. (13.2)), these two legs of the triangle can be related to the hypotenuse (L) by the Pythagorean theorem:

$$L^2 = W^2 + F_c^2. \tag{13.7}$$

Solving for centripetal force and substituting the definition of load factor (13.3), we obtain

$$F_c = W\sqrt{n^2 - 1}. \tag{13.8}$$

Equation (13.8) can be equated with the definition of centripetal force (13.2) to produce a relationship between turn radius, true airspeed, and load factor:

$$R = \frac{V^2}{g\sqrt{n^2 - 1}}. \tag{13.9}$$

In addition, by substitution of (13.9) into $V = \omega R$, the turn rate can be expressed as

$$\omega = \frac{g\sqrt{n^2 - 1}}{V}. \tag{13.10}$$

Equations (13.9) and (13.10) are fundamental relations between the key parameters governing turning flight. From (13.9) we see that turn radius decreases as the true airspeed goes down, and turn radius also decreases for higher load factors. Recall from (13.4) that load factor is inversely proportional to the bank angle, so a steeper turn will naturally lead to a higher load factor and a tighter (smaller radius) turn. Equation (13.10) shows that the turn rate increases as the load factor goes up and decreases as airspeed increases.

A good illustration of the relationship between these factors is a comparison of the time and turn radius required to complete a 180° turn in two different aircraft at the same bank angle (load factor). A Cessna 172 traveling at 100 knots true airspeed (KTAS) in a 30° bank will have a turn radius of 1500 ft (about 0.25 nmi) and complete a 180° turn within 28.5 seconds. On the other hand, an SR-71 traveling at 1720 KTAS (Mach 3) in a 30° bank will have a turn radius of 4.5×10^5 ft (nearly 75 nmi) and take over 8 minutes to complete the same 180° turn.

In certain instances, particularly under instrument conditions, a pilot may wish to make a slow turn in order to stay well clear of stall and to minimize the physiological impact of a sustained turn without visual reference. It can also be useful to estimate the heading change in a turn by timing the duration of a turn with a known turn rate. Thus, a "standard rate turn" has been defined, which is a rate of 3° of heading change per second (52.3×10^{-3} rad/s), meaning that a 180° turn at this rate takes one minute. For a standard rate turn, we can determine what bank angle is needed at a particular true airspeed. Starting with Eq. (13.10) and solving for load factor (n), we obtain

$$n = \sqrt{1 + \left(\frac{\omega V}{g}\right)^2}. \tag{13.11}$$

Equating this to Eq. (13.4) and solving for bank angle we have,

$$\phi = \cos^{-1}\left\{\left[1 + \left(\frac{\omega V}{g}\right)^2\right]^{-1/2}\right\}. \tag{13.12}$$

Following the comparative example given earlier, contrasting the flight performance characteristics of a C-172 at 100 KTAS with the SR-71 at Mach 3, we see that the Cessna requires a bank angle of 15° for a standard rate turn, while the SR-71 would require a bank angle of about 78°. Clearly, as airspeed goes up, the bank angle for a standard rate turn also goes up. Thus, faster aircraft (typically above 250 KTAS) will often fly a half-standard rate turn (1.5°/s), which leads to a bank angle of 19° instead of 34° at an airspeed of 250 KTAS. In fact, an aircraft can fly as fast as 710 KTAS in a half-standard rate turn without exceeding a comfortable bank angle of 30°.

Equations (13.3), (13.4), (13.9), and (13.10) are the critical relationships that describe turn performance. However, they do not yet form a complete picture of an aircraft's turn performance. There are two critical limits on how high the load factor can go, which then limit the turn radius and turn rate of the aircraft. These two critical limits are aerodynamic stall and the maximum structural loading (the number of g's the structure can sustain). We'll look at both of these limits in detail as follows.

From the definition of load factor (13.3) we can write

$$L = nW = \frac{1}{2}\rho V^2 SC_L. \tag{13.13}$$

We'll then solve (13.13) for airspeed (V) and substitute into Eqs. (13.9) and (13.10). We now have expressions for turn radius and turn rate:

$$R = \frac{2(W/S)}{g\rho}\frac{1}{C_L}\frac{n}{\sqrt{n^2 - 1}} \tag{13.14}$$

and

$$\omega = \left[\frac{g^2\rho}{2(W/S)}C_L\frac{n^2 - 1}{n}\right]^{1/2}. \tag{13.15}$$

These expressions illustrate the fundamental limits on turn performance (where we wish to minimize R and maximize ω). We can see that wing loading (W/S) must be minimized in order to maximize turn performance. In fact, this is one of the most important reasons why fighter aircraft have a relatively small wing area. Second, we note that turn performance improves at lower altitudes (higher density), so a fighter aircraft will be more maneuverable at low altitude. Within a turn, we see that a pilot will want to increase both C_L and n as high as possible in order to optimize turn performance. However, both of these factors have inherent limits. The lift coefficient has an aerodynamic limit based on stall, while the load factor has a structural loading limit based on certification standards related to the structural design of the wing. The turn performance of an aircraft will be limited by whichever parameter reaches a maximum first. At low airspeeds, the wing is working harder (higher C_L for a given amount of lift), and the wing will stall before reaching the structural load limit. At high airspeeds, on the other hand, the required lift coefficient for a given flight condition is much lower than the stall limit, and the aircraft turn performance is limited by load factor. Typical load factor limits for general aviation aircraft are $n_{max} = 3.8$ and $n_{min} = -1.52$ (40% of n_{max}), see Federal Aviation Administration (FAA) (2002b). While $C_{L,\,max}$ and n_{max} limit the turn performance, recall that Eqs. (13.9) and (13.10) show that airspeed should be minimized as well. Thus, in order to maximize turn performance for a given aircraft at a given altitude, the pilot should fly at the highest possible load factor and the minimum airspeed possible without stalling the aircraft.

These limits on turn performance are best illustrated on a simplified V–n diagram (Figure 13.3), which plots load factor versus equivalent airspeed and defines the performance envelope of the aircraft. The V–n diagram shown in Figure 13.3 is for a general aviation aircraft with a wing loading of 21 lb/ft^2, a maximum lift coefficient of 1.385, and a never-exceed speed of 200 knots equivalent airspeed (KEAS). The positive and negative load limits, $n_{max} = 3.8$ and $n_{min} = -1.52$, are indicated by horizontal lines, and the maximum dynamic pressure limit is indicated by a vertical line. The stall limit line is found by solving Eq. (13.13) for load factor and evaluating the expression at $C_{L,max}$. At low speeds, aerodynamic stall limits the maximum load factor that can be achieved, thus limiting the turn performance of the aircraft.

Recalling that maximum turn performance occurs when load factor is maximized and airspeed is minimized (subjected to the stall limit), we see that optimum turning performance occurs at the intersection of the stall line with the maximum structural limit. This intersection defines the maneuver speed for the aircraft (V_A), which is 131 KEAS for this aircraft. This is the point on the diagram where the aircraft will be most maneuverable (highest ω, lowest R). It also represents a safe maximum airspeed for flight in turbulent air. Turbulent gusts (updrafts or downdrafts) produce sudden changes in angle of attack of the wing, leading to impulsive loading and elevated load factor. It's possible for a turbulent gust to produce a load factor that exceeds the structural load limit. However, if an aircraft is flying at a speed at or below the maneuver speed, and encounters such a gust, the wing will stall before loading up to the point of structural failure. Thus, flight in anything greater than light turbulence should be flown at a speed no greater than the maneuver speed.

(Note that the V–n diagram shown in Figure 13.3 is a simplified version relative to the definition of the performance envelope established in FAA (2002b). Here, the gust lines are ignored, and the notch in the envelope at negative load factors and high speeds for normal-category aircraft

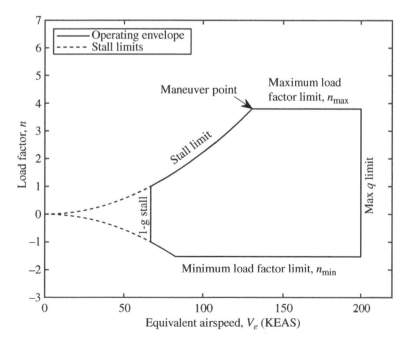

Figure 13.3 Simplified V–n diagram for a typical, normal category general aviation aircraft certified under 14 CFR §23.

is omitted. Detailed information on the exact performance envelope that must be demonstrated through flight testing of normal category aircraft is available in FAA (2002a).

Figure 13.4 shows $V-n$ diagrams for the same aircraft, but at different operating weights. The solid black lines represent the operating envelope for the aircraft at maximum takeoff weight (3050 lb), while the gray lines are the envelope for the same aircraft at 70% of maximum weight (2135 lb). Several interesting changes take place when the operating weight changes. First, the 1 g stall line moves to a lower airspeed (from 67 KEAS down to 56 KEAS) – this is because the aircraft doesn't need to produce as much lift in order to sustain the weight, so the aircraft can fly more slowly before $C_{L,max}$ is reached. Similarly, when in a turn the lighter aircraft can reach a higher stall-limited load factor for a given airspeed, because the necessary lift coefficient for a given flight condition is lower. At higher airspeeds, the structural load factor limits remain unchanged. Since the maneuver point is defined as the intersection of the stall-limited load factor line with the structure-limited load factor line, the maneuver speed decreases when aircraft weight decreases (in this case, from 131 KEAS down to 109 KEAS).

This trend of the maneuver speed can seem counterintuitive. Pilots often think of the maneuver speed as a speed limit: in turbulent air, or when maneuvering, the aircraft must be slowed down to the maneuver speed in order to avoid the possibility of structural damage. (Recall that at speeds below the maneuver point, the wings will stall before being loaded to the point of structural failure – an aerodynamic safety mechanism.) Since the maneuver "speed limit" is being driven by structural concerns, it can be counterintuitive that a *lighter* aircraft will have a *lower* speed limit (one might initially think that a lighter aircraft is less susceptible to structural damage). But if the structural limit remains constant (at a positive load factor limit of +3.8 g), then the maneuver speed will decrease for a lighter aircraft (see Figure 13.4), since the safety relief mechanism of aerodynamic stall has changed, becoming less conservative. With a lighter aircraft requiring less lift, the aircraft can sustain a higher load factor before reaching stall. Thus, the stall line moves toward

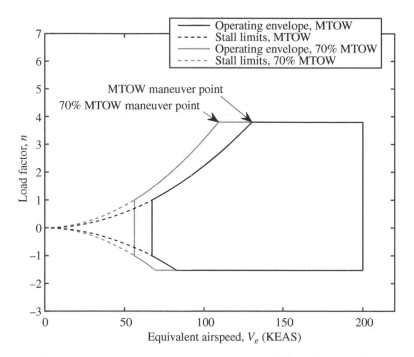

Figure 13.4 $V-n$ diagrams for the same aircraft at maximum takeoff weight and at 70% of maximum takeoff weight.

lower airspeeds, leading to a lower speed where the safety mechanism kicks in. (Note that there is not consensus about maintaining the same load factor limit for a given aircraft design that is lightly loaded. For example, Phillips (2004) allows for the load limit to increase as the weight decreases, keeping the maneuver speed the same. On the other hand, the Pilot Operating Handbooks for the Piper Arrow (1995) and the Cirrus SR20G6 (2017) recommend a lower maneuver speed when the aircraft is lightly loaded.)

Now, when we consider steady, *sustained* turning flight, there is another physical limitation on the turn performance other than stall and the load factor limit. The power available from the engine often defines the limitation on *sustained* turning flight performance. As the load factor increases, more lift is generated, leading to an increase in the induced drag. This added drag can increase to a point where the power required for flight reaches the power available from the engine, and specific excess power drops to zero.

Figure 13.5 illustrates the significance of power limits on turn performance by overlaying contours of specific excess power (defined as the weight-normalized difference between power available and power required) on a $V{-}n$ diagram that includes shaded contours of turn rate. (Recall from Eq. (13.10) that turn rate is a function of only load factor and airspeed, resulting in a single set of contours overlaid on the $V{-}n$ diagram). For this aircraft, we assume that the standard level-flight drag polar of the form

$$C_D = C_{D_0} + \frac{C_L^2}{\pi\,\text{AR}\,e} \tag{13.16}$$

is the same for turning flight, with $C_{D_0} = 0.0260$, $e = 0.76$, and $\text{AR} = 10.12$. Power required is determined by the product of airspeed with drag, and power available is estimated by the shaft horsepower of the engine (200 hp here) multiplied with the propeller efficiency (assumed constant at 0.85 for this example). The conditions for best sustained turn performance (maximum turn rate and minimum turn radius) can be found by following the contour for zero specific excess power (the power limit) to the highest turn rate along this contour. For this particular drag polar and stall speed, stall occurs before the back side of the power curve develops, so the intersection of the stall line with the contour of zero specific excess power defines the maximum sustained maneuver point.

The point for maximum sustained turn rate can also be found analytically as follows. In the power-limited turning performance case, power available is equal to power required:

$$P_A = P_R = T_R V = DV = qSC_D V. \tag{13.17}$$

Substituting in the drag polar (13.16), we have

$$P_A = qSV\left(C_{D_0} + \frac{C_L^2}{\pi\,\text{AR}\,e}\right). \tag{13.18}$$

In the stall-limited case, the optimum airspeed and maximum load factor occur when the lift coefficient is maximized,

$$P_A = \frac{1}{2}\rho V_{\text{opt}}^3 S\left(C_{D_0} + \frac{C_{L,\text{max}}^2}{\pi\,\text{AR}\,e}\right). \tag{13.19}$$

Solving (13.19) for the airspeed for optimum turn performance, we have

$$V_{\text{opt}} = \left[\frac{2P_A}{\rho S(C_{D_0} + C_{L,\text{max}}^2/\pi\,\text{AR}\,e)}\right]^{1/3}. \tag{13.20}$$

The corresponding load factor for this optimum condition is

$$n_{\text{opt}} = \frac{1}{2W}\rho V_{\text{opt}}^2 S C_{L,\text{max}}. \tag{13.21}$$

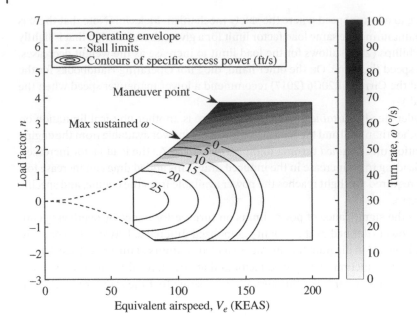

Figure 13.5 *V–n* diagram with overlaid contours of specific excess power (ft/s).

For the aircraft illustrated in the *V–n* diagram of Figure 13.5, the optimum turn performance condition (power limited) is at an airspeed of 102.3 KEAS and a load factor of 2.33, occurring right at the stall limit with a bank angle of 64°, a turn rate of 22 ° /sec, and a turn radius of 445 ft. (Note that the FAA requires that all occupants of the aircraft be wearing parachutes for maneuvers at sustained bank angles of 60° or higher (FAA 2019). Thus, flight at this condition – particularly near the stall limit – is *not* recommended unless conducted by a professional flight test pilot.)

13.2 Flight Testing Procedures

13.2.1 Airworthiness Certification

Flight testing for FAA airworthiness certification involves demonstration of turning stall and ensuring benign post-stall flight characteristics (FAA 2011). The Federal Aviation Regulations (2002c) mention two different types of turning stalls to be evaluated, delineated by the rate of airspeed decay when entering the stall. First is a quasi-steady turning stall, where the deceleration rate must be constant and less than or equal to 1 knot per second, as specified for level stalls (see Chapter 12). The second test, specifically termed an "accelerated turning stall," involves a much faster deceleration rate, specified as between 3 and 5 knots per second with a steadily increasing normal acceleration. In either case, the aircraft is placed in a sustained 30° coordinated turn and airspeed is steadily decreased by increasing back pressure on the control stick to progressively tighten the turn. Stall is defined in the same way as it is for level flight: as an uncontrolled pitch-down of the nose, activation of a stick pusher (if the aircraft is so equipped), or the elevator reaching the aft control limit. Ideally, airspeed is measured through an auxiliary system such as a boom-mounted pitot probe and/or a trailing bomb for static pressure.

Following the stall event, but before application of power, the aircraft response is monitored (see FAA, 2011). It must be possible to recover from the stall without adding power and with normal use of the flight controls. Throughout the recovery process, the altitude loss must be reasonable (in the test pilot's opinion), there must not be undue pitch-up, and no uncontrollable tendency to spin. Furthermore, the aircraft attitude must remain within a limit of 60° bank angle in the original direction of the turn or within 30° in the opposite direction (for turning stalls); or for accelerated turning stalls, within 90° bank angle in the original direction of the turn and within 60° in the opposite direction. Throughout the recovery process, the airspeed must not exceed the maximum speed or the maximum structural load limit. The aircraft should be configured in all relevant configurations for various phases of flight: flaps deployed or retracted at all possible settings, power off and power on (75% maximum), landing gear retracted and extended, and spoilers or speed brakes retracted and extended (if the aircraft is so equipped, and if these have a measurable impact on stall). Any cowl flaps on the engine should be set for the appropriate flight condition, the aircraft should be trimmed for flight at an airspeed 50% greater than the clean aircraft stall speed, and a variable-pitch propeller should be set at the maximum RPM setting for the power-off stall.

13.2.2 Educational Flight Testing

For flight testing in an educational environment, there are two primary flight maneuvers that can be conducted for turning flight. The first entails a series of turning stalls, flown in a conservative manner in the same vein as the FAA procedures described earlier. As with all flight testing procedures conducted for educational purposes, standard piloting procedures must be followed with conservative safety estimates, and the performance envelope for the aircraft must never be exceeded. The second type of experiment for turning flight involves verification of the theory presented in this chapter. These two flight tests are described in detail as follows, and both may be accomplished in a carefully planned 1-hour test flight.

The intent of turning stall flight test is to record the stall speed at various bank angles (load factors) in a sustained turn. A student-focused flight test for turning stall can include several stalls at various bank angles (load factors) in order to define the lower portion of the stall boundary on the V–n diagram. In order to ensure safe flight conditions, the bank angle should not exceed 30°, and coordinated flight must be maintained (more details on proper piloting practice are provided in the next subsection). The deceleration rate in the turn leading up to stall should be steady and no more than 1 knot per second. Turning stalls can be conducted in both directions (right and left turns) in order to assess any effects from the swirling propeller wake on the stall characteristics. (For a clockwise-rotating propeller, from the pilot's perspective, the swirling wake will induce downwash on the right wing and upwash on the left wing. This could lead to earlier stall on the left wing since the upwash increases the local angle of attack, while the downwash on the right wing decreases the local angle of attack.) A curve fit to the turning stall data, based on Eq. (13.13), allows for definition of the stall line on the V–n diagram and determination of the maneuver speed if the structural load limit is specified (typically $+3.8\,g$ for light general aviation aircraft). Additions to this basic flight test could include accelerated turning stalls, where the deceleration rate is higher (between 3 and 5 knots per second), to explore dynamic effects on stall performance.

Verification of the turning flight performance theory presented here involves simply piloting the aircraft through a series of steep turns at airspeeds high enough such that stall is not concern. The airspeed, bank angle, load factor, turn rate, and even turn radius (in zero-wind conditions) can all be directly measured in flight and mutually compared to validate the theory. All turning maneuvers should be done with coordinated aileron and rudder usage.

13.2.3 Piloting

Since the amount of lift must increase with increasing load factor in a banked turn, back pressure on the control stick is required to perform a level turn (back pressure increases the angle of attack, and lift increases to equal nW in a level turn). It is important for the pilot conducting the flight test to maintain coordinated flight – that is, appropriate use of the rudder in conjunction with the ailerons to have the longitudinal axis of the aircraft following the flight path (the nose and tail following the same path). If the turn is not coordinated, the aircraft will either be slipping or skidding through the turn. Both skidding and slipping turns can be felt in the seat of the pants through a side force and indicated by the position of the ball on the turn/slip indicator in the aircraft (toward the inside for a slipping turn, toward the outside for a skidding turn). The consequence of conducting stall in an uncoordinated turn is asymmetric development of stall on the wings, which can easily lead to incipient spin and departure.

A slipping turn is defined as one where not enough rudder is used for the amount of roll input, the nose points away from the turn, and the bank angle will be too large relative to the rate of turn. This leads to an imbalance between the turning force and the centrifugal force, such that the horizontal component of lift is excessive. If the aircraft stalls in a slipping turn, the wing on the outside of the turn will tend to stall first, resulting in a rolling moment to the outside of the turn.

In contrast, a skidding turn is one where too much rudder is employed for the set amount of bank angle, resulting in the nose pointing toward the inside of the flight path. In this case, the centrifugal force is larger than the horizontal component of lift and the aircraft skids toward the outside of the turn. If stall is encountered in a skidding turn, the wing on the inside of the turn will stall first, leading to a roll toward the inside of the turn and possible incipient spin.

The asymmetric stall development in slipping or skidding turns is due to a relative velocity difference on either side of the wing. For a skidding turn to the right, there is a sustained yaw rate of the aircraft toward the right as long as the skidding turn is maintained. This yaw rate creates higher velocity on the outside wing and lower velocity on the inside wing. The lower relative velocity on the inside wing, for a given angle of attack, leads to stall happening first on the inside wing.

In a coordinated turn, the stall behaves very much the same as it does in level flight – the nose drops. Recovery from a coordinated turning stall involves simply relieving back pressure to lower the nose and then level the wings. As airspeed builds back up, the aircraft returns to normal straight-and-level flight.

13.2.4 Instrumentation and Data Recording

Most of the information needed for turning flight test can be read directly from the cockpit display and additional rudimentary tools. However, further insight can be obtained through use of digital data acquisition devices. As always, the airspeed can be read directly from the airspeed indicator. Constant altitude may be maintained by monitoring the altimeter and the vertical speed indicator. The turn rate may be measured by marking the time on a stopwatch as the aircraft passes through an arbitrary pair of headings. Turning flight through at least two full turns (720°) is recommended for averaging data. The bank angle may be read from the attitude indicator, which has tick marks for 10°, 20°, 30°, and 60° bank angles, with 45° bank also marked on some attitude indicators. Setting and reading bank angle from these tick marks can be somewhat crude; a more precise method of measuring bank angle could involve a digital protractor mounted on a level surface in the cockpit or a protractor with a plumb line attached. Load factor can be measured by a spring scale with a known weight (for example, a fish scale or a luggage scale). The range and resolution of the scale, along with the magnitude of the selected weight, should be appropriate to ensure a sufficiently accurate measurement for the intended load factor. (Load factor can be determined by dividing the measured weight in the turn by the measured weight in level flight.)

For turning stall flight tests, the same parameters must be recorded as for level stall, and the same corrections made (see Chapter 12 for further detail). In particular, pressure altitude, outside air temperature, and fuel burn should be recorded, as these values are needed for determining airspeed corrections and weight corrections.

If digital data acquisition techniques are used, a wealth of data is available for analysis. Data acquisition provides a direct record of the heading as a function of time from Global Positioning System (GPS) (or the magnetometers), which can be used to determine turn rate. Further, the bank angle may be inferred by integrating the rate signals from the gyros. Load factor may be determined from the accelerometers by performing a ratio of z-axis acceleration in the turn with z-axis acceleration in level flight. GPS data provides position and heading information, along with groundspeed. If wind effects are carefully treated, the GPS data can provide airspeed, turn radius, and turn rate.

13.3   Flight Test Example: Diamond DA40

One way to evaluate the turn performance data is to check for self-consistency of measured values of turn rate, turn radius, bank angle, and load factor. As discussed earlier, the bank angle and load factor can be measured in the aircraft using a protractor and fish scale. Sample data is presented here for a Diamond DA40 aircraft in level turning flight at 115 knots indicated airspeed, where the measured bank angle was approximately 40° and the load factor was 1.3. (From Eq. (13.4) we see that these two values of bank angle and load factor are self-consistent.) GPS flight data of turning flight through a total heading change of about 600° is shown in Figure 13.6(a). Assuming constant turning flight conditions with zero wind, we would expect the flight path to form a perfect circle. However, the ground track data clearly shows elongated loops, which are indicators that the wind aloft is pushing the aircraft in the positive y-direction. Wind effects make direct determination of the turn radius from the ground track challenging – the radius cannot be directly inferred from the raw data alone.

In order to find the turn radius from ground track data, a set of equations representing the ground track subjected to wind effects can be derived by parameterizing equations for a circle. The following development assumes steady wind speed and direction, as well as steady turn rate and turn radius. If the x-axis is aligned East–West, and the y-axis is North–South, we can write the ground track in Cartesian coordinates as

$$x(t) = a + tV_{\text{wind},x} + R\cos(\omega t)$$
$$y(t) = b + tV_{\text{wind},y} + R\sin(\omega t), \tag{13.22}$$

where (a, b) are the coordinates of the center of the circle, t is the elapsed time since the beginning of the steady turn, and $V_{\text{wind},x}$ and $V_{\text{wind},y}$ are the x- and y-components of the wind speed at altitude. The wind terms in the equations effectively describe how the center of the circle is being shifted as time progresses. For a given set of GPS data, estimates for the wind vector, turn radius, and turn rate can be found through nonlinear least squares curve fitting of Eq. (13.22) to the data. For the example shown in Figure 13.6(a), the resulting curve fit is for a wind speed of 6.6 knots from 170°, a turn rate of 8.2° /s, and a turn radius of 1348 ft. These values all show good consistency with theory and with hand-recorded data from flight. Figure 13.6(b) shows the same data and curve fit, but with the wind effects removed. As we would anticipate, the fit becomes a perfect circle centered on the origin. The GPS data collapses nicely to this circle, providing an indication of the quality of the fit.

The data shown in Figure 13.6 was acquired via a smartphone placed on top of the dashboard of the DA40, with a reasonably clear view of the sky. The data shows some interesting artifacts that are inherent to the data acquisition device: the arrows indicate data points that depart from the curve fit and exhibit a discontinuity in the data set. Steady conditions were maintained throughout the turn, so the discontinuity is not a result of suddenly shifting winds or pilot

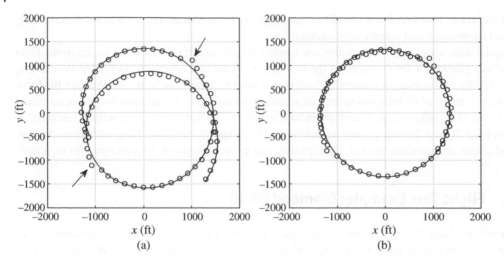

Figure 13.6 (a) GPS flight test data from a DA40 in turning flight. Symbols represent GPS data, while the solid line is a curve fit that accounts for wind. (b) Corrected GPS data and fit, with wind effects removed.

technique. Rather, the departure of these data points from the rest of the series is likely an error in the GPS data acquisition. While the aircraft is in turning flight, the GPS unit's view of the sky is continually changing, with satellites coming in and out of view of the GPS antenna. Inspection of the corresponding GPS accuracy record (a measure of the number of satellite signals used for the computation) revealed large jumps in the number of satellites used, and these jumps coincide with the two discontinuities indicated by the arrows in Figure 13.6(a). It is also interesting to note that the discontinuities are approximately 180° apart along the circular path. Thus, the shifting number of satellites used in the computation of position, resulting from the rapidly changing view of the sky, is likely the cause of the errors in the data set. This is a commonly encountered error when using consumer-grade GPS units in a maneuvering aircraft.

Nomenclature

a	x-axis coordinate for the center of the turn
AR	aspect ratio
b	y-axis coordinate for the center of the turn
C_D	drag coefficient
C_{D_0}	parasite drag coefficient
C_L	lift coefficient
D	drag
e	Oswald efficiency factor
F_c	centripetal force
F_{cfg}	centrifugal force
g	gravitational acceleration
L	lift
n	load factor, L/W
P_A	power available
P_R	power required
q	dynamic pressure
R	turn radius
S	wing area

t	time
T_R	thrust required
V	airspeed
V_A	maneuver speed
V_e	equivalent airspeed
W	aircraft weight
x	horizontal direction (East–West, inertial axes)
y	horizontal direction (North–South, inertial axes)
y'	lateral direction (body axes)
z	vertical direction (inertial axes)
z'	vertical direction (body axes)
ϕ	bank angle
θ	heading
ρ	density
ω	turn rate

Subscripts

max	maximum
min	minimum
opt	optimum
wind	wind speed

Acronyms and Abbreviations

DAQ	data acquisition
FAA	Federal Aviation Administration
KEAS	knots equivalent airspeed
KIAS	knots indicated airspeed
KTAS	knots true airspeed
NACA	National Advisory Committee for Aeronautics
GPS	Global Positioning System

References

Cirrus Design Corporation. (2017). Pilot's Operating Handbook and FAA Approved Airplane Flight Manual for the Cirrus SR20. P/N 11934-005, Duluth, MN.

Federal Aviation Administration. (2002a). Flight Envelope. 14 CFR §23.333

Federal Aviation Administration. (2002b). Limit Maneuvering Load Factors. 14 CFR §23.337.

Federal Aviation Administration. (2002c). Turning Flight and Accelerated Turning Stalls. 14 CFR §23.203

Federal Aviation Administration. (2011). Flight Test Guide for Certification of Part 23 Airplanes. Advisory Circular 23-8C, U.S. Department of Transportation, Federal Aviation Administration.

Federal Aviation Administration. (2019). Parachutes and parachuting. 14 CFR §91.307.

New Piper Aircraft Inc. (1995). Arrow PA-28R-201 SN 2844001 and up Pilot's Operating Handbook and FAA Approved Airplane Flight Manual. Report Number VB-1612, Vero Beach, FL.

Phillips, W.F. (2004). *Mechanics of Flight*. Hoboken, NJ: John Wiley & Sons.

14

Longitudinal Stability

Stability is the characteristic of an aircraft that determines its response to a perturbation such as a wind gust, determining whether the aircraft will naturally return to a stable equilibrium position. Restoration of the aircraft to stable flight requires the generation of restoring moments when the aircraft has been perturbed. For example, if the aircraft pitches up in response to a wind gust, a stable aircraft will generate a nose-down pitching moment that will restore the aircraft orientation to its original stable configuration. On the other hand, an unstable aircraft will generate a nose-up pitching moment in response to an increase in angle of attack, which further exacerbates the pitch-up.

Generally speaking, the stability characteristics are coupled across all three axes. In this chapter, however, we'll focus on longitudinal (pitch) stability only, since its coupling with the lateral and directional axes is negligible. Thus, we'll focus on the pitch attitude of the aircraft, the required forces and moments generated by the horizontal tail, and the impact of elevator deflection on the static and dynamic longitudinal stability of the aircraft.

Flight testing of static longitudinal stability involves determination of the static margin and neutral point for the aircraft under evaluation, which are essentially measures of how stable an aircraft is with its center of gravity (CG) at a given location. Dynamic longitudinal stability, on the other hand, assesses the unsteady motion of the aircraft in response to a perturbation, including the frequency of oscillatory motion and the damping of those oscillations. The primary dynamic longitudinal stability mode of interest here is the long-period (phugoid) mode.

14.1 Static Longitudinal Stability

14.1.1 Theory

We'll begin our discussion of stick-fixed static longitudinal stability with a brief review of longitudinal stability theory, followed by derivation of a specific expression for finding the neutral point, which we'll use in flight testing. Let's begin with a refresher on key definitions related to longitudinal stability. This discussion parallels the description provided by Smith (1981). Other helpful references include Anderson (2012), Phillips (2004), and Kimberlin (2003).

Figure 14.1 illustrates the forces and moments acting on an aircraft in flight. Key forces of interest include the lift produced by the wing–body combination, the lift on the tail, and a moment acting about the aerodynamic center. The *aerodynamic center* of the aircraft is the point along the longitudinal axis about which the moment does not change with angle of attack. The moment about the aerodynamic center (m_{ac}) can be expressed in nondimensional form as:

$$C_{m,ac} = \frac{m_{ac}}{q S \bar{c}} = C_{m,L=0}, \tag{14.1}$$

Introduction to Flight Testing, First Edition. James W. Gregory and Tianshu Liu.
© 2021 John Wiley & Sons Ltd. Published 2021 by John Wiley & Sons Ltd.
Companion website: https://www.wiley.com/go/flighttesting

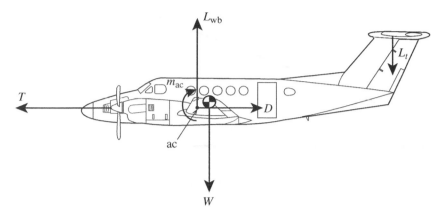

Figure 14.1 Forces and moments acting on an aircraft.

where $\bar{c}$ is the mean aerodynamic chord of the wing, q is the dynamic pressure, and S is the wing area. The moment about the aerodynamic center is equal to the pitch moment when no lift is produced ($C_{m,L=0}$) since $C_{m,ac}$ does not change with angle of attack (by definition).

Now, all of these forces and moments work in conjunction to produce a net moment about the center of gravity, which we'll refer to as m_{CG}, or in coefficient form as $C_{m,CG}$. In order for the aircraft to maintain a fixed pitch attitude, all of these moments must be in balance – in other words, it must be possible to trim the aircraft such that $C_{m,CG}$ is equal to zero for a particular flight condition. This is the first constraint on aircraft stability – a pilot must be able to trim the aircraft such that all moments are balanced and the aircraft pitch attitude is fixed.

The other constraint on aircraft stability is related to how the aircraft responds to a disturbance. If the aircraft were to momentarily pitch up in response to a perturbation such as a wind gust, the aircraft must produce a restoring downward moment that would lower the nose back to its equilibrium position. Since the pitch moment is positive pitch-up (by convention), then a negative pitching moment must be produced when the nose is high. Similarly, if a disturbance causes the nose to pitch down, then a positive pitching moment must be produced to bring the nose back up to equilibrium.

For analysis of aircraft stability, it's helpful to consider a plot of the pitching moment coefficient about the center of gravity versus the aircraft angle of attack (see Figure 14.2). Three curves are illustrated for stable, neutrally stable, and unstable conditions. The constraint of negative pitching moment for a pitch-up disturbance results in a negative slope of the stable C_m-α curve, or $\partial C_{m,CG}/\partial \alpha < 0$. And, the need to trim the aircraft dictates that the curve must cross the x-axis at some point, making it possible to have zero net moment about the center of gravity. Thus, at zero angle of attack for a stable aircraft, the pitching moment coefficient must be positive.

Let's now turn our attention to Figure 14.3, which depicts the forces and moments acting about the aircraft center of gravity. This depiction assumes that there is no contribution to the moments from the thrust or drag – the line of action of these forces is assumed to pass through the CG. Also, the lift produced by the wing and tail is assumed to be oriented perpendicular to the longitudinal axis, such that the freestream velocity vector is aligned with the longitudinal axis, and presuming that the downwash on the tail is small. The net pitching moment about the center of gravity can be expressed as:

$$m_{CG} = m_{ac} + L_{wb}X_{ac} - L_t l_t, \tag{14.2}$$

where m_{ac} is the moment about the aerodynamic center; the second term is the moment induced by the lift of the wing–body combination (L_{wb}) acting at the aerodynamic center, with X_{ac} being the

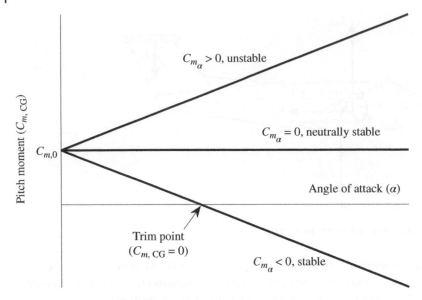

Figure 14.2 Illustration of pitch moment coefficient versus angle of attack curves for stable, neutrally stable, and unstable conditions.

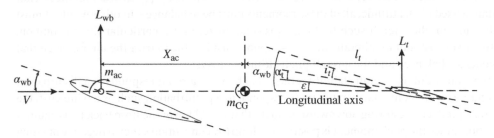

Figure 14.3 Forces and moments acting about the aircraft center of gravity.

distance between the aerodynamic center and the center of gravity; and the last term is the moment due to the tail lift vector (L_t), with l_t being the moment arm between the center of gravity and the aerodynamic center of the tail. We can express Eq. (14.2) in nondimensional form as:

$$C_{m,\text{CG}} = C_{m,\text{ac}} + C_{L,\text{wb}}\frac{X_{\text{ac}}}{\bar{c}} - C_{L,t}\frac{S_t l_t}{S_{\text{wb}}\bar{c}}\frac{q_t}{q}. \tag{14.3}$$

Note that the lift coefficient of the tail is defined relative to the local dynamic pressure and local tail area, leading to the appearance of S_t/S_{wb} and q_t/q in Eq. (14.3). We can define a tail volume coefficient as:

$$\overline{V}_H = \frac{S_t l_t}{S_{\text{wb}}\bar{c}}, \tag{14.4}$$

and express the ratio of dynamic pressures as a tail efficiency factor:

$$\eta_t = \frac{q_t}{q}. \tag{14.5}$$

If we assume for a moment that the zero-lift angle of attack is equal to zero degrees, we can rewrite Eq. (14.3) as:

$$C_{m,\text{CG}} = C_{m,\text{ac}} + a_{\text{wb}}\alpha_{\text{wb}}\frac{X_{\text{ac}}}{\bar{c}} - a_t\alpha_t\overline{V}_H\eta_t, \tag{14.6}$$

where the wing–body lift and tail lift have each been expressed as a product of the local angle of attack (α_{wb} or α_t) and the lift curve slope (a_{wb} or a_t). The relationship between tail angle of attack and wing–body angle of attack (illustrated in Figure 14.3) allows us to write

$$\alpha_t = \alpha_{wb} - i_t - \varepsilon, \tag{14.7}$$

where i_t is the tail incidence angle and ε is the downwash angle at the tail. The downwash angle itself depends on the angle of attack of the aircraft, so the tail angle of attack can be expanded as:

$$\alpha_t = \alpha_{wb} - i_t - \varepsilon_0 - \frac{\partial\varepsilon}{\partial\alpha_{wb}}\alpha_{wb}. \tag{14.8}$$

Substituting Eq. (14.8) into (14.6), we can write

$$C_{m,\text{CG}} = C_{m,\text{ac}} + a_{wb}\alpha_{wb}\frac{X_{ac}}{\bar{c}} - a_t\overline{V}_H\eta_t\left(\alpha_{wb} - i_t - \varepsilon_0 - \frac{\partial\varepsilon}{\partial\alpha_{wb}}\alpha_{wb}\right) \tag{14.9}$$

Finally, we can differentiate (14.9) with respect to the wing–body angle of attack to obtain

$$\frac{\partial C_{m,\text{CG}}}{\partial\alpha_{wb}} = a_{wb}\frac{X_{ac}}{\bar{c}} - a_t\overline{V}_H\eta_t\left(1 - \frac{\partial\varepsilon}{\partial\alpha_{wb}}\right) \tag{14.10}$$

Recall that, in order to ensure longitudinal stability, the change in pitching moment with an increase in angle of attack must be negative. When we consider (14.10), this condition stipulates that the slope of the curve must be negative: $\partial C_{m,\,\text{CG}}/\partial\alpha_{wb} \leq 0$. The most significant observation we can make is that the first term in (14.10) must be smaller in magnitude than the second term in order for the stability criterion to be satisfied. All of the parameters appearing in the second term such as tail volume coefficient, tail efficiency factor, and tail lift curve slope are set in the aircraft design process, so the second term is fixed. The magnitude of the first term, however, depends on the distance of the aerodynamic center from the aircraft center of gravity. The location of the aerodynamic center is fixed in the design process, but the CG location can change substantially in routine operation of an aircraft due to variations in loading (fuel, cargo, and passengers). Thus, stability places a limit on how far aft the center of gravity may be moved: X_{ac} can only be so large. In fact, if we consider the special case of neutral stability, where $\partial C_{m,\,\text{CG}}/\partial\alpha_{wb} = 0$, we can write

$$\frac{X_{ac}}{\bar{c}} = \frac{a_t}{a_{wb}}\overline{V}_H\eta_t\left(1 - \frac{\partial\varepsilon}{\partial\alpha_{wb}}\right) \equiv \frac{X_n}{\bar{c}}, \tag{14.11}$$

This particular condition, defined as the neutral point ($X_n/\bar{c}$) and shown in Figure 14.4, expresses the distance from the center of gravity to the aerodynamic center when the aircraft is neutrally stable. If the CG is moved further aft from the neutral point (X_{ac} is increased beyond X_n), the aircraft will become unstable. Conversely, if the CG is moved forward from the neutral point (X_{ac}

Figure 14.4 Depiction of the neutral point.

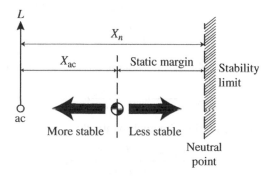

is decreased), the aircraft becomes more stable. We can take this definition of the neutral point, Eq. (14.11), and rewrite (14.10) such that:

$$
\begin{aligned}
\frac{\partial C_{m,\text{CG}}}{\partial \alpha_{\text{wb}}} &= a_{\text{wb}} \left[\frac{X_{\text{ac}}}{\bar{c}} - \frac{a_t}{a_{\text{wb}}} \overline{V}_H \eta_t \left(1 - \frac{\partial \varepsilon}{\partial \alpha_{\text{wb}}} \right) \right] \\
&= a_{\text{wb}} \left[\frac{X_{\text{ac}} - X_n}{\bar{c}} \right].
\end{aligned}
\tag{14.12}
$$

Equation (14.12) is an expression of the *static margin* of an aircraft, which is essentially a measure of how stable it is. Once the neutral point is known, the static margin depends only on the location of the aircraft center of gravity.

14.1.2 Trim Condition

Now that we have established a criterion for ensuring negative slope of the C_m-α curve, let's turn our attention to the stipulation that the aircraft must be trimmed. Trim is achieved by setting the elevator at a desired position such that there is no net moment about the center of gravity, which ensures that the aircraft attitude remains fixed. Referring to Figure 14.2, the trim condition occurs when the aircraft moment coefficient crosses the x-axis (denoted by the trim point on the diagram). In order for it to be possible to trim the aircraft, the zero-lift moment coefficient must be positive if the C_m-α slope is negative.

Analysis of the trim condition can be achieved by rewriting Eq. (14.9) as:

$$
C_{m,\text{CG}} = C_{m,\text{ac}} + a_t \overline{V}_H \eta_t (i_t + \varepsilon_0) + \alpha_{\text{wb}} \left[a_{\text{wb}} \frac{X_{\text{ac}}}{\bar{c}} - a_t \overline{V}_H \eta_t \left(1 - \frac{\partial \varepsilon}{\partial \alpha_{\text{wb}}} \right) \right],
\tag{14.13}
$$

where the first two terms are constant, and the third term alone varies with angle of attack. The first two terms can be grouped together and labeled as the pitch moment about the center of gravity at zero angle of attack ($C_{m,0}$), and the term in brackets is $\partial C_{m,\text{CG}}/\partial \alpha_{\text{wb}}$ (Eq. (14.10)). At this point, we'll now include the effects of the elevator on the horizontal tail to write Eq. (14.13) as:

$$
C_{m,\text{CG}} = C_{m,0} + \alpha_{\text{wb}} \frac{\partial C_{m,\text{CG}}}{\partial \alpha_{\text{wb}}} + \delta \frac{\partial C_{m,\text{CG}}}{\partial \delta},
\tag{14.14}
$$

where δ is the elevator deflection angle, with a downward deflection having a positive sign.

Definition of the trim condition involves finding the elevator deflection angle necessary to achieve $C_{m,\text{CG}} = 0$, so we can write

$$
\delta_{\text{trim}} = -\frac{C_{m,0} + \alpha_{\text{trim}}(\partial C_{m,\text{CG}}/\partial \alpha_{\text{wb}})}{\partial C_{m,\text{CG}}/\partial \delta}.
\tag{14.15}
$$

Here, α_{trim} is the wing–body angle of attack associated with the trim condition, which can be expanded out as follows. The lift coefficient for the entire aircraft can be written as:

$$
C_L = \frac{\partial C_L}{\partial \alpha_{\text{wb}}} \alpha_{\text{wb}} + \frac{\partial C_L}{\partial \delta} \delta,
\tag{14.16}
$$

to account for the effect of the tail and elevator. Similarly, the trim condition can be expressed as:

$$
C_{L,\text{trim}} = \frac{\partial C_L}{\partial \alpha_{\text{wb}}} \alpha_{\text{trim}} + \frac{\partial C_L}{\partial \delta} \delta_{\text{trim}}.
\tag{14.17}
$$

Solving for the trim angle of attack:

$$
\alpha_{\text{trim}} = \frac{C_{L,\text{trim}} - (\partial C_L/\partial \delta)\delta_{\text{trim}}}{\partial C_L/\partial \alpha_{\text{wb}}},
\tag{14.18}
$$

we can substitute into Eq. (14.15):

$$\delta_{\text{trim}} = -\frac{C_{m,0} + \left\{\left[C_{L,\text{trim}} - (\partial C_L/\partial\delta)\delta_{\text{trim}}\right] / (\partial C_L/\partial\alpha_{\text{wb}})\right\} (\partial C_{m,\text{CG}}/\partial\alpha_{\text{wb}})}{\partial C_{m,\text{CG}}/\partial\delta}. \tag{14.19}$$

Now that our equations are becoming unwieldy, and let's introduce some compact notation such that Eq. (14.19) can be rewritten as:

$$\delta_{\text{trim}} = -\frac{1}{C_{m_\delta}} \left(C_{m,0} + \frac{C_{m_\alpha}}{C_{L_\alpha}} C_{L,\text{trim}} - \frac{C_{m_\alpha}}{C_{L_\alpha}} C_{L_\delta} \delta_{\text{trim}} \right). \tag{14.20}$$

Solving for the elevator trim angle, we obtain

$$\delta_{\text{trim}} = \frac{-C_{m,0} C_{L_\alpha} - C_{m_\alpha} C_{L,\text{trim}}}{C_{m_\delta} C_{L_\alpha} - C_{m_\alpha} C_{L_\delta}}. \tag{14.21}$$

We can differentiate Eq. (14.21) with respect to the trim lift coefficient to get

$$\frac{d\delta_{\text{trim}}}{dC_{L,\text{trim}}} = -\frac{C_{m_\alpha}}{C_{m_\delta} C_{L_\alpha} - C_{m_\alpha} C_{L_\delta}}. \tag{14.22}$$

This equation is particularly revealing of the longitudinal stability characteristics: when C_{m_α} goes to zero, the amount of elevator change required for a change in lift coefficient also decreases and goes to zero. In other words, elevator control becomes ineffective for changing lift coefficient when $C_{m_\alpha} = 0$. We can see this by considering Figure 14.5, where the results of Eq. (14.22) are notionally plotted for various CG locations (after USN TPS 1997; Kimberlin 2003). The CG location at which a change in lift coefficient (dC_L or $d\alpha$) produces no change in moment ($dC_m = 0$) is the neutral point, by definition. This is because the aerodynamic center of the entire aircraft is collocated with the CG at the neutral point, and no moment is produced by changing angle of attack if the CG is at the neutral point.

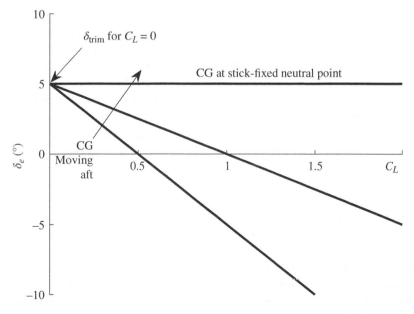

Figure 14.5 Notional data of required elevator position (trim angle) as a function of lift coefficient for various CG positions.

Since C_{m_α} is proportional to the static margin (Eq. (14.12)), we can also write

$$\frac{d\delta_{\text{trim}}}{dC_{L,\text{trim}}} = \frac{-a_{\text{wb}}(X_{\text{ac}} - X_n)/\bar{c}}{C_{m_\delta}C_{L_\alpha} - C_{m_\alpha}C_{L_\delta}} = K\left(\frac{X_n - X_{\text{ac}}}{\bar{c}}\right), \tag{14.23}$$

where K is a constant. Now this is where our discussion starts to converge on a strategy for flight test determination of static longitudinal stability. Equation (14.23) shows that the amount of elevator deflection needed to trim the aircraft at a particular lift coefficient is proportional to the static margin! In other words, if the CG is located farther forward (X_{ac} decreases, and the static margin increases – see Figure 14.4), then greater amounts of elevator deflection are required to trim the aircraft across a range of lift coefficients. The relationship shown in Eq. (14.23) suggests that we can measure the static margin by somehow recording the elevator deflection across a range of flight conditions (lift coefficient and center of gravity).

14.1.3 Flight Testing Procedures

Direct measurement of elevator deflection in flight is not feasible for most flight test programs (unless you employ a "wing walker" with a protractor!). Instead, we can rely on the fact that control stick or yoke position is directly proportional to the elevator deflection, so we can write

$$K\left(\frac{X_n - X_{\text{ac}}}{\bar{c}}\right) = \frac{d\delta_{\text{trim}}}{dC_{L,\text{trim}}} \propto \frac{ds_{\text{trim}}}{dC_{L,\text{trim}}}, \tag{14.24}$$

where s is the position of the control stick or yoke. The equivalence between stick position and elevator deflection is due to the control input being connected to the control surfaces via fixed-length linkages or push rods in nearly all general aviation (GA) aircraft and unmanned aerial vehicles (UAVs). The position of the control yoke (s) can be conveniently measured relative to any fixed reference point in the cockpit using a measuring tape.

Flight test determination of the neutral point involves finding the static margin at several CG locations, and extrapolating the elevator trim sensitivity (Eq. (14.24)) to the CG location where the slope goes to zero. This method is detailed as follows. First, a number of practical CG locations are selected for flight testing, each of which must be within the fore and aft boundaries established in the aircraft limitations (detailed in the Pilot Operating Handbook, or POH, for that aircraft). If the reader is unfamiliar with methods for calculating CG, it may be helpful at this point to read or review the discussion on weight and balance in Chapter 6. Center of gravity location can be set by moving passengers or cargo around inside the cabin, as long as aircraft operating limitations are met. In order to obtain the best extrapolation to find the neutral point, the range of CG locations should be as wide as possible, while still remaining within the aircraft operating limitations. One straightforward way of moving the CG is to carry sandbags as cargo and then move the sandbags around the aircraft.

Fuel burn during flight will also have an impact on the CG location, since the fuel's center of gravity is not necessarily collocated with the aircraft's center of gravity. For example, on the Cirrus SR20, for every 10 gallons of fuel burned the CG will move forward by approximately two-tenths of an inch. Thus, the aft-most CG location should be tested at the beginning of a flight, and the forward-most CG location should be tested at the end of the flight, in order to maximize the range of CG locations evaluated. Measurement of fuel burn must be done by some other method than reading the fuel gauges, since they are notoriously inaccurate and lack sufficient precision. Instead, fuel burn can be recorded by taking note of the engine power settings and duration of time at those settings, and then calculating fuel burn via the specified fuel burn rates at those power settings (referring to the POH). Modern aircraft make this task straightforward through direct measurement of fuel burn via a fuel totalizer that directly measures the instantaneous fuel flow rate and integrates this parameter over time.

Now that we've formed a strategy for controlling and measuring the CG location, let's turn our attention to measuring the required stick position for trim across a range of lift coefficients. For each CG location, the pilot must fly a range of stabilized airspeeds (recall that lift coefficient is inversely proportional to the square of airspeed). At each airspeed condition, a measurement of the stick position is made relative to some convenient reference location using a tape measure or something similar. Note that in some aircraft such as the Cirrus SR20, the amount of stick deflection is very small (fractions of an inch) across a range of lift conditions, so care must be taken to read the deflection as accurately as possible (measurements to the nearest 1/32 of an inch were required for reasonable accuracy on the SR20).

The flight profiles should be conducted as follows. First, the pilot should select a stable flight altitude and airspeed, trim the aircraft at that condition via the trim wheel to obtain zero stick force, and record the baseline stick position for the first data point. The aircraft is then flown at that condition with a constant altitude, but at airspeed test points above and below by ±5 KIAS and ±10 KIAS (or greater excursions for greater stick deflection), with the stick position recorded each time. Airspeeds are flown at alternately higher and lower set points in order to facilitate maintaining constant altitude during the test. It is very important to note that the trim wheel should not be adjusted during this range of test speeds! Elevator angle to trim should be controlled via stick position alone. As a check on the integrity of the data set, the concluding point in the set should be a repeat of the initial trim condition and airspeed, to ensure that the same stick position is obtained.

For each test point (defined by CG location, altitude, and airspeed), the following parameters should be recorded: aircraft loading configuration, fuel burn, stick position, outside air temperature, and pressure altitude. Data analysis proceeds by converting indicated airspeed to true airspeed, and obtaining lift coefficient from true airspeed, air density, and instantaneous aircraft weight at that test condition. Following this, stick position should be plotted against lift coefficient for all of the test points (airspeeds and CG locations), with a linear fit applied to each CG location set of airspeeds.

14.1.4   Flight Test Example: Cirrus SR20

Typical flight test results are shown in Figures 14.6 and 14.7 for a Cirrus SR20 flown with three different CG locations. The first CG location was 145.9 in. aft of the datum, with full fuel, pilot and co-pilot in the front seats, a passenger in the back seat, and sandbags in the baggage area. After testing the first CG location, the pilot landed the aircraft to remove the sandbags, and the second CG location of 143.7 in. was flown. The pilot landed again to drop off the passenger, and the third CG location of 141.7 in. was flown. (Note that with fuel burn and removal of payload and passengers, the aircraft weight changes as well.) Indicated airspeeds of 110 ± 20 and ± 10 KIAS were flown using a two-axis autopilot in altitude and heading mode.

Figure 14.6 plots the stick deflection at each of these airspeeds (note that the entire y-axis range is only half an inch!) as a function of lift coefficient at each point. The slope of each line in Figure 14.6 is the value of the derivative in Eq. (14.24) for that CG location. We can plot each slope value versus CG location, as shown in Figure 14.7. Applying a linear fit to the slope values, we can extrapolate the trend down to the x-axis to find the location of the neutral stability point (see Eq. (14.24)). For the data sets shown in Figures 14.6 and 14.7, the calculated neutral point is at 151.9 in.. This value may be compared with the aft CG limit of 148.1 in. specified in the POH. Thus, the static margin at the aft CG limit is 3.8 in., or 8.3% of the mean aerodynamic chord (45.7 in.). Aircraft designs generally maintain a static margin of at least 5% (Phillips 2004), so these values of the neutral point and minimum static margin are reasonable.

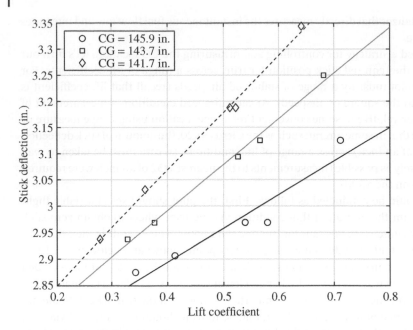

Figure 14.6 Measured stick deflection at various airspeed and CG test points.

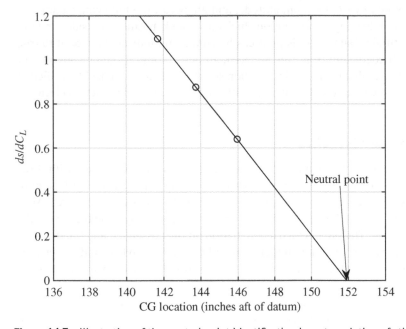

Figure 14.7 Illustration of the neutral point identification by extrapolation of stick deflection slope data.

14.2 Dynamic Longitudinal Stability

14.2.1 Theory

Dynamic stability differs from static stability due to the time-varying nature of the forces and moments involved. While dynamic stability about all three axes is important, we'll focus our attention here on longitudinal stability about the pitch axis. There are two characteristic modes that are typically present – the short-period and long-period modes – and pilots are immediately concerned about whether the amplitude of those modes will grow or decay in time. The damping characteristics (underdamped leads to growth; damped leads to decay) directly affect the handling qualities of the aircraft – if insufficiently damped, these modes can make the aircraft difficult to fly (for example, when tracking the inbound course and glideslope on an instrument approach to a runway). The short-period mode generally has a characteristic period of a few seconds and represents an exchange in energy between rotational kinetic energy and potential energy (altitude). It manifests as excursions in angle of attack and altitude, with only small changes in velocity. On most aircraft, the short-period mode is significantly damped such that it is not a concern for pilots. The long-period mode, however, is often only very lightly damped and is readily apparent to pilots. Since the period of this mode tends to be very long (on the order of 30 seconds or longer), it is typically not objectionable to most pilots. The long-period mode represents an interchange between translational kinetic energy (airspeed) and potential energy (altitude). Since this mode is present on most aircraft and is readily observable, the long-period mode will be the focus of the dynamic longitudinal flight test described here.

We'll begin by defining the equations of motion that describe the longitudinal flight characteristics. These equations will be defined relative to the wind axes (rather than body-fixed or Earth-fixed coordinate axes), where the x-axis is aligned with the freestream velocity vector. These coupled equations will describe the aircraft angle of attack, pitch attitude, and velocity in terms of stability derivatives and can be solved simultaneously for an analytical prediction of the longitudinal stability characteristics (both short- and long-period modes). The derivation below generally follows the discussion of Smith (1981) and will regularly refer to Figure 14.8, which depicts the relevant axes, angles, and forces acting on the aircraft.

Starting with the vertical (z) axis, a summation of the forces acting perpendicular to the freestream is

$$\sum F_z = L - W \cos \gamma - F_c = 0 \tag{14.25}$$

for quasi-steady conditions (acceleration perpendicular to the freestream direction is negligible). Here, γ is the angle between the freestream and the horizon (see Figure 14.8), and F_c is the apparent centrifugal force resulting from motion along a curved flight path. (This equation of motion is not in an inertial frame, but in the aircraft-centered wind axes frame.) If we assume that γ is small and

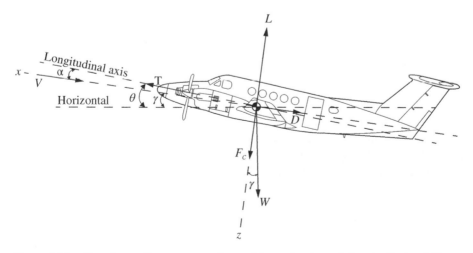

Figure 14.8 Aircraft coordinate axes, angles, and forces for dynamic longitudinal stability.

that the lift can be expressed as the sum of the weight and an additional (time-varying) component of lift (L'), we can solve Eq. (14.25) for centrifugal force:

$$F_c = (W + L') - W = L'. \tag{14.26}$$

We'll now expand out definitions for the centrifugal force and the additional lift. From classical mechanics, the centrifugal force acting on a body following a curved path of radius R can be expressed as:

$$F_c = M\omega^2 R = MV\omega, \tag{14.27}$$

where ω is the same as $\dot{\gamma}$, M is the mass of the flight vehicle, and V is the freestream velocity. From Figure 14.8, we can also express the angular rate as $\omega = \dot{\theta} - \dot{\alpha}$, giving

$$F_c = \frac{W}{g}V(\dot{\theta} - \dot{\alpha}). \tag{14.28}$$

Additional lift can be expanded out from the lift equation ($L = 1/2\rho V^2 S C_L$) via partial derivatives as:

$$L' = \frac{\partial L}{\partial \alpha}\alpha' + \frac{\partial L}{\partial V}V', \tag{14.29}$$

since all terms in the lift equation are constant except for α and V, and there is a linear relationship between α and C_L. The prime notation (α' and V') denotes that these quantities represent changes from the steady-state trimmed values of angle of attack and airspeed. We can expand the first partial derivative in (14.29) as:

$$\frac{\partial L}{\partial \alpha} = \frac{\partial L}{\partial C_L}\frac{\partial C_L}{\partial \alpha} = \frac{1}{2}\rho V^2 S\frac{\partial C_L}{\partial \alpha} = \frac{L}{C_L}C_{L_\alpha}, \tag{14.30}$$

using shorthand notation for the lift curve slope. The second partial derivative in (14.29) can be similarly expanded as:

$$\frac{\partial L}{\partial V} = \rho V S C_L = \frac{1/2\rho V^2 S C_L}{1/2 V} = \frac{2L}{V}. \tag{14.31}$$

Combining (14.30) and (14.31) with (14.29), we get

$$L' = \frac{L}{C_L}C_{L_\alpha}\alpha' + \frac{2L}{V}V'. \tag{14.32}$$

Now, we're in a position to relate the centrifugal force to the additional lift, substituting (14.28) and (14.32) into (14.26) to get

$$\frac{W}{g}V(\dot{\theta} - \dot{\alpha}) = \frac{L}{C_L}C_{L_\alpha}\alpha' + \frac{2L}{V}V', \tag{14.33}$$

which can be solved for pitch rate and rewritten as:

$$\dot{\alpha}' = \dot{\theta}' - \frac{2g}{V^2}V' - \frac{g}{VC_L}C_{L_\alpha}\alpha', \tag{14.34}$$

using the fact that $L = W$, and that the time derivative of the mean angles is zero (*i.e.*, $\dot{\alpha} = \dot{\bar{\alpha}} + \dot{\alpha}' = \dot{\alpha}'$). This is the first longitudinal equation of motion.

We'll now proceed by considering the forces acting in the freestream (x) direction. Let's first consider the steady condition, where the summation of forces parallel to the freestream is

$$\sum F_x = T\cos\alpha - D - W\sin\gamma = 0. \tag{14.35}$$

Assuming that α and γ are small, we obtain

$$T = D + W\gamma. \tag{14.36}$$

Let's now consider the accelerated case, since the accelerations in the freestream direction are non-negligible. This equation of motion can be written as:

$$\sum F_x = T - (D + D') - W(\gamma + \gamma') = Ma_x,$$ (14.37)

where we've assumed that drag and flight path angle (γ) change in time, but that thrust is constant. Substituting (14.36) into (14.37) gives

$$-D' - W\gamma' = Ma_x.$$ (14.38)

Since $a_x = \dot{V}'$, we can then write

$$\frac{W}{g}\dot{V}' = -D' - W\gamma'.$$ (14.39)

We'll now expand the two terms on the right-hand side in a manner similar to what we did with the first equation of motion. The change in drag can be expanded from the drag equation ($D = 1/2\rho V^2 S C_D$) via partial derivatives as:

$$D' = \frac{\partial D}{\partial \alpha}\alpha' + \frac{\partial D}{\partial V}V',$$ (14.40)

since drag only varies with changes in angle of attack and airspeed in this case. The first partial derivative is subsequently expanded as:

$$\frac{\partial D}{\partial \alpha} = \frac{\partial D}{\partial C_D}\frac{\partial C_D}{\partial \alpha} = \frac{1}{2}\rho V^2 S\frac{\partial C_D}{\partial \alpha} = \frac{L}{C_L}C_{D_\alpha},$$ (14.41)

and the second partial derivative is

$$\frac{\partial D}{\partial V} = \rho V S C_D = \frac{1/2\rho V^2 S C_D}{1/2V} = \frac{2D}{V}.$$ (14.42)

Substituting (14.41) and (14.42) into (14.40) and recalling that $L = W$ gives

$$D' = \frac{W}{C_L}C_{D_\alpha}\alpha' + \frac{2D}{V}V'.$$ (14.43)

Since $\gamma' = \theta' - \alpha'$ (see Figure 14.8), we can substitute (14.43) into (14.39) to get

$$\frac{W}{g}\dot{V}' = -\frac{W}{C_L}C_{D_\alpha}\alpha' - \frac{2D}{V}V' - W(\theta' - \alpha').$$ (14.44)

We can rearrange this equation, considering that $D/W = C_D/C_L$, to obtain our second equation of motion:

$$\dot{V}' = -g(\theta' - \alpha') - \frac{g}{C_L}C_{D_\alpha}\alpha' - \frac{2g}{V}\frac{C_D}{C_L}V'.$$ (14.45)

The third and final equation of motion comes from a summation of the moments acting about the center of gravity and recalling from dynamics that:

$$\sum m' = I_{yy}\ddot{\theta}',$$ (14.46)

where I_{yy} is the moment of inertia about the lateral axis and $\ddot{\theta}'$ is the perturbation to the angular acceleration about the lateral axis. The moment disturbance about the center of gravity can be expressed as:

$$m' = \frac{1}{2}\rho V^2 S\bar{c}C_m',$$ (14.47)

where $\bar{c}$ is the mean chord. The moment disturbance is only appreciably sensitive to α' and θ', so we can neglect any impact due to V'. The moment coefficient perturbation can be expanded as:

$$C_m' = C_{m_\alpha}\alpha' + C_{m_q}q',$$ (14.48)

where $q = \dot{\theta}$ is the pitch rate (not to be confused with dynamic pressure), and q' is the pitch rate perturbation, which is equal to $\dot{\theta}'\bar{c}/V$ for dimensional consistency. Combining (14.47) and (14.48) with (14.46), we obtain our third equation of motion:

$$\ddot{\theta}' = \frac{\rho V^2 S\bar{c}}{2I_{yy}}\left(C_{m_\alpha}\alpha' + C_{m_q}\dot{\theta}'\frac{\bar{c}}{V}\right). \tag{14.49}$$

We now have a coupled set of three equations of motion (Eqs. (14.34), (14.45), and (14.49)), which are collected and summarized here:

$$\dot{\alpha}' = \dot{\theta}' - \frac{2g}{V^2}V' - \frac{g}{VC_L}C_{L_\alpha}\alpha'$$

$$\dot{V}' = -g(\theta' - \alpha') - \frac{g}{C_L}C_{D_\alpha}\alpha' - \frac{2g}{V}\frac{C_D}{C_L}V'$$

$$\ddot{\theta}' = \frac{\rho V^2 S\bar{c}}{2I_{yy}}\left(C_{m_\alpha}\alpha' + C_{m_q}\dot{\theta}'\frac{\bar{c}}{V}\right). \tag{14.50}$$

This set of equations may be solved numerically (for example, using differential equation solvers in MATLAB). Numerical solution of these governing equations for longitudinal motion will provide time-resolved values of angle of attack ($\alpha(t)$), airspeed ($V(t)$), and aircraft pitch attitude ($\theta(t)$), along with their derivatives. The analytical estimates of angle of attack and airspeed are particularly useful for comparison with flight test data.

Now that we have defined the governing equations, all that we need to do at this point is to determine values for all of the input parameters in order to solve for α, V, and θ. Most of the input parameters are straightforward, but we'll need to devote a little more attention to the lift curve slope for the airplane (C_{L_α}), the slope of the pitch moment/angle of attack relationship (C_{m_α}), the sensitivity of drag to the aircraft angle of attack (C_{D_α}), and the pitch damping term (C_{m_q}). Collectively, these are referred to as the aircraft's stability derivatives. Verification of the stability derivatives is one of the central aims of aircraft flight testing!

Let's now provide details on how to estimate the stability derivatives for an aircraft. We'll begin with the typical definitions of the aircraft lift and drag coefficients:

$$C_L = \frac{L}{1/2\rho V^2 S} = \frac{2W}{\rho V^2 S}$$

$$C_D = C_{D_0} + \frac{C_L^2}{\pi\,\text{AR}\,e}. \tag{14.51}$$

The lift curve slope for the entire aircraft (C_{L_α}) can be found by considering that overall lift is a combination of the wing–body lift and the tail lift (neglecting any contributions from any other components):

$$L = L_{wb} + L_t, \tag{14.52}$$

which can be expanded out as:

$$L = a_{wb}\alpha_{wb}\frac{1}{2}\rho V^2 S + a_t\alpha_t\eta_t\frac{1}{2}\rho V^2 S_t, \tag{14.53}$$

where a is the lift curve slope, α is the local angle of attack, and the subscripts "wb" and "t" denote the wing–body and tail, respectively. Note that the ratio of the local dynamic pressure between the tail and the wing is still defined as η_t (recall Eq. (14.5)). The definition of the tail angle of attack is the same as it was for static longitudinal stability, so Eq. (14.8) can be used here. If we substitute Eq. (14.8) into Eq. (14.53) and differentiate with respect to α_{wb}, we get

$$\frac{\partial L}{\partial \alpha_{wb}} = a_{wb}\frac{1}{2}\rho V^2 S + a_t\left(1 - \frac{\partial\varepsilon}{\partial\alpha_{wb}}\right)\eta_t\frac{1}{2}\rho V^2 S_t. \tag{14.54}$$

If we assume that the angle of attack for the entire aircraft is approximately the same as the wing–body angle of attack, the lift curve slope for the entire aircraft (C_{L_α}) can then be expanded from (14.54) via the chain rule:

$$C_{L_\alpha} = \frac{\partial C_L}{\partial \alpha_{wb}} = \frac{\partial C_L}{\partial L} \frac{\partial L}{\partial \alpha_{wb}} = \frac{2}{\rho V^2 S} \left[a_{wb} \frac{1}{2} \rho V^2 S + a_t \left(1 - \frac{\partial \varepsilon}{\partial \alpha_{wb}} \right) \eta_t \frac{1}{2} \rho V^2 S_t \right],$$ (14.55)

which simplifies to

$$C_{L_\alpha} = a_{wb} + a_t \eta_t \left(1 - \frac{\partial \varepsilon}{\partial \alpha_{wb}} \right) \frac{S_t}{S}.$$ (14.56)

The sensitivity of the drag coefficient to the aircraft angle of attack (C_{D_α}) can be determined in a similar manner, using the chain rule:

$$C_{D_\alpha} = \frac{\partial C_D}{\partial D} \frac{\partial D}{\partial C_L} \frac{\partial C_L}{\partial \alpha}.$$ (14.57)

Since drag coefficient is defined as $C_D = 2D/\rho V^2 S$ and can also be expressed as a drag polar in Eq. (14.51), we can evaluate the first two derivatives in (14.57):

$$\frac{\partial C_D}{\partial D} = \frac{2}{\rho V^2 S}$$
$$\frac{\partial D}{\partial C_L} = \rho V^2 S \frac{C_L}{\pi \, AR \, e}.$$ (14.58)

Substituting (14.58) into (14.57) we get

$$C_{D_\alpha} = \frac{2C_L}{\pi \, AR \, e} C_{L_\alpha}$$ (14.59)

for the drag sensitivity.

The pitch damping term (C_{m_q}) may be found by considering that the pitching moment about the aircraft center of gravity is dominated by the tail lift. Under this assumption, we can write

$$m_t = -l_t L_t,$$ (14.60)

where l_t is the distance between the aerodynamic center of the tail and the center of gravity for the aircraft, and the tail lift force is defined as positive in the upward direction (see Figure 14.3). The pitching moment can then be expanded out as:

$$m_t = -l_t \eta_t \frac{1}{2} \rho V^2 S_t a_t \alpha_t.$$ (14.61)

The tail angle of attack may be approximated by relating it to the pitch rate of the aircraft:

$$\alpha_t \cong \frac{\dot{\theta} l_t}{V}.$$ (14.62)

Expressing (14.61) in nondimensional form, and substituting (14.62), we get

$$C_{m,t} = -\frac{l_t}{\bar{c}} \frac{S_t}{S} \eta_t a_t \frac{\dot{\theta} l_t}{V}.$$ (14.63)

Since pitch rate can also be expressed as:

$$q = \frac{\dot{\theta} \bar{c}}{V},$$ (14.64)

and the tail volume coefficient is

$$\overline{V}_H = \frac{l_t}{\bar{c}} \frac{S_t}{S},$$ (14.65)

the pitching moment due to the tail can be written as:

$$C_{m,t} = -\overline{V}_H \frac{l_t}{\overline{c}} \eta_t a_t q.$$
(14.66)

The partial derivative of (14.66) with respect to pitch rate (q) then is

$$C_{m_q,t} = \frac{\partial C_{m,t}}{\partial q} = -\overline{V}_H \frac{l_t}{\overline{c}} \eta_t a_t.$$
(14.67)

According to Smith (1981), the pitch damping term for the entire aircraft is proportional to the pitch damping term of the tail such that:

$$C_{m_q} = -1.1 \overline{V}_H \frac{l_t}{\overline{c}} a_t.$$
(14.68)

The final stability derivative to be determined is the slope of the pitch moment with respect to angle of attack (C_{m_α}), which is found via Eq. (14.12). We now have expressions for the four stability derivatives: C_{L_α} in Eq. (14.56), C_{D_α} in Eq. (14.59), C_{m_q} in Eq. (14.68), and C_{m_α} in Eq. (14.12).

Other parameters that must be determined as input values to the stability derivatives and/or the longitudinal equations of motion include the following. Flight conditions lead to definition of gravitational acceleration (g, which is a function of altitude), mean freestream airspeed (V), and density (ρ). The geometry, weight, and balance of the aircraft will determine values for wing area (S), mean aerodynamic chord ($\overline{c}$), tail moment arm (l_t), tail planform area (S_t), wing aspect ratio ($AR = b^2/S$), and aircraft weight (W). The mass moment of inertia about the pitch axis (I_{yy}) can be found from

$$I_{yy} = \frac{\ell^2 M \overline{R}_y}{4},$$
(14.69)

where M is the vehicle mass, ℓ is the length of the air vehicle, and $\overline{R}_y$ is the nondimensional radius of gyration. A typical value of $\overline{R}_y$ for a single-engine propeller aircraft is 0.38 (more values are provided by Roskam (2003) or Raymer (2006)). For reference, mass moment of inertia values for various aircraft were measured and reported by Gracey (1940).

Other aerodynamic parameters such as the lift curve slope of the wing–body combination (a_{wb}), lift curve slope of the tail (a_t), sensitivity of downwash angle to angle of attack ($\partial \varepsilon/\partial \alpha$), and the drag polar ($C_{D_0}$ and e) can be estimated via standard methods used in aircraft design. Raymer (2006) estimates the lift curve slope (for incompressible flow, with no or little leading edge sweep, which is typical of most GA aircraft) as:

$$a = \frac{2\pi\, AR}{2 + \sqrt{4 + AR^2}} \left(\frac{S_{exposed}}{S} \right) f.$$
(14.70)

Here, the factor f is defined as:

$$f = 1.07(1 + d/b)^2,$$
(14.71)

where d is the fuselage diameter. If the product of ($S_{exposed}/S$) and f is greater than 1, Raymer indicates that the product should be truncated to some value slightly lower than 1 (say 0.98). Alternatively, Anderson (2012) gives a simpler form of the three-dimensional lift curve slope (neglecting the fuselage) as:

$$a = \frac{a_0}{1 + a_0/\pi e_1\, AR},$$
(14.72)

where the units of the lift curve slope are per radian, and e_1 is approximately equal to the Oswald efficiency factor (e). These expressions for lift curve slope can be used to find either a_{wb} or a_t, using the appropriate input values. Downwash sensitivity ($\partial\varepsilon/\partial\alpha$) and the drag polar can be estimated by comparison with empirical data and established estimation methods (*e.g.*, see Raymer 2006 or Williams and Vukelich 1979a,1979b,1979c).

Bringing the analysis to a conclusion, Eq. (14.50) are modeled using a numerical solver (ode45 in MATLAB). Note that this set of differential equations must be represented as a series of four coupled first-order equations (*i.e.*, the second-order equation for pitch attitude, Eq. (14.49), must be split up into two first-order equations for MATLAB to solve). Input values are based on a notional aircraft similar to the Cirrus SR20. The estimated stability derivatives are $C_{L_\alpha} = 5.68\,\text{rad}^{-1}$, $C_{D_\alpha} = 0.242\,\text{rad}^{-1}$, $C_{m_q} = -13.2\,\text{rad}^{-1}$, and $C_{m_\alpha} = -1.03\,\text{rad}^{-1}$. Other input parameters are $\bar{c} = 1.15\,\text{m}$, $S = 13.5\,\text{m}^2$, $I_{yy} = 3136\,\text{kg/m}^2$, AR $= 10.1$, $C_{D_0} = 0.0275$, and $e = 0.80$. The mean freestream velocity is 112 KIAS at an altitude of 3000 ft ($\rho = 1.121\,\text{kg/m}^3$). Initial conditions for the simulation are a perturbation of –10° in pitch attitude and +20 kts in indicated airspeed. Results from the simulation are shown in Figure 14.9 for time histories of angle of attack, airspeed, pitch angle, vertical acceleration, flight path angle, and vertical acceleration for body-fixed axes (what would be measured on the aircraft). Note that the body-fixed vertical acceleration is not the same as the vertical acceleration, since the aircraft is pitching in time. It's important to recognize that both of the contributions to acceleration (changes in pitch attitude as well as acceleration of the flight vehicle) need to be mapped into body coordinates: changes in the orientation of the gravity vector relative to the aircraft body axes are related by the pitch angle (θ), while flow accelerations relative to the body axes are related by the angle of attack (α). These contributions can be expressed as:

$$a_z = g\cos\theta + \dot{V}'\sin\alpha \qquad (14.73)$$

Since α is small, changes in acceleration in the body axes are dominated by the pitch attitude of the aircraft (θ). Thus, Eq. (14.73) can be used with analytical results from Eq. (14.50) to directly compare with flight test results.

Results in Figure 14.9 indicate that the period of the phugoid mode is approximately half a minute and that the response is lightly damped (underdamped), which are typical characteristics for a general aviation aircraft. The system of equations in (14.50) may be thought of as a second-order system, allowing us to use common methods (Ogata 1992; Phillips 2004) to analyze the impulse response of this dynamic system. The eigenvalues of the matrix constructed for solution of (14.50) can be used to directly find the damped natural frequency (ω_d), undamped natural frequency (ω_n), and the damping ratio (ζ). Of the two sets of eigenvalues for the response matrix in our aircraft example here, the eigenvalues for the long-period mode are the smaller set ($\lambda_{1,2} = -0.009 \pm 0.1989i$). The damped natural frequency for the underdamped second-order system is found from the imaginary part of the eigenvalue:

$$\omega_d = |\text{imag}(\lambda)|, \qquad (14.74)$$

giving a frequency of 0.1989 rad/s in this case. The undamped natural frequency, which is the value of the natural frequency if no damping is present, is found from the eigenvalue pair:

$$\omega_n = \sqrt{\lambda_1\lambda_2}, \qquad (14.75)$$

which in this case is 0.1991 rad/s. The period of oscillations may be found by

$$t_{\text{period}} = \frac{2\pi}{\omega_d}, \qquad (14.76)$$

giving a period of 31.59 seconds. The damping ratio is found from

$$\zeta = -\frac{\lambda_1 + \lambda_2}{2\sqrt{\lambda_1\lambda_2}}, \qquad (14.77)$$

giving a value of 0.0453.

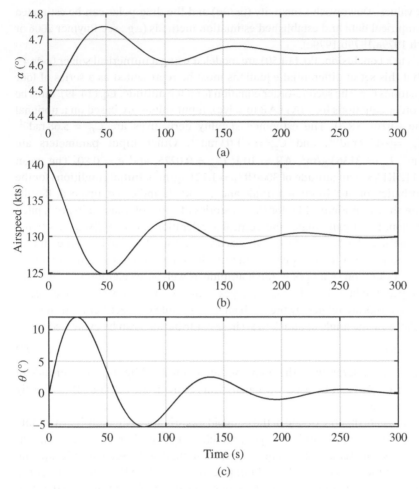

Figure 14.9 Analytical predictions of the long-period mode based on estimated SR20 stability derivatives and numerical simulation of Eq. (14.50), providing time histories of (a) angle of attack, (b) airspeed, and (c) pitch angle.

14.2.2 Flight Testing Procedures

Flight test assessment of the long-period (phugoid) mode is fairly straightforward. From an initial condition of trimmed, steady, level flight, the pilot provides an impulse pitch input to pitch the air-craft nose-up or nose-down. The airspeed is allowed to change by about 10 or 15 kts above or below the trimmed airspeed. After the perturbation airspeed is reached, the pitch input is then brought back to the neutral position and the aircraft motion is allowed to dynamically pitch up and down. Measurements of the aircraft response could possibly be made via a pencil-and-paper approach, although the flight test engineer would need to read and record values in rapid succession! A much more straightforward approach is to record the data using a digital data acquisition device. The relevant sensors to monitor include pressure (altitude), ground speed, and z-axis acceleration. All three of these will exhibit oscillations that indicate the damped oscillation period and the damping ratio. The z-axis acceleration, however, will be a combination of the aircraft acceleration with the pitch attitude of the aircraft, as illustrated by Eq. (14.73).

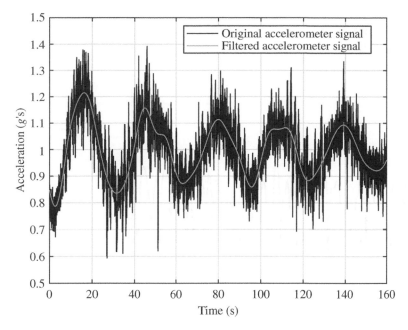

Figure 14.10 *Z*-Axis acceleration time record of the Cirrus SR20 phugoid mode.

14.2.3 Flight Test Example: Cirrus SR20

Dynamic flight test results can be directly analyzed to find the longitudinal stability characteristics and compared with analytical predictions. First, the experimental data should be filtered using the techniques described in Chapter 4, where we initially presented phugoid flight test data as an example of data analysis techniques. Acceleration data is plotted in Figure 14.10, showing both the original noisy data set and the filtered data after an eighth-order low pass IIR filter with a cutoff of 0.1 Hz and passband ripple of 0.01 dB has been applied. Removal of the noise greatly assists with identification of peaks and their amplitudes in the response.

The damped natural frequency (ω_d) of the long period may be determined in one of two ways. First, spectral analysis can be performed on the time record of z-axis acceleration to identify the frequency of the predominant peak and find $\omega_d = 2\pi f_{\text{peak}}$. For our flight test example here, the power spectrum of the z-axis accelerometer data is shown in Figure 14.11, where the frequency peak is 0.032 Hz, corresponding to a damped natural frequency of 0.2013 rad/s and a period of 31.2 seconds. The alternate method is to find the mean period of the oscillations directly from the time record by identifying the time interval between convenient points in the waveform (*e.g.*, successive maxima). Note that this procedure must be done on filtered data in order to make the points readily identifiable. The measured period, t_{period}, can then be related to the damped natural frequency by $\omega_d = 2\pi/t_{\text{period}}$.

The damping ratio of the long-period mode can be found from the ratio of succeeding amplitudes of the peaks in whatever quantity (acceleration, airspeed, etc.) is selected for analysis. The log of this ratio is referred to as the logarithmic decrement for a second order system,

$$\ln\left(\frac{x_1}{x_2}\right) = \zeta\omega_n t_{\text{period}}, \tag{14.78}$$

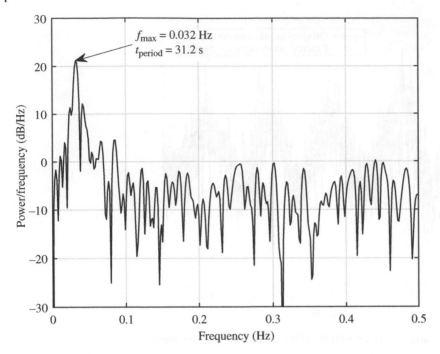

Figure 14.11 Spectral analysis of the time record shown in Figure 14.10 indicating a phugoid mode period of 31.2 seconds for the Cirrus SR20.

where ω_n is the *undamped* natural frequency and x_1 and x_2 are the amplitudes of the quantity being measured (successive peaks). Note that ω_n is not the same as ω_d, but the two are related by

$$\omega_n = \frac{\omega_d}{\sqrt{1 - \zeta^2}}. \tag{14.79}$$

If we generalize Eq. (14.78) such that the selection of peaks is not restricted to subsequent peaks in the oscillation, and substitute Eqs. (14.76) and (14.79) we have

$$\frac{1}{n-1} \ln \left(\frac{x_1}{x_n} \right) = \frac{2\pi\zeta}{\sqrt{1 - \zeta^2}}, \tag{14.80}$$

where n is the number of the peak. Equation (14.80) can then be solved for damping ratio to get

$$\zeta = \frac{\frac{1}{n-1} \ln \left(\frac{x_1}{x_n} \right)}{\sqrt{4\pi^2 + \left[\frac{1}{n-1} \ln \left(\frac{x_1}{x_n} \right) \right]^2}}. \tag{14.81}$$

Thus, the damping ratio can be found from the ratio of the amplitude of any two peaks in the oscillatory decay of the long-period response. Using the data shown in Figure 14.10, the measured damping ratio from the logarithmic decrement method is 0.0496.

Both the flight test results and the analytical predictions for the Cirrus SR20 are in good agreement. Table 14.1 shows a summary of the results from both. In both cases, the analytical predictions are slightly lower than the flight test results, but the percentage difference is relatively low.

Table 14.1 Summary of long-period dynamic longitudinal stability results.

	Flight test	Analytical prediction	Percent difference (%)
Damped natural frequency	0.2013 rad/s	0.1989 rad/s	1.2
Damping ratio	0.0496	0.0453	8.7

Nomenclature

a	lift curve slope for the aircraft
a_t	lift curve slope at tail
a_{wb}	lift curve slope at wing–body
a_x	acceleration in freestream direction
a_z	acceleration in vertical direction
a_0	two-dimensional (airfoil) lift curve slope
AR	aspect ratio of wing
$\bar{c}$	mean aerodynamic chord
C_D	drag coefficient
C_{D_0}	parasitic drag coefficient
C_{D_α}	$\partial C_D / \partial \alpha_{\mathrm{wb}}$
C_L	lift coefficient
$C_{L,t}$	lift coefficient for the tail
$C_{L,\mathrm{trim}}$	lift coefficient at trim
$C_{L,\mathrm{wb}}$	lift coefficient for the wing–body lift
C_{L_α}	$\partial C_L / \partial \alpha_{\mathrm{wb}} = a_{\mathrm{wb}}$
C_{L_δ}	$\partial C_L / \partial \delta$
C_m	pitch moment coefficient
C_m'	perturbation of pitch moment coefficient
$C_{m,\mathrm{ac}}$	pitch moment coefficient about the aerodynamic center
$C_{m,\mathrm{CG}}$	pitch moment coefficient about the center of gravity
$C_{m,L=0}$	pitch moment coefficient when no lift is produced
$C_{m,t}$	pitch moment coefficient about the center of gravity due to the tail
$C_{m,0}$	pitch moment coefficient at zero angle of attack
C_{m_q}	pitch damping, $\partial C_{m,\mathrm{CG}} / \partial q = \partial C_{m,\mathrm{CG}} / \partial \dot{\theta}$
$C_{m_q,t}$	pitch damping due to tail moment, $\partial C_{m,t} / \partial q = \partial C_{m,t} / \partial \dot{\theta}$
C_{m_α}	$\partial C_{m,\mathrm{CG}} / \partial \alpha_{\mathrm{wb}}$
C_{m_δ}	$\partial C_{m,\mathrm{CG}} / \partial \delta$
d	fuselage diameter
D	drag
D'	change in drag from steady-state value
e	Oswald efficiency factor
e_1	efficiency factor, assumed equal to e
f	factor used in estimating lift curve slope for an aircraft
f_{peak}	peak frequency
F	force
F_c	centrifugal force

g	gravitational acceleration
i_t	tail incidence angle
I_{yy}	moment of inertia about the lateral axis
K	constant
l_t	tail moment arm, distance between CG and the tail ac
ℓ	length of the aircraft
L	lift
L'	additional lift (time-varying component of lift)
L_t	lift produced by the tail
L_{wb}	lift of the wing–body combination
M	aircraft mass
m	moment
m'	change in moment from steady-state value
m_{ac}	pitch moment about the aerodynamic center
m_{CG}	pitch moment about the center of gravity
m_t	pitch moment about the center of gravity, produced by the tail
n	peak number in an oscillatory waveform
q	dynamic pressure
q	pitch rate, $\dot{\theta}$
q'	perturbation to the pitch rate
q_t	dynamic pressure at the tail
R	radius of curvature of flight path
$\overline{R}_y$	radius of gyration
s	position of the control stick (control yoke)
s_{trim}	trim position of the control stick (control yoke)
S	wing area
S_{exposed}	exposed wing area (outside the fuselage)
S_t	horizontal tail area
S_{wb}	planform area of the wing
t_{period}	period of oscillations
T	thrust
V	freestream velocity
V'	change in velocity from steady-state value
$\overline{V}_H$	horizontal tail volume coefficient
W	aircraft weight
x	distance along the freestream direction (wind axes)
$x_{1, 2, \ldots, n}$	amplitude of the 1st, 2nd, or nth peak of a quantity being measured
X_{ac}	distance from aerodynamic center to center of gravity
X_n	neutral point
y	distance along the horizontal direction (wind axes)
z	distance along the vertical direction (wind axes)
α	angle of attack
α'	change in angle of attack from steady-state value
α_t	angle of attack at tail
α_{trim}	angle of attack at trim
α_{wb}	angle of attack at wing–body
$\dot{\alpha}$	rate of change of angle of attack
$\dot{\alpha}'$	perturbation of rate of change of angle of attack

γ	angle between freestream and the horizon
γ'	change in angle from steady-state value
δ	elevator deflection angle
δ_{trim}	elevator deflection angle at trim
ε	downwash angle at tail
ε_0	downwash angle at tail, component insensitive to α
ζ	damping ratio
η_t	tail efficiency factor
θ	angle between longitudinal axis and horizontal
θ'	change in angle from steady-state value
$\dot{\theta}$	pitch rate
$\dot{\theta}'$	perturbation to pitch rate
$\ddot{\theta}$	angular acceleration about lateral axis
$\ddot{\theta}'$	perturbation to angular acceleration about lateral axis
λ	eigenvalue
ρ	density
ω	angular rate, $\dot{\gamma}$
ω_d	damped natural frequency
ω_n	natural frequency (undamped)

Subscripts

ac	aerodynamic center
CG	center of gravity
t	tail
trim	trim
wb	wing–body combination

Acronyms and Abbreviations

ac	aerodynamic center
CG	center of gravity
GA	general aviation
IIR	infinite impulse response
KIAS	knots indicated airspeed
UAVs	unmanned aerial vehicles

References

Anderson, J.D. Jr., (2012). *Introduction to Flight*, 7e. New York: McGraw-Hill.

Gracey, W. (1940). Measured Moments of Inertia for 32 Airplanes. *National Advisory Committee for Aeronautics* TN-780.

Kimberlin, R.D. (2003). *Flight Testing of Fixed-Wing Aircraft*. Reston, VA: American Institute of Aeronautics and Astronautics.

Ogata, K. (1992). *System Dynamics*, 2e. Englewood Cliffs, NJ: Prentice-Hall.

Phillips, W.F. (2004). *Mechanics of Flight*. Hoboken, NJ: Wiley.

Raymer, D.P. (2006). *Aircraft Design: A Conceptual Approach*, 4e. Reston, VA: American Institute of Aeronautics and Astronautics.

Roskam, J. (2003). *Airplane Design, Part V: Component Weight Estimation*. Lawrence, KS: Design, Analysis, and Research Corporation.

Smith, H.C. (1981). *Introduction to Aircraft Flight Test Engineering*. Basin, WY: Aviation Maintenance Publishers.

US Naval Test Pilot School (1997). *Fixed Wing Stability and Control: Theory and Flight Test Techniques*, U.S. Naval Test Pilot School Flight Test Manual, USNTPS-FTM-No. 103. Patuxent River, MD: Naval Air Warfare Center, Aircraft Division.

Williams, J.E. and Vukelich, S.R. (1979a). The USAF Stability and Control Digital DATCOM. Volume I. Users Manual. AFFDL-TR-79-3032 Volume I, DTIC accession number ADA086557.

Williams, J.E. and Vukelich, S.R. (1979b). The USAF Stability and Control Digital DATCOM. Volume II. Implementation of Datcom Methods. AFFDL-TR-79-3032 Volume II, DTIC accession number ADA086558.

Williams, J.E. and Vukelich, S.R. (1979c). "The USAF Stability and Control Digital DATCOM. Volume III. Plot Module," AFFDL-TR-79-3032 Volume III, DTIC accession number ADA086559.

15

Lateral-Directional Stability

Lateral-directional stability is different from longitudinal stability in that the stability characteristics of the roll and yaw axes are coupled, whereas the pitch (longitudinal) stability could easily be isolated. Stability about the lateral and directional axes is coupled because the kinematics of the aircraft produce coupling (*i.e.*, the cross-product of inertia, I_{xz}), aerodynamic forces and moments affect both axes simultaneously, and rudder or aileron control inputs applied in isolation will produce *both* roll and yaw moments. This inherent coupling makes the flight testing of the lateral-directional stability more complicated.

Directional stability (yaw about the vertical axis) is often referred to as weathercock stability (as in a weathervane) – a directionally stable aircraft will point the nose back into the wind when subjected to a disturbance in sideslip angle. In this way, directional stability is analogous to pitch stability, and a stable trim condition is easily identified. Lateral stability (roll about the longitudinal axis), on the other hand, does not have a stable trim condition that can be identified. There are many stable bank angle positions, making it difficult to define a single stable condition. Instead, lateral stability is identified by a wing-leveling tendency when an aircraft is subjected to a bank angle disturbance.

Dynamic lateral-directional stability deals with the temporal response of the aircraft after it is disturbed from an equilibrium condition. The dynamic response can be either first order, with an exponential response, or second order, with an oscillatory response. For the lateral-directional axes, the dominant modes are the Dutch roll (second order), roll (first order), or spiral (first order) modes. Stable modes will return to the original trim condition, while neutral dynamic stability neither converges nor diverges from the trim condition, and an unstable mode will diverge from the trim condition. The damping characteristics and frequencies of these modes determine how much pilot input is required, and how difficult the aircraft is to fly, particularly for course tracking missions or formation flying.

15.1 Static Lateral-Directional Stability

15.1.1 Theory

We'll start our discussion of static lateral-directional stability with an overview of the nomenclature for the forces, moments, angles, stability derivatives, and sign conventions (see Figures 15.1 and 15.2). For roll motion, a positive rolling moment (ℓ) induces a downward motion of the right wing (counter-clockwise rotation in the front view of Figure 15.1) and an increase in the bank angle, ϕ. Negative deflection of the ailerons, δ_a, is shown in the front view of Figure 15.1, where the right aileron (from the pilot's perspective) is up and the left aileron is deflected down. Note that a

Introduction to Flight Testing, First Edition. James W. Gregory and Tianshu Liu.
© 2021 John Wiley & Sons Ltd. Published 2021 by John Wiley & Sons Ltd.
Companion website: https://www.wiley.com/go/flighttesting

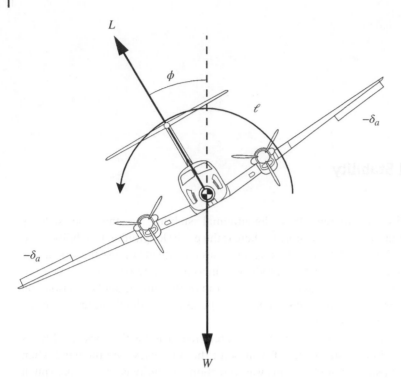

Figure 15.1 Definition of forces and moments for lateral-directional stability (front view).

negative aileron deflection leads to a positive rolling moment, as shown in Figure 15.1. The roll rate is defined as $p = \dot{\phi}$, which is the time rate of change of bank angle. For yawing motion, a positive yawing moment (n) turns the nose to the right (Figure 15.2). The angle between the freestream velocity vector and the aircraft longitudinal axis is the sideslip angle, β. Yaw rate, $r = \dot{\beta}$, is the time rate of change of sideslip angle. Side force, Y, is positive acting toward the right. Rudder deflection angle, δ_r, is defined negative when the trailing edge is deflected to the right, as shown in Figure 15.2. For a stable response to a positive sideslip angle, a side force L_V produced by the vertical tail must act in the negative y-direction (as shown in Figure 15.2), through moment arm l_V, in order to produce a positive yaw moment (n) that turns the nose back into the freestream direction.

The governing equations for lateral-directional stability are a coupled set of two moment equations (about the longitudinal and directional axes) and one force equation (in the lateral direction). The moment equations are

$$C_\ell = C_{\ell_\beta}\beta + C_{\ell_p}p + C_{\ell_r}r + C_{\ell_{\delta_a}}\delta_a + C_{\ell_{\delta_r}}\delta_r \tag{15.1}$$

and

$$C_n = C_{n_\beta}\beta + C_{n_p}p + C_{n_r}r + C_{n_{\delta_a}}\delta_a + C_{n_{\delta_r}}\delta_r, \tag{15.2}$$

where C_ℓ and C_n are the total roll and yaw moment coefficients (respectively). The derivatives C_{ℓ_β} and C_{n_β} (shorthand for $\partial C_\ell/\partial\beta$ and $\partial C_n/\partial\beta$, respectively) are the rolling and yawing moment coefficient due to sideslip, which are functions of the aircraft geometry (tail, wing, etc.) and are independent of control inputs (δ_a and δ_r). Similarly, the derivatives C_{ℓ_p} and C_{n_p} are the rolling and yawing moment coefficient due to roll rate (p), and the derivatives C_{ℓ_r} and C_{n_r} are the rolling and yawing moment coefficient due to yaw rate (r). Likewise, $C_{\ell_{\delta_a}}$ and $C_{n_{\delta_a}}$ are the roll and yaw moment

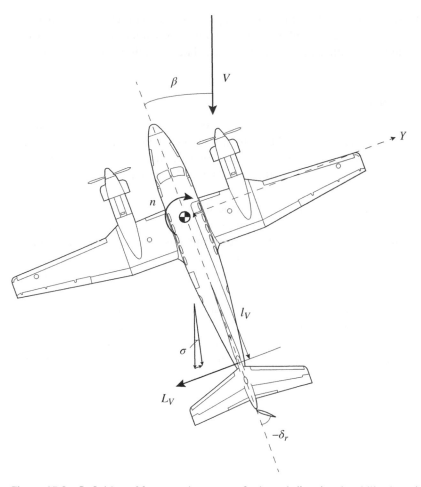

Figure 15.2   Definition of forces and moments for lateral-directional stability (top view).

coefficients due to aileron deflection, and $C_{\ell_{\delta_r}}$ and $C_{n_{\delta_r}}$ are the roll and yaw moment coefficients due to rudder deflection, which describe the moments due to the control inputs.

For our purposes here, the yaw rate and roll rate are assumed to be zero, which reduces the equations to

$$C_\ell = C_{\ell_\beta}\beta + C_{\ell_{\delta_a}}\delta_a + C_{\ell_{\delta_r}}\delta_r \tag{15.3}$$

and

$$C_n = C_{n_\beta}\beta + C_{n_{\delta_a}}\delta_a + C_{n_{\delta_r}}\delta_r. \tag{15.4}$$

Note that (15.3) and (15.4) are coupled equations, where both roll and yaw moment are impacted by both aileron and rudder inputs. Similar to the moment equations, the steady side force equation is

$$C_{Y_\beta}\beta + C_{Y_{\delta_r}}\delta_r + C_{Y_{\delta_a}}\delta_a + C_L\phi = 0, \tag{15.5}$$

where C_Y is the side force coefficient, C_L is the lift coefficient, and C_{Y_β}, $C_{Y_{\delta_r}}$, and $C_{Y_{\delta_a}}$ are corresponding derivatives.

Even though lateral-directional stability is inherently coupled, as seen in the cross-coupled terms in the above equations, we'll isolate each stability axis for the purposes of our discussion here. We'll start with directional stability ("weathercock") and move on to lateral stability ("wing leveling").

15.1.2 Directional Stability

Static stability about the directional axis requires that the nose be pointed back into the oncoming wind after the aircraft is subjected to a sideslip disturbance. The required moment (C_n) must be positive (tending to rotate the nose to the right) when the sideslip angle (β) is also positive (oncoming wind deflected to the right relative to the nose). Thus, the stability derivative C_{n_β} must be positive to ensure directional stability.

The predominant contributor to directional stability is the side force produced by the vertical tail and rudder, which acts through a moment arm between the center of gravity and the center of lift of the vertical tail. The impact of this side force, and the resulting moment, is directly analogous to the impact of the horizontal tail and elevator on longitudinal (pitch) stability. Thus, we'll develop the governing equations for directional stability in a similar manner to the development of the longitudinal stability equations (see Chapter 14).

Since the vertical tail is the predominant factor that impacts the yawing moment due to sideslip, we'll neglect any other factors such as the fuselage or engine nacelles, which tend to have a small destabilizing effect. With the effects of all other aircraft components neglected, we'll express this stability derivative, $C_{n_{\beta_V}}$, as a function of the side force coefficient and the tail moment arm:

$$C_{n_{\beta_V}} = -C_{Y_{\beta_V}} \frac{l_V}{b}, \tag{15.6}$$

where l_V is the tail moment arm and b is the wing span. The minus sign in Eq. (15.6) is due to the fact that side force (Y) is defined as positive to the right, by convention, but yaw moment is defined as positive when the nose moves to the right. The side force coefficient due to the vertical tail, $C_{Y_{\beta_V}}$, is given by

$$C_{Y_{\beta_V}} = -a_V \left(1 - \frac{\partial \sigma}{\partial \beta} \right) \eta_V \frac{S_V}{S}, \tag{15.7}$$

where the minus sign is again employed due to the fact that the side force is defined as positive to the right, but a positive sideslip angle produces a side force to the left. Here, a_V is the lift curve slope of the vertical tail, σ is the sidewash (defined as positive to the right when viewed from the top), η_V is the tail efficiency factor (the ratio of the local dynamic pressure at the tail relative to the freestream dynamic pressure), S_V is the vertical tail area, and S is the wing area. The derivative of sidewash with respect to side slip ($\partial \sigma / \partial \beta$) is due to interference with other components of the aircraft, such as the fuselage, affecting the local sideslip angle at the tail. Combining (15.6) and (15.7) results in

$$C_{n_{\beta_V}} = a_V \left(1 - \frac{\partial \sigma}{\partial \beta} \right) \eta_V \frac{S_V}{S} \frac{l_V}{b}, \tag{15.8}$$

which illustrates that the key factors affecting the directional stability derivative are the lift curve slope of the tail (a_V), the area of the tail (S_V), and the moment arm of the tail (l_V). Since each variable in Eq. (15.8) is positive with a vertical tail, then C_{n_β} is positive as long as the positive contribution to stability from the vertical tail dominates the destabilizing effects of the fuselage and nacelles on directional stability.

Returning to the equation of motion for yaw, we'll evaluate the directional stability while considering the impact of the vertical tail and fixed rudder deflection. Solving Eq. (15.4) for rudder deflection angle, we obtain

$$\delta_r = \frac{1}{C_{n_{\delta_r}}} \left(C_n - C_{n_\beta} \beta - C_{n_{\delta_a}} \delta_a \right). \tag{15.9}$$

Taking the derivative of (15.9) with respect to sideslip angle we find

$$\frac{\partial \delta_r}{\partial \beta} = -\frac{C_{n_\beta}}{C_{n_{\delta_r}}}. \tag{15.10}$$

Since the sign convention of rudder deflection is positive for trailing-edge-left, the sign of $C_{n_{\delta_r}}$ for a typical vertical tail and rudder configuration is negative (positive rudder deflection produces a negative yaw moment). And, as we saw earlier, C_{n_β} must be positive in order to have positive directional stability. Thus, the derivative of rudder angular position with respect to sideslip angle must be positive for directional stability ($\partial \delta_r / \partial \beta > 0$). In general, higher values of C_{n_β} are desired, so larger values of $\partial \delta_r / \partial \beta$ are generally favorable.

In summary, the value of C_{n_β}, which must be positive, is impacted by many aspects of the aircraft geometry and control surfaces. The fuselage and nacelles are generally destabilizing influences, so the vertical tail must be sized to produce sufficiently positive C_{n_β} in order to ensure directional stability. If inadequate directional stability is anticipated, the aircraft might be modified by adding dorsal or ventral fins to the top or bottom of the fuselage, respectively. A dorsal fin will reduce the tail aspect ratio and increase the tail stall angle to improve the directional stability, while a ventral fin will increase the vertical tail aspect ratio and steepen the tail lift curve slope (a_V). A free-floating rudder tends to decrease the directional stability, in the same way that a free-floating elevator decreases the longitudinal stability. And, one of the most significant ways that the lateral and directional stability are coupled is exemplified by the impact of the ailerons on the directional stability. For a right turn of the aircraft, the left aileron is deflected trailing-edge-down, while the right aileron is deflected trailing-edge-up, in order to produce the lift differential that results in a rolling moment to the right. But, with this lift differential, there is a difference in induced drag on either wing. The left wing, with slightly higher lift, will also have slightly higher induced drag than the right wing. This leads to a yaw moment to the left, which is opposite of the intended direction of the turn. This tendency is known as adverse yaw, which works against positive directional stability. One final flight characteristic to consider for directional stability is high flight speeds – as compressibility effects increase, the lift curve slope of the vertical tail will reduce, which in turn reduces the directional stability.

15.1.3 Lateral Stability

Roll stability is distinctly different from longitudinal or directional stability because no clear definition exists for a trimmed roll condition. The forces and moments acting on an aircraft are only indirectly dependent on the bank angle of the aircraft – so, unlike the effect of the horizontal or vertical tail, there is no part of the aircraft that produces restorative forces and moments that would return the aircraft to a stable equilibrium condition.

Instead, lateral stability can be thought of as a wing-leveling tendency. That is, if an aircraft experiences a disturbance in roll, then the aircraft should tend to roll back to a wing-level condition. The mechanism for this depends on sideslip. When an aircraft encounters a bank angle disturbance, the lift vector is tilted to one side, which induces a sideslip. If the aircraft has positive lateral stability, then a positive sideslip angle will produce a negative rolling moment, which will tend to restore the aircraft to a wings-level condition. Likewise, a negative sideslip angle should induce a positive roll moment. While positive stability is desired, only weak positive stability should be present. Otherwise, certain aircraft maneuvers such as crosswind landings (where sustained flight in sideslip is required) become more difficult. Also, if the roll-yaw coupling becomes too strong, then it is difficult for the pilot to maintain a ground track, such as tracking the instrument landing system localizer on an instrument approach, or strafing runs for ground-attack aircraft.

There are a number of geometrical features of the aircraft that will promote positive roll stability. Wing dihedral has the most significant impact – positive dihedral (an upward tilt to the wings) increases the lateral stability of the aircraft. When the aircraft is subjected to a roll disturbance (and the resulting sideslip), the wing on the side of the sideslip will have less downwash than the opposing wing, giving a higher angle of attack on the wing facing the sideslip. Thus, the difference in angle of attack on the two wings will create a lift differential that rolls the aircraft back to a wings-level condition. Wing sweep also has a stabilizing effect for lateral stability. A swept wing faces more directly into the wind when the aircraft is in a sideslip, which causes an effective reduction in sweep on that side and an increase in lift that creates a restorative moment. The stabilizing effect of wing sweep actually increases as the angle of attack increases, leading to strong roll-yaw coupling at high angles of attack for highly swept wings. The location of the wing on the fuselage will also have an effect, with high-mounted wings possessing positive lateral stability, and low-mounted straight wings having negative lateral stability. (Thus, low-wing aircraft need some level of dihedral in order to maintain lateral stability.) Finally, a vertical tail positioned above the longitudinal axis of the aircraft will induce positive lateral stability due to the location of the net force produced by the tail above the center of gravity.

15.1.4 Flight Testing Procedures

The main purpose of lateral-directional stability flight testing is to verify the stability of the aircraft and the degree of roll-yaw coupling. Flight tests are conducted by piloting a series of steady-heading sideslips and measuring various control inputs. Starting with a completely trimmed aircraft, the pilot establishes a steady heading sideslip using rudder in one direction, and data are recorded. At each test condition, the bank angle, rudder deflection angle, rudder stick force, aileron deflection angle, aileron stick force, and sideslip angle are all recorded. In some situations, it may not be feasible to record stick forces, so these can be neglected if necessary. Bank angle is measured with an inclinometer or can be inferred from measurements of the gravitational acceleration vector relative to the body axes of the aircraft. (The data acquisition unit should be fixed and aligned with the aircraft. A bank angle will shift a portion of the measured gravitational acceleration from the vertical axis to the lateral component of a three-axis accelerometer. Basic vector addition can be used to find the resultant gravity vector and its angle from the vertical, which then is the bank angle.) Aileron deflection angles may be found by measuring stick position using a ruler and calibrating the relationship between stick position and aileron surface deflection via pre-flight measurements. Aileron deflection angle can be measured by recording the vertical displacement of the aileron trailing edge, and using trigonometry with the control surface dimensions to find the deflection angle. Rudder deflection angle may be found by calibrating rudder pedal input to the angular deflection of the rudder. This ground-based calibration is done in the same way that the aileron calibration is done. Rudder input is inherently an imprecise measurement, since it is cumbersome to physically reach the rudder pedals with a ruler; instead, the flight test engineer must rely on the pilot's ability to reliably and repeatably provide the desired input. (This should be practiced several times before flight.) Finally, the sideslip angle is measured in flight for each steady-heading sideslip. The amount of sideslip can be found by recording the difference between aircraft heading (read from the directional gyro) and the aircraft course (recorded from the GPS ground track). This relationship inherently assumes that there is no wind altering the aircraft's course. If winds aloft are nonzero, then the maneuver can be flown on opposite headings in order to cancel the effect. Alternatively, the winds aloft can be directly measured through cloverleaf flight tests (see Chapter 8), or by relying on the wind vector reported by modern avionics on the aircraft. A series of steady-heading sideslips should be flown, corresponding to rudder inputs of $\pm 25\%$, $\pm 50\%$, $\pm 75\%$, and $\pm 100\%$.

Data reduction involves plotting the various parameters – bank angle, rudder deflection angle, and aileron deflection angle as a function of sideslip angle. (The relationship between rudder input and measured sideslip angle effectively serves as a calibration between the two, allowing all other parameters to be plotted as a function of sideslip angle.)

15.1.5 Flight Testing Example: Cirrus SR20

Sample results from a flight test in a Cirrus SR20 are shown in Figures 15.3–15.6. First, the calibration relationship between rudder or aileron input and the corresponding control surface deflection is shown in Figure 15.3. There is a reasonably linear fit for these relationships, allowing the test engineer to make a direct measurement of stick or rudder pedal deflection as a proxy for control surface deflection.

Figure 15.4 shows that the slope of the $\partial \delta_r / \partial \beta$ curve is positive, which is the criterion that must be satisfied for Eq. (15.10) to hold true, since $C_{n_{\delta_r}}$ is negative and we want C_{n_β} to be positive for stability. Note that the flight test community often uses a sign convention for rudder deflection that is opposite of what is commonly used in general aerospace engineering. The general definition of rudder deflection follows the convention where a positive deflection is trailing-edge-left, which induces a positive sideslip. However, the flight test community will often use the opposite convention, which is positive trailing-edge-right, which corresponds to the rudder deflection needed to create a positive yawing moment. Thus, some presentations of lateral-directional stability results (*e.g.*, Kimberlin 2003) will have a negative slope for $\partial \delta_r / \partial \beta$ due to the flight test definition of rudder

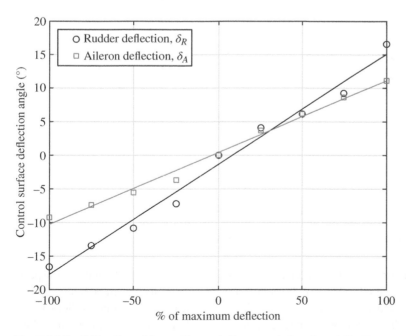

Figure 15.3 Calibration of the rudder and aileron control surface deflections.

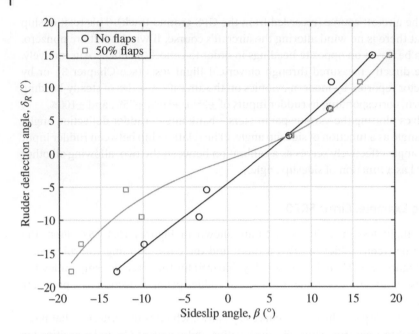

Figure 15.4 Static directional stability characteristics from steady heading sideslip data.

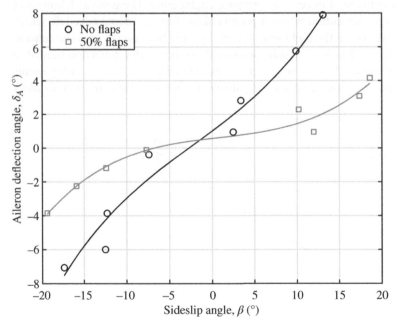

Figure 15.5 Static lateral stability characteristics from steady heading sideslip data.

deflection angle. For consistency with the standard sign convention generally adopted in aerospace engineering, the trailing-edge-left positive definition is used for all results presented here.

For each of the data sets in Figures 15.4–15.6, the data are typically not symmetric (left-right), and there is nonlinearity in the response. These phenomena are largely due to separation of flow over the fuselage while the aircraft is in a sideslip. Aircraft with a "boxy" fuselage (such as the Piper

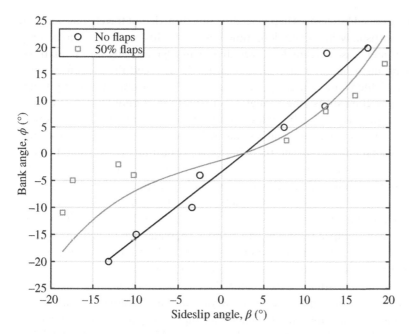

Figure 15.6 Side force characteristics from steady heading sideslip data.

Arrow) will tend to have more significant flow separation than what occurs over the Cirrus SR20 shown here. The left–right asymmetry of these plots is due to the effects of the propeller slipstream causing asymmetric flow over the fuselage. In all cases, deployment of the flaps had a significant impact on the slope, nonlinearity, and asymmetry of the data. This behavior is not unexpected due to the increased complexity of the flow and potential for separation of flow over the flaps while the aircraft is in a sideslip.

15.2   Dynamic Lateral-Directional Stability

15.2.1   Theory

The lateral-directional dynamic response characteristics of an aircraft are primarily composed of three characteristic modes: roll, spiral, and Dutch roll. A full theoretical development of these dynamic modes is beyond the scope of this book, but the interested reader can consult standard texts such as Yechout et al. (2003), Phillips (2004), or Nelson (1998). However, we'll briefly discuss the main characteristics of each mode and focus our attention for flight testing on the Dutch roll mode since it often has the largest bearing on a pilot's opinion of an aircraft's handling qualities.

The roll mode is typically a heavily damped, convergent, first-order response to an aileron input. This mode can be excited by inputting a step change in aileron deflection and observing the roll rate of the aircraft. The roll rate quickly converges to a steady-state value, where the aircraft will increase bank angle at a constant rate. This mode of the aircraft's dynamic response largely influences a pilots' opinion of the roll responsiveness of the aircraft for maneuvering flight, and a pilot's work-load in maintaining level flight in turbulent conditions. The key parameters of interest to a test pilot are the time constant of the roll mode, the steady-state roll rate that can be achieved with varying control inputs, and the amount of adverse yaw generated by the rolling maneuver (USN TPS 1997).

The spiral mode is characterized by changes in the heading and direction of travel of the aircraft, has first-order behavior, and usually has a very long time constant. This mode tends to be only slightly convergent, but in many cases the mode is even divergent (dynamically unstable). If the aircraft encounters a lateral disturbance, a divergent spiral mode will lead the aircraft to enter a spiraling flight path that tightens into a high-speed spiral dive. This divergent property is not a concern as long as the time constant is long enough for a well-trained pilot to correct for the disturbance. However, the spiral mode often becomes a challenge for less experienced pilots who inadvertently fly into a cloud. For flight in clouds, visual reference to the horizon is lost, and the pilot no longer has a visual cue to indicate the wings-level condition. Thus, some form of disturbance is inevitably encountered, and the aircraft will tend to enter a spiral dive. The most significant characteristics of the spiral mode are the nature of the mode (convergent, divergent, or neutral), and the time required for the amplitude of the motion to double or halve (USN TPS 1997).

The Dutch roll mode is the short-period mode for lateral-directional dynamic stability, and typically a second-order, heavily damped response. The mode is characterized by coupled yawing and rolling motions that are out of phase with one another, which can be quite uncomfortable for the pilots and passengers. The name of the mode dates back at least a century, when the similarity of the mode with the rolling and twisting motion an ice skater makes was noted by Hunsaker (1916). The Dutch roll emerges due to coupling between the yaw and roll dynamic response of the aircraft – the stabilizing dihedral effect for roll (augmented for high sweep and/or high dihedral) tends to be more strongly stable than the directional stability provided by the vertical tail. Higher lateral static stability tends to exacerbate the Dutch roll mode due to roll-yaw coupling. The Dutch roll is characterized by a 90° phase difference between the roll and yaw dynamic response, making it feel as though the aircraft is wagging its tail, causing an unpleasant ride for the passengers. Critical parameters that influence how objectionable the Dutch roll mode is are the ratio of the roll-to-yaw motion, expressed as ϕ/β, and the amount of damping present. A high roll-to-yaw ratio that is lightly damped will be highly objectionable. The ϕ/β ratio can be conceptualized by considering an observer who is in an aircraft adjacent to the test aircraft, and viewing the position light of the test aircraft while the Dutch roll mode of the test aircraft is excited. A mode with low roll-to-yaw ratio (less than 1) will trace out an ellipse with a major axis that is horizontal, since the yaw dominates the motion. On the other hand, a Dutch roll mode with high roll-to-yaw ratio (greater than 1) will trace out an ellipse with a vertical major axis. If the roll and yaw motions are approximately equal in magnitude ($\phi/\beta \approx 1$), then the wing tip light will trace out a pattern that is close to circular. Again, high values of the roll-to-yaw ratio are most objectionable. Yechout (2003) provides an estimate for the roll-to-yaw ratio as:

$$\left|\frac{\phi}{\beta}\right| \approx \frac{C_{\ell_\beta}}{C_{n_\beta}} \frac{I_{zz}}{I_{xx}} \frac{1}{\rho V}, \tag{15.11}$$

where C_{ℓ_β} is the lateral static stability (dihedral effect) and C_{n_β} is the directional static stability.

Typical values for the period of the Dutch roll are 3 to 15 seconds (Corda 2017), which is short enough that it is difficult for the pilot to compensate in real time. In fact, if a pilot attempts to actively correct for a short-period dynamic instability such as this one, the phasing of the input is delayed due to human limitations of sensing and response. This out-of-phase input often ends up exacerbating the instability, resulting in a common problem referred to as pilot-induced oscillation (PIO).

An accurate prediction of the Dutch roll characteristics should involve analysis of the fully coupled, three degrees of freedom (3-DOF) equations of motion. However, various simplifications can be made to reduce the analysis to a 2-DOF system, where first-order approximations of the Dutch roll characteristics may be made (*e.g.*, see Philips (2004), Yechout (2003), Nelson (1998), or

USN TPS (1997)). The full analysis of these approaches is beyond the scope of this book, but we'll provide final results for two approaches.

Most Dutch roll mode approximations begin with the assumption that the roll-to-yaw amplitude ratio is small. The linearized 3-DOF equations of motion are reduced to 2-DOF expressions by neglecting the roll equation by assuming $\phi/\beta \ll 1$, *i.e.*, that the Dutch roll mode is dominated by sideslip and yaw. In practice, this is a dubious approximation since the Dutch roll mode is inherently a coupling of roll and yaw. However, the resulting approximation provides useful insight into the physical properties that influence the Dutch roll mode and is much easier to analyze than a full simulation from the 3-DOF system of equations. Based on these approximations, Yechout (2003) provides the following approximations for the natural frequency and damping ratio of the Dutch roll mode:

$$\omega_{n_{\mathrm{DR}}} \approx \sqrt{n_\beta + \frac{1}{V}(Y_\beta n_r - n_\beta Y_r)}. \tag{15.12}$$

$$\zeta_{\mathrm{DR}} \approx \frac{-(n_r + Y_\beta/V_\infty)}{2\omega_{n_{\mathrm{DR}}}}. \tag{15.13}$$

Here, the lateral-directional stability parameters n_β (yaw moment due to sideslip), Y_β (side force due to sideslip), n_r (yaw moment due to yaw rate), and Y_r (side force due to yaw rate) are defined as follows:

$$n_\beta = \frac{qSb}{I_{zz}}C_{n_\beta}$$

$$Y_\beta = \frac{qS}{(W/g)}C_{Y_\beta}$$

$$n_r = \frac{qSb^2}{2I_{zz}V}C_{n_r}$$

$$Y_r = \frac{qSb}{2(W/g)V}C_{Y_r}. \tag{15.14}$$

where V and q are the time-averaged values of freestream velocity and dynamic pressure, S is the wing area, b is the wing span, W is the aircraft weight, g is gravitational acceleration, and I_{zz} is the moment of inertia about the vertical (yaw) axis. Note that this approximation for the Dutch roll mode is inherently limited, since the contribution of roll is entirely eliminated from the physics – in other words, we've assumed infinite roll damping for these approximations. For a much improved approximation of the Dutch roll mode, Phillips (2000) retained the roll contribution and derived approximations with much better fidelity and only a modest increase in the complexity of the resulting approximation (see Phillips (2000) or Phillips (2004) for details).

For purposes of quick assessment, we can further simplify the approximations given in (15.12) and (15.13). Considering typical values of stability derivatives for general aviation aircraft (*e.g.*, Phillips 2000, Nelson 1998, or the values tabulated in Appendix C), we can neglect the second term on the right-hand side of (15.12) and assume that Y_β/V is small relative to n_r in Eq. (15.13). Then, substituting (15.14) into (15.12) and (15.13), we obtain the following simplified (albeit somewhat crude) approximations (Yechout (2003), Kimberlin (2003), or USN TPS (1997)):

$$\omega_{n_{\mathrm{DR}}} \approx V\sqrt{\frac{\rho Sb}{2I_{zz}}C_{n_\beta}}. \tag{15.15}$$

and

$$\zeta_{\mathrm{DR}} \approx -C_{n_r}\sqrt{\frac{1}{32}\frac{\rho Sb^3}{I_{zz}C_{n_\beta}}}. \tag{15.16}$$

Here, we see that the directional stability (C_{n_β}) and yaw damping (C_{n_r}) have a significant role in determining the Dutch roll characteristics. It's interesting to note that the frequency of the Dutch roll mode scales linearly with the airspeed, but that the Dutch roll damping is independent of airspeed. Both the frequency and the damping of the Dutch roll mode decrease with altitude (decreasing density).

The dynamic stability of aircraft with objectionable Dutch roll characteristics can be improved most readily by increasing the yaw damping term (C_{n_r}), which is commonly done through the addition of an active-feedback yaw damper. The yaw damper takes input from the yaw rate gyros and provides a direct input to the rudder to counteract the yaw instability. Yaw dampers are fairly common on aircraft; in fact, many large commercial aircraft rely upon a yaw damper to maintain adequate Dutch roll characteristics, since the wing sweep leads to high lateral stability and exacerbates the Dutch roll mode.

Determination of the set of stability derivatives for an aircraft is the subject of extensive analysis done in the design phase of an aircraft. This can be done using the USAF Stability and Control DATCOM software (Williams and Vukelich 1979a, 1979b, 1979c), if the geometry and moments of inertia of the aircraft are known. Otherwise, one can refer to values of the stability derivatives of a typical general aviation aircraft, which are available from Teper (1969) for the Navion or a generic general aviation aircraft in Appendix C (from Example 8.2.1 of Phillips 2004).

15.2.2 Flight Testing Procedures

For flight testing, the critical parameters that must be characterized for the Dutch roll mode are the damping, the frequency or period of the mode, and the ratio of the roll-to-yaw motion (ϕ/β). It's important to begin the flight test maneuver with the aircraft in a trimmed condition – take time to carefully establish trimmed and equilibrium flight (the "trim shot"). The Dutch roll mode is most cleanly excited by pulsing the rudder in what is known as a "doublet." The rudder pedals are alternately depressed by a sufficient amount to induce some sideslip (but not too much!), and at a frequency that approximately corresponds with the expected natural frequency of the mode (such that the natural instability is reinforced). An example of a typical doublet actuation schedule is shown in Figure 15.7, with a frequency of $2.24\,\mathrm{s}^{-1}$ ($T_2 = 1.4\,\mathrm{s}$). Several doublets can be performed in succession, if needed to sufficiently excite the Dutch roll mode.

Most test pilots will use equal duration and amplitude on the rudder pedals for a doublet, but in some cases it may be beneficial to use the "3-2-1-1" input as described by Morelli (1998) or Jategaonkar (2015) in order to excite a broader range of frequencies. The time history of rudder deflection for the 3-2-1-1 input is shown in Figure 15.8, where the half-period of the second pulse is set equal to the expected natural frequency, $T_2 = 1/(2\omega_{n_{DR}})$. The half-period of the longest (first) pulse is $T_3 = 1.5\,T_2$ and the half-period of the third and fourth (shortest) pulses is $T_1 = 0.5\,T_2$.

Once the excitation (either doublet or 3-2-1-1) is applied, the pilot should terminate the excitation maneuver such that the wings are level and the spiral mode is not excited. Following the conclusion of the input, the controls should be released or held in a neutral condition. Sufficient time should elapse for the natural mode to decay and a number of oscillations can be recorded.

15.2.3 Flight Test Example: Cirrus SR20

Manual recording of the flight data by pencil and paper may be attempted, but digital data acquisition techniques will provide a much higher fidelity data record for analysis. Sample data from a Cirrus SR20 flight are shown in Figure 15.9, which is a plot of the y-axis acceleration measured by a flight data recorder. The methods employed in the analysis of dynamic longitudinal

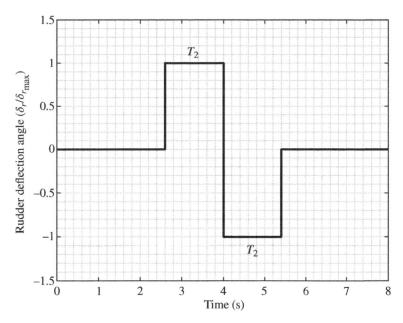

Figure 15.7 Doublet rudder input to excite lateral-directional dynamic response.

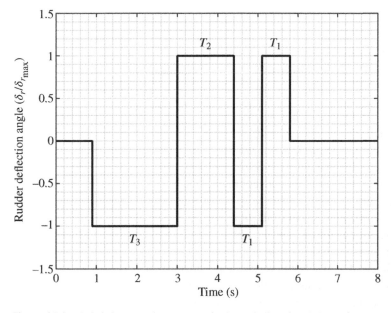

Figure 15.8 3-2-1-1 rudder input to excite lateral-directional dynamic response.

stability data may be used here (refer to the peak identification and log decrement method described in Chapter 14). For this data set, the measured natural frequency is $\omega_{n_{DR}} = 2.3\,\mathrm{s}^{-1}$ and the damping is $\zeta_{DR} = 0.088$. Based on the data from the rate gyros, the magnetometers, and the accelerometers, the amplitude of the roll motion (ϕ) and of the yaw motion (β) can be determined and the roll-to-yaw ratio found.

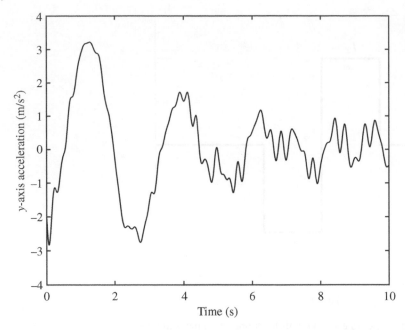

Figure 15.9 *Y*-axis acceleration (low-pass filtered at 4 Hz) for a SR20 in Dutch roll.

Nomenclature

a_V	lift curve slope of the vertical tail
b	wing span
C_L	lift coefficient
C_ℓ	roll moment coefficient
C_{ℓ_p}	roll moment coefficient due to roll rate, $\partial C_\ell / \partial p$
C_{ℓ_r}	roll moment coefficient due to yaw rate, $\partial C_\ell / \partial r$
C_{ℓ_β}	roll moment coefficient due to sideslip, $\partial C_\ell / \partial \beta$
$C_{\ell_{\delta_a}}$	roll moment coefficient due to aileron deflection, $\partial C_\ell / \partial \delta_a$
$C_{\ell_{\delta_r}}$	roll moment coefficient due to rudder deflection, $\partial C_\ell / \partial \delta_r$
C_n	yaw moment coefficient
C_{n_p}	yaw moment coefficient due to roll rate, $\partial C_n / \partial p$
C_{n_r}	yaw moment coefficient due to yaw rate, $\partial C_n / \partial r$
C_{n_β}	yaw moment coefficient due to sideslip, $\partial C_n / \partial \beta$
$C_{n_{\beta_V}}$	yaw moment coefficient due to sideslip, vertical tail contribution
$C_{n_{\delta_a}}$	yaw moment coefficient due to aileron deflection, $\partial C_n / \partial \delta_a$
$C_{n_{\delta_r}}$	yaw moment coefficient due to rudder deflection, $\partial C_n / \partial \delta_r$
C_Y	side force coefficient
C_{Y_r}	side force coefficient due to yaw rate, $\partial C_Y / \partial r$
C_{Y_β}	side force coefficient due to sideslip, $\partial C_Y / \partial \beta$
$C_{Y_{\beta_V}}$	side force coefficient due to sideslip, vertical tail contribution
$C_{Y_{\delta_a}}$	side force coefficient due to aileron deflection, $\partial C_Y / \partial \delta_a$
$C_{Y_{\delta_r}}$	side force coefficient due to rudder deflection, $\partial C_Y / \partial \delta_r$
g	gravitational acceleration
I_{xx}	moment of inertia about x-axis when rotated about x-axis

I_{xz}	moment of inertia about z-axis when rotated about x-axis
I_{zz}	moment of inertia about z-axis when rotated about z-axis
ℓ	roll moment
l_V	moment arm between CG and the vertical tail ac
L_V	side force on vertical tail
n	yaw moment
n_r	yaw moment due to yaw rate
n_β	yaw moment due to sideslip
p	roll rate, $\dot{\phi}$
q	freestream dynamic pressure
r	yaw rate, $\dot{\beta}$
S	wing area
S_V	vertical tail planform area
T_1	half-period of the third and fourth pulses of a 3-2-1-1 input
T_2	half-period of a doublet, or the second pulse of a 3-2-1-1 input
T_3	half-period of the first pulse of a 3-2-1-1 input
V	freestream velocity
W	aircraft weight
x	direction along the longitudinal axis of the aircraft (body axes)
y	direction along the lateral axis of the aircraft (body axes)
Y	side force
Y_r	side force due to yaw rate
Y_β	side force due to sideslip
z	direction along the vertical axis of the aircraft (body axes)
β	sideslip angle
δ_a	aileron deflection angle
δ_r	rudder deflection angle
ζ_{DR}	damping for the Dutch roll mode
η_V	vertical tail efficiency factor
ρ	density of air
σ	sidewash angle
ϕ	bank angle
$\omega_{n_{DR}}$	undamped natural frequency for the Dutch roll mode

Acronyms and Abbreviations

ac	aerodynamic center
CG	center of gravity
DOF	degrees of freedom
PIO	pilot induced oscillation

References

Corda, S. (2017). *Introduction to Aerospace Engineering with a Flight Test Perspective*. Chichester, West Sussex, UK: Wiley.

Hunsaker, J.C. (1916). Dynamical stability of aeroplanes. *Proceedings of the National Academy of Sciences of the United States of America* 2 (5): 278–283.

Jategaonkar, R.V. (2015). *Flight Vehicle System Identification: A Time-Domain Methodology*, 2e, Chapter 2. Reston, VA: American Institute of Aeronautics and Astronautics.

Kimberlin, R.D. (2003). *Flight Testing of Fixed-Wing Aircraft*. Reston, VA: American Institute of Aeronautics and Astronautics.

Nelson, R.C. (1998). *Flight Stability and Automatic Control*, 2e. New York: McGraw-Hill.

Morelli, E.A. (1998). In-Flight System Identification," AIAA-98-4261, 23rd AIAA Atmospheric Flight Mechanics Conference, Boston, MA. http://hdl.handle.net/2060/20040090518

Phillips, W.F. (2000). Improved Closed-Form Approximation for Dutch Roll. *Journal of Aircraft* 37 (3): 484–490.

Phillips, W.F. (2004). *Mechanics of Flight*. Hoboken, NJ: Wiley.

Teper, G. (1969). "Aircraft Stability and Control Data," NASA-CR-96008, STI-TR-176-1, Section X (pp. 99–103). http://hdl.handle.net/2060/19690022405.

U.S. Naval Test Pilot School (1997). *Fixed Wing Stability and Control: Theory and Flight Test Techniques*, USNTPS-FTM-No. 103. Patuxent River, MD: Naval Air Warfare Center.

Williams, J. E. and Vukelich, S. R., 1979a, "The USAF Stability and Control Digital DATCOM. Volume I. Users Manual," AFFDL-TR-79-3032 Volume I, DTIC accession number ADA086557.

Williams, J. E. and Vukelich, S. R., 1979b, "The USAF Stability and Control Digital DATCOM. Volume II. Implementation of Datcom Methods," AFFDL-TR-79-3032 Volume II, DTIC accession number ADA086558.

Williams, J. E. and Vukelich, S. R., 1979c, "The USAF Stability and Control Digital DATCOM. Volume III. Plot Module," AFFDL-TR-79-3032 Volume III, DTIC accession number ADA086559.

Yechout, T.R., Morris, S.L., Bossert, D.E., and Hallgren, W.F. (2003). *Introduction to Aircraft Flight Mechanics: Performance, Static Stability, Dynamic Stability, and Classical Feedback Control*. Reston, VA: American Institute of Aeronautics and Astronautics.

16

UAV Flight Testing[1]

Flight testing of unmanned aerial vehicles (UAVs), or drones as they are colloquially known, is a fascinating, multifaceted, and rapidly changing contemporary topic. With the explosion of interest in drones, there is a surging demand for flight testing guidelines for assessing vehicle and system performance. This chapter addresses the critical features of these flight test programs and maps out possible future developments in this dynamic field.

16.1 Overview of Unmanned Aircraft

First, it is helpful for us to clearly define what a drone or UAV is. Many specialists have taken exception to the term "drone" since it conjures up specific negative mental images – such as Predator MQ-9 military drones striking a target – which are not representative of the full breadth of technologies, systems, and uses for these unmanned aircraft. However, we think that the term "drone" will have permanence in the aerospace lexicology, perhaps most significantly due to the monosyllabic simplicity of the word. (UAV is too unwieldy to say, and many in the public do not know what a UAV is.) However, both "UAV" and "drone" are still lacking in their comprehensiveness, since both terms refer to only the vehicle. However, the full system – including the vehicle, ground control station (GCS), control link antennas, and ground support infrastructure – is critical for the safe and successful deployment of an unmanned aircraft. Thus, "unmanned aircraft system," or UAS, is a common term that encompasses not only the aircraft but also all of the required supporting infrastructure. Many people prefer this term, since it encompasses all of the hardware, software, and people involved with successful flight operations – whether those assets are on the ground or in the air. Other terms that are used to describe unmanned aircraft include "remotely piloted aircraft" (RPA) and "remotely piloted aircraft system" (RPAS), in order to denote that a pilot is very much engaged with safe flight of the vehicle, albeit not onboard. In this text, we will most commonly and interchangeably use the terms "drone," "UAS," or "UAV" – and we will consider them to be roughly synonymous.

UAVs can vary in size from extremely small ("micro scale") aircraft that can fit in the palm of your hand (Figure 16.1), all the way up to some of the largest aircraft built (e.g., the Global Hawk has a wingspan larger than that of a Boeing 737!) as shown in Figure 16.2. Finally, the adjective "small" is often used as a prefix to UAV, UAS, or RPA and has a specific definition. Small unmanned aircraft systems (sUASs) are defined by the U.S. Federal Aviation Administration (FAA) as those having a maximum take-off weight of <55 lb. The U.S. Department of Defense uses a classification system

---

1 This chapter was co-authored with Dr. Matthew H. McCrink, The Ohio State University.

Introduction to Flight Testing, First Edition. James W. Gregory and Tianshu Liu.
© 2021 John Wiley & Sons Ltd. Published 2021 by John Wiley & Sons Ltd.
Companion website: https://www.wiley.com/go/flighttesting

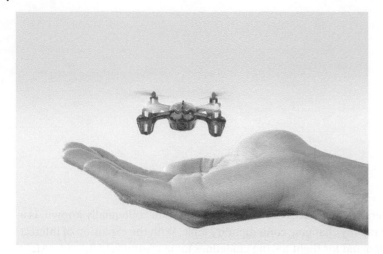

Figure 16.1 Example of a palm-sized quad-rotor drone. Source: Christine Daniloff, MIT.

Figure 16.2 An RQ-4 Global Hawk drone. Source: U.S. Air Force.

with five groups based on vehicle weight, operating altitude, and speed. These groups are detailed in Table 16.1 (UAS Task Force Airspace Integration Integrated Product Team 2011).

There is a wide and ever-increasing array of use cases for drones or UAVs. Current or envisioned missions include providing persistent communications or internet connectivity over vast remote regions; inspection of linear infrastructure such as power lines, pipelines, or railroad tracks; inspection of hazardous or difficult-to-access infrastructure such as bridges or gas plume stacks; various forms of package delivery, including medical supplies; search and rescue; and many, many other applications. In short, any mission that is "dull, dirty, or dangerous" is an ideal candidate for an unmanned aircraft instead of a manned aircraft – with obvious cost savings and possible improvements in system safety and efficiency. With the rapid evolution of the field of UAVs, it has been difficult for textbooks to keep pace with the developments in the field. As of the writing of this text, other relevant books on UAV design and operation include Beard and McLain (2012), Fahlstrom and Gleason (2012), Gundlach (2014), Marshall et al. (2016), Atkins et al. (2016), and Marqués and

Table 16.1 U.S. Department of Defense UAS classifications.

UAS group	Maximum weight (lb)	Nominal operating altitude (ft)	Speed (kts)
Group 1	0–20	<1200 AGL	100
Group 2	21–55	<3500 AGL	<250
Group 3	<1320	<FL 180	<250
Group 4	>1320	<FL 180	Any
Group 5	>1320	>FL 180	Any

Source: UAS Task Force Airspace Integration Integrated Product Team (2011).

Da Ronch (2017). These references provide insight into the design principles and operation of UAVs under various scenarios.

The explosion of interest and capability in UAVs can be traced to the evolution of the same technology that drove the advancement of mobile computing and communications (*e.g.*, smartphones). As we saw earlier in this book (Chapter 3), the sensors used on smartphones are the same sensors needed for maneuvering and navigation of unmanned aircraft. The critical advancement was the creation of small, lightweight, and low-power sensor systems that could be integrated into even the smallest of unmanned aircraft. These sensors include three-axis rate gyroscopes, three-axis accelerometers, three-axis magnetometers, and a global navigation satellite system (GNSS) (*e.g.*, Global Positioning System (GPS)) receiver. The low cost and widespread availability of these fundamental sensors is the key enabling feature for the proliferation of drones.

At the outset, given the context of the rest of this textbook, we must consider what it is that makes flight testing of UAVs different from traditional manned aircraft flight testing. Most significantly (and obviously): there is no pilot on board the aircraft. There are clear safety implications resulting from the absence of a human pilot on board: in some ways, personal safety is enhanced, since the risk of injury or death to the pilot drops to near zero. However, removing the pilot from the aircraft also elevates the risk, since UAS pilots can no longer rely on their senses to detect and interpret anomalous flight behavior. Even with all of the modern sensors available for flight testing, the pilot's presence in the cockpit provides critical input to safety decisions. A clear example of the role of the pilot is in flutter flight testing – the pilot senses vibrations and hears the sound emitted from aeroelastic phenomena and combines this sensory input with information from the cockpit displays to provide input to decision-making on the safety of flight. Thus, while the risk posed to the pilot is dramatically reduced for UAV flight testing, the risk to the flight vehicle is elevated.

The second most significant difference between flight testing of manned and unmanned aircraft is that there must be extra systems carried on board the UAV in order to offset the lack of a pilot on board. The two most significant additional systems are a control link between the pilot and the aircraft, and a system for detecting and avoiding collisions with other aircraft or ground-based obstacles. Each of these systems is essential for the safety of UAS flight.

16.2 UAV Design Principles and Features

We will now discuss a few of the basic principles of UAV design, and what makes unmanned aircraft different from manned aircraft. We will first consider the different configurations of the most common flight vehicles in operation today. We will then dwell on the overall system architecture of unmanned aircraft, followed by discussions of the various unique systems on board. Systems and architectures found on unmanned aircraft include[2] (but are not limited to) electric propulsion, the control and communications link, and advanced levels of automation and autonomy.

16.2.1 Types of Airframes

Unmanned aircraft can take on the form of traditional fixed-wing aircraft such as the turbofan-powered Global Hawk shown in Figure 16.2. The configuration and operation of these aircraft is very similar to traditional manned aircraft, but without a traditional cockpit. Pilotless aircraft can also be vertical take-off and landing (VTOL) vehicles. These may be traditional helicopters such as the MQ-8B Fire Scout (Figure 16.3), which has a single main rotor and a tail rotor to counter the reaction torque exerted by the main rotor. Or, much more commonly, VTOL drones take the form of quadcopters, such as the drone depicted in Figure 16.4. These multi-rotor vehicles achieve flight control by independently adjusting the rotational speed of each of the rotors, in order to control the net forces and moments acting on the flight vehicle. For example, to transition a quadcopter into forward flight, the revolutions per minute (RPM) of the front two rotors is lowered, while the RPM of the rear rotors is increased. This produces more thrust on the rear rotors, causing the vehicle to tilt forward and orient a component of the thrust vector in the desired direction of flight. Multi-rotor drones are not limited to just the quadcopter configuration. Vehicles with six or even eight rotors, such as Ohio State University's Octavian (Figure 16.5), are common. These multi-rotor vehicles with more rotors can often carry heavier payloads and are more robust toward in-flight failure of a single rotor. Finally, hybrid configurations such as NASA's Greased Lightning (Figure 16.6) are also becoming popular. These vehicles may not only take off and land vertically but can also transition into fixed wing flight for higher efficiency in cruise. These vehicles may have an array of multiple rotors, such as the distributed electric propulsion configuration depicted in Figure 16.6, where the entire main wing assembly pivots upward for take-off and landing, or rotates down for cruise conditions. One interesting challenge of hybrid aircraft is the complex control laws that must be developed to handle the transitions between

Figure 16.3 The MQ-8B Fire Scout, an example of an unmanned aircraft with a traditional rotorcraft configuration with a main rotor and a tail-rotor. Source: U.S. Navy.

---

2 One other key feature found on advanced UAVs is a detect-and-avoid (DAA) system. DAA systems are beyond the current scope of this book; however, a brief overview of DAA is provided as an online supplement 5.

Figure 16.4 A typical quadrotor drone, the most common configuration of a UAV. Source: Photo by Josh Sorenson on Unsplash, https://unsplash.com/photos/ouuigywbXlI.

Figure 16.5 The OSU Octavian, an eight-rotor VTOL drone carrying a mast-mounted sonic anemometer for local wind measurements. Source: Courtesy of Holly Henley, The Ohio State University.

vertical flight and cruise. These control laws are often determined and refined in flight testing due to the difficult-to-predict interaction effects between multiple rotors, the wing, and the fuselage.

16.2.2 UAV System Architecture

In order to understand the most critical aspects of drone flight testing, we have to develop an understanding of the typical system architecture of small UAS. The ground-based pilot interfaces with the GCS, which provides a display of flight-critical data. This display often includes the same information displayed to a pilot via the primary flight display (PFD) and multifunction display (MFD) on a manned aircraft with glass panel avionics. A typical GCS display, based on the open source QGroundControl package, is shown in Figure 16.7. Flight-critical information includes airspeed, altitude, heading, and attitude information via an artificial horizon and turn/bank coordinator. In addition, the lateral and vertical position of the aircraft is conveyed on a map of the area, with

Figure 16.6 The NASA "Greased Lightning" GL-10 prototype UAV, an example of a tilt-wing vehicle with distributed electric propulsion. Source: NASA Langley Research Center/David C. Bowman.

a tag for the aircraft displaying altitude. Other traffic in the same area is often displayed on the same map, providing the operator with situational awareness not unlike that provided to an air traffic controller. The UAS operator provides control inputs to the vehicle through either waypoint programming, or via direct control input (*e.g.*, through joystick control). This categorization of control input is similar to the differences between programming avionics on a manned aircraft, versus direct stick and rudder control inputs.

The control inputs provided by the pilot are uplinked to the vehicle through the command and control (C2) link. The C2 link is physically established by communications radio(s) and antennas on both the GCS and on board the aircraft. Control information is fed up to the vehicle by the C2 link, and received telemetry data from the vehicle are passed through the C2 link to the GCS for monitoring. This two-way communication of data is critical for safe and robust functioning of the flight vehicle. The presence of a C2 link is one of the most significant differences between manned and unmanned aircraft and is part of the flight vehicle system that must be evaluated in flight testing.

The avionics system architecture for the flight vehicle itself is depicted in the block diagram in Figure 16.8 (this representation is tailored to the Avanti aircraft described at the end of this chapter, but is also generally applicable to many high-end UAS). On board the vehicle, the flight management computer (FMC) serves as the central location for the control logic and executive functioning of the aircraft. It should be noted that the FMC is distinctly different from the autopilot, which provides automatic actuation of the flight control surface servos in order to stabilize flight or achieve a desired aircraft state (the autopilot can be considered a subservient system of the FMC). The FMC is a higher-level function, more analogous to the role of the pilot on board a manned aircraft. The function of the FMC includes parsing and processing control input data, determining the exact flight path to be flown in order to reach desired waypoints, monitoring the health of all systems, executing a lost link function when needed, providing flyaway protection, and implementing appropriate collision avoidance maneuvering.

Data received by the C2 link from the GCS are conveyed to the FMC via a common bus. Note that multiple control links may be present in order to provide redundancy and fault tolerance. Figure 16.8 depicts triple-redundant C2 links, requiring the FMC to also incorporate link integrity

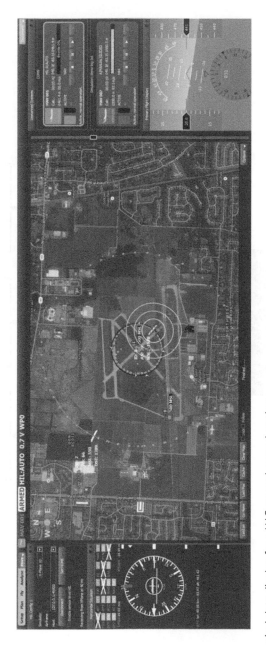

Figure 16.7 Typical data display for a UAS ground control station.

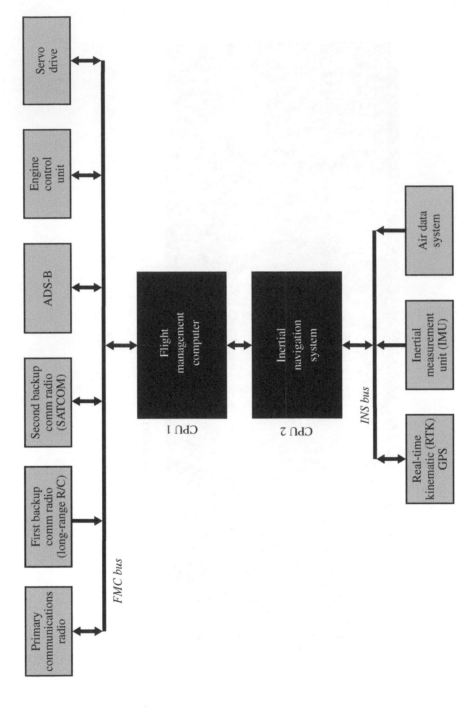

Figure 16.8 Typical avionics system architecture for an unmanned aircraft.

monitoring and C2 link source selection logic if multiple links are employed. Also feeding into the FMC is transponder data from automatic dependent surveillance-broadcast (ADS-B) system, which can provide data directly to the aircraft for autonomous resolution of flight conflicts (*e.g.*, performing a maneuver to remain well clear of another aircraft on a collision course with the UAV). The FMC is directly connected to the servos to actuate the various flight control surfaces (*e.g.*, ailerons, rudder, elevator, flaps, landing gear, etc. for a traditional fixed-wing aircraft) and to the engine control unit.

From the vehicle side, feeding into the FMC is information from the inertial navigation system (INS), which provides a real-time estimate of the vehicle state (position, velocity, acceleration, orientation, and angular rate of change). This information is based on data from a satellite-based navigation system (*e.g.*, GPS), air data (*e.g.*, pitot-static information, angle of attack, sideslip, and pressure altitude), and data from the inertial measurement unit (IMU) comprised sensors such as the accelerometers, gyroscopes, and magnetometers discussed in Chapter 3. This data – not all of which may be available on a given UAS – are fused together through techniques such as the extended Kalman filter (EKF), which is an elegant method for handling dropout of GPS data, integration error of the IMU data, or other deficiencies. The EKF essentially determines the most reliable estimate of vehicle state based on the available data. Also, it should be noted that the underlying sensors of the IMU are the same as those described in Chapter 3 for digital data acquisition (DAQ) of flight test data (*e.g.*, smartphone-based DAQ).

16.2.3 Electric Propulsion

Electric propulsion is becoming increasingly common in aviation, particularly on UAVs and quadrotor drones. Multi-rotor vehicles take advantage of an electric motor's precise speed (RPM) control and rapid response to commanded speed changes in order to actively stabilize the vehicle and achieve the desired control response. Electric motors also have a power output that is independent of altitude (*i.e.*, pressure) since no oxygen is needed (in contrast to internal combustion engines which must have air for the combustion process). Another advantage of electric propulsion is the improved environmental impact, with lower noise output, and zero emissions in the airport environment (opening the possibility of greener aviation if electricity is derived from renewable sources). Finally, electric motors have much higher efficiency (80–90%), compared to the thermal efficiency of an internal combustion engine (25–50%). However, a comparison of motor efficiency does not convey the full picture: the advantage of improved electric motor efficiency is more than offset by the vastly superior energy density of aviation fuel, compared to batteries or other electrical energy storage methods. Aviation fuel has a specific energy on the order of 45 MJ/kg. However, typical lithium-ion batteries might have a specific energy of only about 0.72 MJ/kg (200 Wh/kg). At a system level – accounting for specific energy of the fuel storage as well as motor efficiency – propulsion based on fossil fuels still has much better performance than an all-electric propulsion system, since the required weight for energy (fuel) storage on board the vehicle is so much less for fossil fuels. The current compromise on emerging aircraft is for hybrid-electric systems to be used, where fuel and a gas generator are coupled with batteries in order to provide sufficient power for flight while maintaining propulsion system mass at a reasonable level.

Despite these tradeoffs with electric propulsion, there is an increasing use of all-electric aircraft – both for small-scale drones and for larger aircraft. As energy storage technology continues to improve (higher energy density), there is likely to be a significant trend toward more all-electric aircraft. Since airworthiness certification standards for propulsion systems have historically been centered around internal combustion engines, there is substantial work that needs to be done to develop certification plans to make a safety case for electric propulsion. At the time of writing this book, manned electric aircraft are being certified under the rewritten Part 23 (14 CFR 23), or under the light sport category (14 CFR 103), but airworthiness certification standards for UAS are still pending.

16.2.4 Command and Control (C2) Link

The command and control (C2) link for the aircraft is a critical differentiating feature relative to manned aircraft. The control link provides the physical interface between the pilot and the aircraft's flight controls, which allows the operator to maintain positive control of the aircraft at all times. If the C2 link was to fail, this would be somewhat analogous to the pilot losing flight control authority on a manned aircraft. Thus, the C2 link is a safety-critical tether that allows the pilot to maintain control of the vehicle. The C2 link is also sometimes referred to as the control and non-payload communications (CNPC) link, implying that the link is limited to critical functions only.

The C2 link is typically established through two-way radio communications, where essential flight telemetry data are streamed down to the GCS, and flight control commands are streamed up to the vehicle. The type of control can range from direct inputs from a control stick and throttle for real-time direct control or can involve a higher level of automation where the autopilot handles basic "stick-and-rudder" flying, while the pilot controls the flight path by setting heading, altitude, and/or sequencing a series of waypoints to be flown. As higher levels of autonomy emerge, more functionality and responsibility may be shifted from the pilot to the flight control system. For example, a pilot may direct the aircraft at the mission tasking level, and the aircraft could make all subsequent decisions. In any case, however, the pilot is in command of the aircraft and is responsible for all airmanship decisions. This requirement for direct control and/or supervision demands that a real-time link be maintained between vehicle and pilot at all times.

The physical infrastructure required for a control link centers on at least one radio and antenna pair on board the vehicle, along with a radio and antenna connected to the GCS. If digital data are being communicated, then a modem is also involved on each end, for modulating and demodulating the signal to convey the data over the radio link. The selection of the operating frequency of the link is driven by technical as well as policy dimensions. Radio spectrum is an increasingly scarce, congested resource with an increasing number of wireless devices vying for spectrum. Also, different frequencies will have different propagation characteristics, due to the effects of the atmosphere on propagation, along with the inherent attenuation of transmitted signals over a free path. Bandwidth is also a concern, with higher data rate messages (more information to convey in a given period of time) generally requiring larger segments of spectrum in order to convey the information. Also, there may be a need to share spectrum with other users, potentially leading to interference from other users. The International Telecommunications Union (ITU) is the body that makes recommendations on spectrum allocations, weighing out the various factors involved and approving these through a World Radiocommunication Conference (WRC) held every 3–4 years.

Most hobbyist grade UAS maintains a control link on an unlicensed band, such as WiFi communications on 2.4 or 5 GHz. Other common control link bands are the 900-MHz unlicensed band, or the 433-MHz (70 cm) amateur radio band, which requires an amateur radio license for operation, with no commercial use allowed. More recently, low-altitude sUASs are increasingly relying on the established cell phone network (LTE/4G/5G) for control link communications. For high-altitude operations in airspace where manned aircraft operate, communications channels for UAS (aeronautical mobile) have been designated by the ITU on the C-band for 5030–5091 MHz.

Antennas can take on many different forms and are designed to efficiently convey the radio frequency (RF) energy from the radio through the transmitting medium. Antennas may be omnidirectional (equal gain in all directions) or directional (enhanced gain in particular directions). High-gain, directional antennas are advantageous for directing the energy where it is desired and suppressing received RF energy from signals other than the intended signal. For optimal energy transmission, the antenna must be cut to a certain fraction of the RF wavelength. Thus, lower-frequency (longer wavelength) emissions incur an additional challenge of larger antennas.

There may also be other, intermediate infrastructure involved in the control link for a UAS. If the direct radio line of sight range is less than the anticipated operating range, then some type of relay mechanism may be needed. This could take the form of a radio repeater, which is an intermediate transceiver that is within range of both the ground station and the UAS, even if the GCS and UAS cannot directly communicate. The repeater can also be a part of a larger communication network. Common examples for UAS control include cell phone towers or satellite communications. The satellite communication platform can be either a constellation of communications satellites in low earth orbit (such as the Iridium network) or a larger platform in geostationary orbit (such as Viasat). Satellite communications offer significant advantages, as well as challenges, for maintaining the UAS control link. On the positive side, satellites provide large-area coverage that is independent of any terrestrial limitations such as coverage in sparsely populated areas, or inhibited propagation due to intermediate terrain. On the negative side, however, the signal levels tend to be weak due to the long propagation distances involved.

In addition to conveying control signals from the GCS to the UAS, and down-linking telemetry data from the UAS to the GCS, there may be a need to sustain voice communications between the pilot and air traffic control (ATC). For UAS operating in any type of controlled airspace within the national airspace system (see Chapter 6 for descriptions of the various types of controlled airspace), communications between the pilot and various air traffic controllers must be maintained. Normally, when there is a pilot on board the aircraft, this communication is done through very high frequency (VHF) radios and the radio coverage range has been well established over decades of use. However, things are complicated somewhat by the removal of the pilot from the aircraft, and the addition of a GCS in the mix. UAS must maintain not only reliable voice communications between both the aircraft and ATC but also the aircraft and the pilot.

Since the C2 link is so critical to the safety of flight, it must be highly robust no matter what the flight condition is. There are a number of hazards posed to the robustness of the C2 link, including RF fading, interference, jamming, or even hacking. The quality of the RF link can depend significantly on the range and orientation of the flight vehicle, since the antenna gain characteristics are seldom isotropic. Thus, there is an elevated risk of lost link when the vehicle is maneuvering. Flight testing of the control link must assess the maximum radio range over which a reliable link may be maintained, quantified through a measure of the packet correction rate.

16.2.5 Autonomy

As UAV technology develops, there will be an ever-increasing push toward enhanced autonomy (National Research Council 2014). At the time of the writing of this book, the state of the art is an ever-increasing level of automation, most often with direct human control or oversight over the automated behavior. The key difference between automation and autonomy is in the level of human oversight and interaction with the aerial vehicle. For systems with high levels of automation, the vehicle performs according to pre-programmed scripts, executing well-bounded behavior in a deterministic manner. Examples of automated systems in aviation include the autopilot on either manned or unmanned aircraft, with the level of automation ranging from a simple wing leveler to a fully coupled autopilot with autoland features. However, there is always a human in the loop, supervising the behavior of the automated system.

As we move toward autonomy, however, the role of the human is reduced even further. An autonomous vehicle has high-level, adaptive decision-making capability. The vehicle makes decisions based on available information when encountering a new situation that has not been pre-programmed by a human operator. This level of behavior is often achieved by training the autonomous system on a large data set, and then operating the autonomous system under close supervision in a wide range of actual conditions. Over time, the system's artificial intelligence builds a decision-making and classification tree that allows it to make autonomous decisions for safe and efficient flight.

Increasing levels of autonomy pose a significant challenge to flight testing and airworthiness certification. Verification and validation (V&V) of autonomous system performance is critical for safety of flight; however, autonomous systems (by definition) do not exhibit deterministic behavior. Thus, it is difficult to develop and conduct a scripted set of tests that can be employed for assessing the reliability and performance of autonomous systems. Later in this chapter, we will present an example of an emergent (unanticipated) property of an autonomous system. V&V and flight testing programs many need to involve a robust and wide-ranging set of controlled flight conditions designed to elicit and identify these emergent properties. Autonomous systems will also need to have well-defined boundary conditions for system performance (*e.g.*, limits on the flight envelope), and rigorous schemes to evaluate system response near the edges of the flight envelope in as many situations as possible. Since this approach to V&V results in an extremely large test matrix, flight testing programs for autonomous systems may need to be augmented by robust simulation efforts.

16.3 Flight Regulations

The rapid surge in popularity of drones has challenged the application of the existing aviation rules toward this disruptive technology. Many of the regulations developed for manned aviation are inapplicable or inappropriate for drones – some regulations may be overly restrictive, while other new regulations are needed to address deficiencies posed by this new technology. Given the rapid evolution and disruptive nature of the technology, the regulatory framework will be evolving for some time.

In the United States, the first major promulgation of regulations for drones was the publishing of 14 CFR 107 in August 2016 (Federal Aviation Administration 2016). This part of the regulations generally applies to the flight of all UAS, unless the flight is being done as a hobbyist (under an exemption provided by Section 336 of the FAA Modernization and Reform Act of 2012) or as a public aircraft (generally operated under an authorized waiver from the standard aviation rules, through a certificate of authorization/waiver, referred to as a COA). Under Part 107 of the federal aviation regulations, a drone operator must hold a remote pilot certificate, which can be obtained by passing a knowledge test. Vehicles are limited to a maximum weight of <55 lb (25 kg) and must be registered with the FAA. The operating limitations include a maximum speed of 100 mph (87 kts) and a maximum altitude of 400 ft above ground level (AGL) (or within 400 ft of a structure). Furthermore, the remote pilot in command must keep the done within visual line of sight (VLOS) of the operator or the observer, and operate in daytime and visual meteorological conditions (VMC) only. These are just a few of the flight restrictions imposed by Part 107 (see 14 CFR §107 for full details). Any deviation from these regulations requires a waiver, the application for which must be backed up by a robust safety case. Current regulations in other nations have similar limitations on the operation of drones.

16.4 Flight Testing Principles

Flight testing of UAS shares many of the same principles as the procedures and requirements for manned aviation flight testing, as discussed throughout the rest of this book. The fundamental principles governing flight are, of course, the same for both manned and unmanned aircraft. However, there remain some key critical differences in the details.

First, there is the impact of wing loading on a fixed-wing aircraft, which is defined as aircraft weight divided by wing area. Small (<55 lb), fixed-wing UAS will have much lower values of wing loading, compared to manned aircraft. For example, Ohio State's Peregrine UAS, with a cruise speed of 27 knots true airspeed (KTAS), a wing area of 3.83 ft^2, and a max weight of 12 lb, will have a wing loading of 3.13 psf. In contrast, a Cirrus SR20, which cruises at 145 KTAS, has a wing area of 144.9 ft^2 and a maximum take-off weight of 3050 lb, giving a wing loading of 21.0 psf. This illustrates a fundamental principle of flight vehicle design, where faster vehicles tend to have higher values of wing loading (see Tennekes 2009).

The implications of lower wing loading (for fixed-wing unmanned aircraft), or lower disk loading (for rotary wing unmanned aircraft), is that the vehicle is much more susceptible to atmospheric disturbances such as wind gusts. This makes it more difficult to acquire high-quality flight test data, since typical wind gusts will have a more pronounced effect on the measured data. In some cases, the magnitude of a wind gust can be of the same order of magnitude as the flight speed of the UAS. Similarly, thermals in the atmosphere can create significant updrafts or downdrafts that are of the same order of magnitude as the rate of climb or descent for a UAS. This accentuates the need for testing UAS flight vehicle performance parameters under extremely calm conditions.

A second significant difference of UAS flight testing is the repeatability enabled by the onboard autopilot, instead of relying on manual pilot control. Desired flight profiles can be pre-programmed and flown with greater precision and repeatability than what can be done under manual control. This improves the quality of flight test data and offsets the negative impact of sensitivity to wind gusts discussed in the previous paragraph.

The absence of the pilot on board the aircraft also alters the assessment of risk when conducting pre-flight safety reviews. With no pilot on board, the flight test team only needs to worry about the safety of people and property on the ground, and the integrity of the flight vehicle. Also, the flight testing regimen for a small UAV does not have to accommodate g-limits induced by a pilot on board the aircraft, which is helpful for clearing the flight envelope. Thus, slightly elevated risks may become acceptable in UAV flight testing (see Chapter 6 for a detailed discussion of flight test planning and risk assessment).

A fourth significant difference in the details of flight testing is the level of access to flight data. Many small UAS tends to have open architectures, with much more direct access to detailed records of flight data. For example, the Pixhawk flight controller can downlink or save all vehicle state information, including GPS data, inertial data for vehicle attitude, etc. Generally, this level of access to the data is not available for general aviation (GA) aircraft. Similarly, detailed records of the flight control inputs are also available from the flight controller, which is a significant advantage for flight tests such as dynamic stability or control law development. Furthermore, even if access to the flight computer data is not available to the flight test engineer, it can be straightforward to add supplemental instrumentation such as flight data recorders and air data systems (including pitot probes or five-hole probes).

We will now provide detail on some of the specific methods and tests associated with UAV flight testing. We will start with a discussion of instrumentation available for UAV flight testing and then discuss specific flight test procedures for documenting the performance of systems unique to UAVs. We will then turn our attention to best practices for piloting UAVs for high-quality flight test data.

16.4.1 Air Data Instrumentation

For most flight test programs, it is important to have a reliable indication of airspeed. Nearly, all UAVs are equipped with GPS, which provides an indication of ground speed and position. Under calm wind conditions, the GPS-derived ground speed can be equated to true airspeed. However, if any wind is present (even a few knots), then it is important to account for the difference between ground speed and true airspeed. There are two approaches to handling this challenge – installation of an air data probe to directly measure true airspeed, or implementation of some other means of wind measurement to convert between ground speed and true airspeed.

The ideal approach is to install an air data probe on the aircraft. This can be an extended boom with total and static pressure orifices, possibly also with vanes for indicating angle of attack and sideslip. Or, a calibrated multi-hole probe such as a five-hole or seven-hole probe can be used to measure total pressure, static pressure, and flow direction. The challenge with using an air data probe, however, is that it consumes payload weight and power. In some cases, these probes may be too heavy or unwieldy for installation on a small UAS. Also, if the size of the probe is large relative to the flight vehicle, the presence of the probe may alter the drag, stability, or handling characteristics of the vehicle. Thus, many small UAS may require an alternate means of determining true airspeed.

If UAS flight testing is conducted near ground level, then a local wind measurement with a ground-based anemometer may suffice. Time-resolved wind speed and direction should be recorded simultaneously with the flight vehicle performance parameters. Ideally, the flight profiles should be conducted in the same direction relative to the wind for each segment. Following the flight, the time-resolved wind can be subtracted from the ground speed in order to find the true airspeed.

Another approach to wind estimation involves analysis of inertial data. As the aircraft maneuvers in flight, the wind will induce a vehicle trajectory that differs from what would be anticipated in no-wind conditions with the measured rates and accelerations. A Kalman filter is used to infer the required wind conditions to account for this difference between expected and actual trajectory. This approach to wind estimation requires detailed knowledge of the flight vehicle's dynamic model such that the Kalman filter can provide a reasonable wind estimate. The quality of this approach to wind estimation improves as the vehicle continues to maneuver (as more data are acquired) and presumes that the winds are steady.

16.4.2 UAV Flight Test Planning

Details of preparation and planning of flight testing programs are detailed in Chapter 6. For UAV flight testing, it may be tempting to circumvent some of the rigor associated with planning manned flight tests. With no pilot on board, and in some cases with a less expensive airframe involved, the risks to personal health or program finances may be dramatically lower than typical manned flight test programs. However, this lax approach can lead to unanticipated negative consequences, and a permissive, casual culture could end up permeating a flight test program even when subsequent risks are higher. Thus, it is important to maintain rigorous workflows and strong adherence to safety protocols. Many of the same principles for planning manned flight test programs (see Chapter 6) are applicable to unmanned aircraft flight testing. Valasek and Bowden (2019) or Abdulrahim (2019) can be consulted, for examples, on how UAV flight test programs are planned and conducted in an academic context.

16.4.3 Piloting for UAV Flight Testing

Manual control of an unmanned aircraft for flight testing is distinctly different from manned flight testing. The pilot is not able to benefit from visual cues from on board the aircraft. If the pilot is relying solely on visual observation of the aircraft from the ground, it can be difficult to infer vehicle attitude and heading from a distance. This can make it extremely challenging to conduct the precision flying needed for acquiring quality flight test data. Instead, the pilot can achieve better manual flight control by directly referring to the GCS display for the vehicle. Flight testing a UAS from a ground station is analogous to instrument flying – maintaining aircraft attitude and airspeed must be done primarily in reference to the instruments on the ground station, which often emulate traditional glass-panel avionics on a manned aircraft. This is more effective than direct observation of the vehicle by the pilot, but is less desirable than traditional flight testing with a pilot on board the aircraft. This is because the pilot loses the advantage of physical cues, such as minor accelerations due to wind gusts, that would be an indicator of a need to repeat a test point. The ground-based pilot also lacks the sensory cues that would be indicators of flight safety and vehicle performance.

A more structured approach to flight testing with a UAS is to program the entire flight test profile into the flight management system. This approach can work extremely well, as the autopilot can often hold flight test conditions (*e.g.*, altitude and airspeed) with much greater accuracy than could be achieved by manual control. (This approach is even better than what most manned aircraft pilots can achieve.) The challenge, however, is that the complete characteristics of the flight management system and autopilot must be completely known and trustworthy, which often requires an accurate and comprehensive model of the aircraft. This can create a chicken-and-egg situation, since the only way to robustly perform system identification is through flight testing. Modern approaches to flight testing under development include "learn to fly" strategies which essentially perform system identification on-the-go, such that system identification can happen quickly and autonomously through a series of well-planned and executed maneuvers.

To illustrate these flight testing principles, we will now present two example flight test programs conducted at The Ohio State University. These flight testing efforts were done in the context of research and development on new technology systems for UAS, and to enable new use cases such as long endurance persistent remote sensing. We will provide a brief summary of these flight testing programs to illustrate the approach and types of data that can be generated in UAV flight tests. The two aircraft studied are the Peregrine and the Avanti, discussed as follows.

16.5 Flight Testing Examples with the Peregrine UAS

Our first example involves the Peregrine, a fixed-wing UAS designed and built at The Ohio State University. As with larger GA aircraft, it is important to perform a careful ground-based characterization of the propulsion system, in order to have an indication of installed power across a range of operating conditions. The work described here summarizes a wind-tunnel investigation of the propulsion system performance, followed by in-flight measurement of the drag polar by measurements of excess power and glide performance. Further details on these studies are provided by McCrink (2015) and McCrink and Gregory (2017).

16.5.1 Overview of the Peregrine UAS

The fixed wing UAS used in this work was designed at The Ohio State University and named the Peregrine. This vehicle was initially envisaged as an agricultural monitoring system, thus prioritizing endurance and stability as key design considerations. The aircraft is designed to cruise at 27 kts with an endurance of 1 hour. This speed represents the velocity necessary to traverse 1.2 square miles while operating at 400 ft AGL with a visual swath area of 100 ft. It has an empty weight of 4.5 lb with a maximum gross weight of 12 lb and a useful payload weight of 3.5 lb. The stall speed is 15 kts, which corresponds to a typical hand launch velocity for aircraft of similar weight. See Table 16.2 for full specifications of the Peregrine UAS.

Table 16.2 Key characteristics of the Peregrine UAS.

Parameter	Value	Units
Gross weight	12	lb
Empty weight	4.5	lb
Wing span	80.87	in.
Wing area	3.83	ft^2
Aspect ratio	11.85	—
Fuselage length	54	in.
Maximum thrust (take-off)	4.5	lb
Stall speed (clean)	15	kts
Parasitic drag coefficient (C_{D_0}) (predicted)	0.0261	—
Oswald efficiency factor (e) (predicted)	0.90	—

Figure 16.9 The Peregrine UAS, a fixed-wing electric propulsion aircraft for long-endurance remote sensing.

The airframe (shown in Figure 16.9) features a relatively high aspect ratio wing with V-tail stabilizer control surfaces. The center section of the aircraft is an open payload volume which can support various camera systems used for precision agriculture applications. The nose compartment contains the aircraft's power systems including the 0.6-hp (450 W) brushless electric motor, electronic speed controller, and 4.8 A-h lithium polymer flight battery (120 W-h capacity at 25 V). The FMC and INSs are located in the fuselage behind the main wing with airspeed information provided by a wing-mounted pitot-static system. Telemetry radios, which provide in-flight data monitoring and control capabilities, are located in the wing tips to minimize impact of electromagnetic interference (EMI) on the INS sensors. The airframe was manufactured in-house using rapid-prototyped molds created on a small-scale three-axis computer numerical control (CNC) machine. The majority of the airframe is a carbon fiber/Divinycell/carbon fiber sandwich construction, resulting in

Table 16.3 Predicted and measured drag polar for the Peregrine UAS.

Parameter	Analytical prediction	Level acceleration flight test	Rate-of-climb flight test
Parasitic drag coefficient (C_{D_0})	0.0261	0.0220	0.0205
Oswald efficiency factor (e)	0.90	0.85	0.86

a rigid and lightweight airframe with a high degree of dimensional stability. Relevant sizing and performance parameters are outlined in Table 16.3.

16.5.2 Propulsion System Characterization

We will now turn to the first set of experiments to characterize the performance of the propulsion system and prepare the way for in-flight measurement of specific excess power. The propulsion system of the Peregrine is common to many UAS platforms and is composed of a lithium polymer battery to provide electrical power to the motor and onboard systems, and an electronic speed controller driving an electric brushless motor. A small fixed-pitch propeller is attached to the motor with no additional gearing, such that motor RPM is equivalent to propeller RPM. To estimate the power available for flight test, a detailed understanding of the individual components is required to relate the electrical power provided by the battery, which is easily measured, to thrust power produced by the propeller.

For small, fixed-pitch propellers, analytic methods used to estimate the thrust, torque, and corresponding efficiency differ greatly from full-scale models due to Reynolds, or scale effects. To account for these effects, a new analytic tool based on blade element momentum theory (BEMT) was developed which relates the geometric shape and aerodynamic characteristics of the individual propeller blades to the overall change in momentum of the flow passing through the propeller disk. BEMT models work by dividing a single propeller blade into small slices and analyzing the resulting thrust and torque produced by the segment when exposed to a variable inflow profile. The predicted thrust and torque acting on the individual blade elements induces upstream and downstream changes in the free stream flow field, which in turn modifies the inflow profile to the blade element. By iteratively adjusting the inflow profiles and reassessing individual blade forces, a balance is reached where the thrust produced by the sum of individual blade elements is equal to the momentum transferred to the free stream. Scale effects appear in estimating individual blade forces and the coupling between blade elements and are described in detail in McCrink and Gregory (2017).

Validation of the BEMT model was conducted in a wind tunnel facility at the Ohio State University using a specially designed test stand capable of measuring both the thrust and torque generated by the motor/propeller system. The tests also measured the electrical power delivered to the motor and resulting RPM as a function of airspeed (which were used for post-flight estimation of the power available). Results from one such test are shown in Figure 16.10 using an APC 10×7E propeller, compared with BEMT predictions over an order of magnitude change in Reynolds number. In this figure, the propulsive efficiency, defined as the ratio of thrust power to motor power, is shown for a range of Reynolds numbers commonly encountered in flight. Clearly, lowering the Reynolds number (low airspeed and/or low RPM) results in a marked reduction in both propulsive efficiency and maximum advance ratio when compared to the same propeller operating at higher Reynolds numbers. These deleterious effects are most prevalent when operating at reduced airspeed (take-off and landing) and reduced power settings (maximum endurance) and are of first-order importance when evaluating UAS performance.

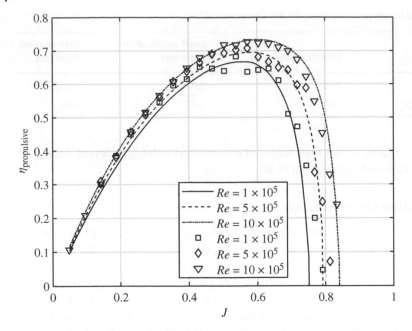

Figure 16.10 Performance data for the APC 10×7E propeller. Note the dependency of efficiency on the Reynolds number at high advance ratios. Source: McCrink and Gregory (2017).

16.5.3 Specific Excess Power: Level Acceleration and Rate of Climb

Now that we have established the baseline performance of the electric propulsion system (full details available from McCrink and Gregory 2017), we can turn our attention to measurements of the specific excess power of the aircraft. Basic principles of flight testing for excess power are covered in Chapter 9, so will not be repeated here. First, we will discuss level acceleration flight tests, followed by climb performance flight tests.

Several modifications to the autopilot system were necessary to conduct level acceleration testing on the Peregrine sUAS. These modifications were driven by the requirement to remain in VLOS of the operator, while being in close proximity to the ground (below 400 ft AGL). The first modification was the inclusion of a feedforward controller to roughly predict the angle of attack for a required airspeed. This is especially critical for level acceleration testing in which the high thrust-to-weight ratio and low wing loading of the aircraft makes maintaining a constant altitude difficult (many small UAS have no difficulty accelerating nearly straight up!). Without a feedforward controller, impulsive application of thrust would lead to significant altitude excursions.

The angle of attack for the feedforward controller is determined by a first-order approximation for the lift curve slope ($\partial C_L / \partial \alpha$) of the aircraft,

$$C_L = \frac{\partial C_L}{\partial \alpha} \alpha + \alpha_{L=0} = \frac{W}{qS}, \tag{16.1}$$

where the lift coefficient (C_L) is found from knowledge of weight (W), wing area (S), and measured dynamic pressure (q). The angle of zero lift, $\alpha_{L=0}$, and the lift curve slope are estimated from analytical predictions. Angle of attack is assumed to be close to the angle between the aircraft's longitudinal axis and the flight path, both of which can be measured. Eq. (16.1) provides a rough estimate of the required angle of attack required for a flight condition, such that the needed angle can be set.

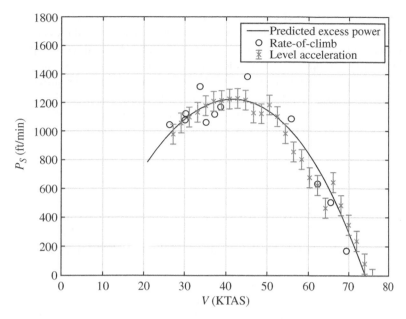

Figure 16.11 Excess power of the Peregrine UAS.

In practice, this estimate of α is sufficient to prevent pitch oscillations or large altitude excursions occurring after full power is commanded. If the flight controller incorporates angle of attack tracking during the first seconds after full throttle is applied, the airspeed and climb rate stabilize quickly. If no attempt is made to initially control the angle of attack, then airspeed and climb rate excursions develop which skew the resulting excess power estimation as additional energy is transferred between the two quantities. Additionally, the autopilot pitch gains must be re-tuned to enable a rapid full throttle command at an airspeed near stall. Rapid application of the throttle in this condition results in a dramatic change in control effectiveness due to the increase in airflow over the stabilizers. In practice, controlling this behavior may be achieved by a reduction in pitch rate damping to prevent the autopilot from commanding extreme or oscillatory pitch changes.

Climb flight tests were conducted in a manner following the methods described in Chapter 9. From a stable, level flight condition at low altitude, full power was added, the trim condition was adjusted and set for climbing flight, and data were acquired in stabilized flight. The key challenges associated with climb flight testing under 14 CFR 107 or a COA were to keep the maximum flight altitude <400 ft AGL and within the VLOS of the operator. At conditions for high rates of climb, the vehicle can quickly reach the altitude limit. Conversely, at high flight speeds (low climb angle), the vehicle can fly beyond VLOS of the operator.

Flight test results for level accelerations and rate-of-climb tests are shown in Figure 16.11. The rate-of-climb data were measured over 15 successive laps of a circuit of climbs and glides, while the level acceleration data was generated in a single run. During the flight test, the electrical power was monitored using onboard voltage and current sensors, while the flight controller logged the vehicle state and air data for post-flight analysis. The vehicle weight for these tests was 6.2 lb. The predicted excess power was estimated during the design phase of the Peregrine along with propulsion characteristics derived from the Reynolds-dependent propulsion model outlined previously. These same model data are used with flight measurements of electrical power and RPM data to approximate the power available. The measured and estimated excess power for the rate-of-climb and level acceleration are shown in Figure 16.11 with the resulting C_{D_0} and e detailed in Table 16.3.

Figure 16.11 demonstrates the close agreement in excess power using the rate-of-climb and level acceleration methods. Of note, while the level acceleration test provides a better overall indication of the trends in the excess power curve, it is a much more difficult flight test to conduct owing to the maneuver-specific tuning of the autopilot system. Additionally, the rate-of-climb tests have a built in advantage in that the data for an entire leg is averaged into a single point, which provides a degree of smoothing to the data. However, if the autopilot can be properly adjusted, the level acceleration may reproduce the entirety of the excess power curve in a single run and allow for more repeatable testing. In practice, this can reduce flight-testing time and cost, yielding benefit to the flight testing program.

16.5.4   Glide Flight Tests

Estimation of the aircraft's lift-to-drag ratio (L/D) requires gliding the aircraft over a range of fixed airspeeds with zero net thrust. The assumption of zero thrust is uniquely met by turning off the motor and allowing the propeller to fold back. For this work, three methods of measuring and analyzing glide performance data are compared. The first is the standard method outlined in Chapter 10, where the rate of descent and the airspeed are used to construct the velocity triangle of a glide. The rate-based calculation for L/D involves comparing the vertical descent rate to the average airspeed. For this test, the aircraft is placed at a predetermined start altitude of 400 ft AGL and commanded to hold a constant airspeed until reaching a hard deck of 50 ft AGL. This test is repeated in low to no-wind conditions over a range of airspeeds corresponding to the flight envelope of the Peregrine (approximately 20–50 kts). The rate method requires airspeed data and works well in steady winds but requires multiple climb/glide patterns to reproduce the entire L/D curve. If no air data are available, the test can be conducted in no-wind conditions using a GPS sensor to estimate an equivalent airspeed. Alternatively, under steady wind conditions, the local wind speed can be measured using the techniques described in Chapter 8 or other direct methods, allowing for conversion between ground speed and true airspeed.

For the second L/D estimation method considered here, the vertical descent rate and airspeed quantities used in the rate method are replaced by the integrated displacements during the unpowered descent. This constructs the velocity triangle based on distances instead of velocities (refer to the velocity triangles in Figure 10.5). Thus, the vertical descent rate is replaced by the vertical displacement (change in altitude) and the average airspeed by the integrated quantity

$$s = \int_{t_0}^{t_1} V \, dt, \tag{16.2}$$

where s is the hypotenuse of the aircraft's velocity triangle and V is the true airspeed. This substitution improves the L/D estimates in gusting wind environments commonly encountered in low-altitude flights and exacerbated by the low wing loading of most sUAS platforms.

Finally, a third method for estimating the L/D for vehicles without an airspeed sensor is demonstrated. While numerous small UASs have air data measurement capabilities, due to cost/space/size constraints, some forgo these sensors in favor of a strictly INS-based attitude and control system. Therefore, it is necessary to evaluate the performance of a strictly INS-based approach for estimating the aircraft's L/D. There are numerous advantages to this method, the first of which is the simplicity of integration into an existing vehicle. Tying into an existing air data computer or installing and calibrating a separate unit can be time-consuming and costly. However, if the GPS or INS data are used directly, the resulting L/D estimates are likely to be highly erroneous. For many sUASs, the wind speed can be a large fraction of the vehicle's airspeed, complicating the determination of horizontal distance traveled. Instead, INS-derived wind speed estimates are used

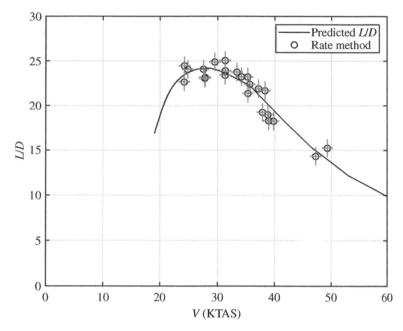

Figure 16.12 Lift-to-drag ratio of the Peregrine from flight testing, compared with predicted results from the drag polar determined by analysis. Testing conducted under calm conditions.

to generate a pseudo airspeed signal which may be used with either of the previously mentioned flight test methods. Briefly, this pseudo airspeed signal is generated by assuming a quasi-steady flight speed and only requires the vehicle to turn. The difference between the computed inertial velocity and GPS velocity on a short time scale (1–2 seconds) provides an indirect indication of the wind vector and can be integrated into the INS filter equations for smoothing.

We will now turn to an evaluation of glide performance data from these three approaches. The L/D resulting from the traditional rate method is shown in Figure 16.12. Error estimates are based on common UAS air data sensor noise characteristics, with the major source of uncertainty being the airspeed measurements (see Chapter 5 for a detailed discussion on uncertainty analysis). Figure 16.12 shows good agreement between the predicted and measured L/D for the aircraft in flights with little to no wind. Flights at higher airspeeds were attempted, but the limited operating ceiling did not allow the aircraft to reach a steady-state descent rate and airspeed before reaching the hard deck. Low speed testing was limited by loss of aircraft control occurring at speeds below 23 kts (stall).

The integration-based L/D estimation method is demonstrated in a flight with high winds (20 kts) and gusts, with results shown in Figure 16.13. The standard rate-based estimates show a larger scatter when compared with the integrated method. In all cases considered, the integration-based method is either equivalent or exceeds the performance of the rate-based method. It should be noted that the efficacy of this method is related to the fidelity and update rate of the airspeed sensors on board the aircraft. If the data rate of the onboard airspeed sensor is decreased from 100 to 10 Hz, then the noise in the L/D estimate increases due to apparent aliasing in the airspeed record. Therefore, careful selection of the air data system and logging hardware should be considered for this flight test.

Figure 16.14 shows data resulting from estimation of L/D using the inertial state estimates to approximate a pseudo airspeed signal. For the case of low wind, the resulting L/D estimates are in reasonable agreement with the rate method previously outlined. If the change in wind speed

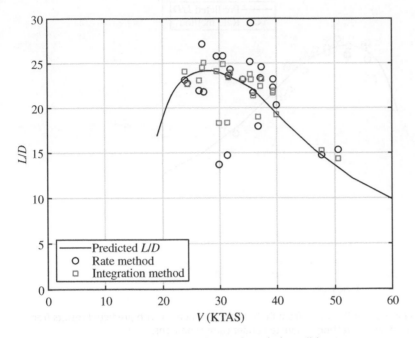

Figure 16.13 Glide flight test results for gusting wind conditions.

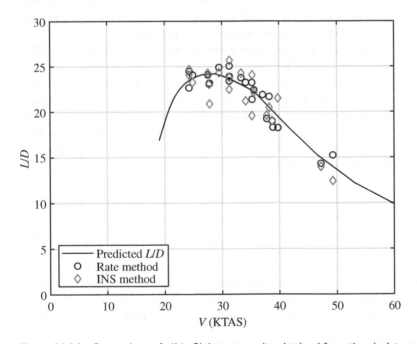

Figure 16.14 Comparison of glide flight test results obtained from the air data system and from inertial measurements.

or direction is relatively slow, compared to the filter time constants in the INS, the method can be applied to flights in wind speeds equivalent to the integration method test case outlined previously. However, the presence of gusts complicates testing as the pseudo airspeed signal is not able to accurately capture the rapid changes in wind speed or direction necessary to utilize either the rate or integration-based methods. Operating within this limitation, however, incorporation of the INS-derived wind speed can dramatically improve the resulting L/D estimates for systems not equipped with air data systems.

16.6 Flight Testing Examples with the Avanti UAS

The goal of the Avanti aircraft development program was to create and demonstrate new systems for robust control links and autonomy. A related goal was to develop a high-speed flight research vehicle that could also be used to set world records for speed and distance of a UAV (see FAI 2017a, 2017b, 2017c; Warwick 2017; or McCrink and Gregory 2021). The flight testing results presented here will provide an assessment of the drag polar by coast-down testing, a critical assessment of radio range for safe flight beyond visual line of sight (BVLOS) of the operator, and a flight demonstration of an emergent hazardous property of an autonomous system.

16.6.1 Overview of the Avanti UAS

The Avanti, depicted in Figure 16.15, is a high-speed research aircraft capable of operating BVLOS. It is powered by a small-scale JetCat P-180 RXi turbojet capable of propelling it to speeds in excess of 140 kts. The airframe is a composite fiberglass/carbon fiber shell produced by Skymaster and was chosen because of its high-speed potential and the large open forward fuselage which allowed for integration of the avionics and telemetry radio systems necessary for BVLOS testing. The onboard avionics included a custom INS and FMC, with an air data system composed of a five-hole probe mounted in the nose of the aircraft providing angle of attack and sideslip information in addition to static and total pressures. The radio systems included redundant command and control links, a satellite communications system, and an ADS-B transceiver. Integration of these radio systems into a composite airframe posed one of the most challenging aspects of the design phase.

Figure 16.15 The Avanti turbojet-powered UAS, depicted in the flare to landing.

Figure 16.16 Testing of the Avanti communications antenna radiation patterns in an RF anechoic chamber at The Ohio State University.

The lack of internal grounding points throughout the airframe increased the interference between the high-powered radio receivers and required extensive testing in an RF anechoic chamber (see Figure 16.16) to ensure sufficient gain margins existed to support long distance flights. Interference issues were exacerbated by broadband EMI generated by the turbojet ignition system and the close proximity of the radio antennas to the highly sensitive satellite communication and GPS receiver antennas. In addition to the electrical issues posed by the propulsion system, the small-scale turbojet used to power the aircraft is much less efficient than large-scale powerplants, necessitating extensive modifications to the internal configuration of the airframe to increase the total fuel capacity to support the range requirements of the aircraft. Extensive ground testing of these systems was performed before the first flight, and refinement of component and systems integration continued throughout the flight-testing campaign. Basic geometric and aerodynamic properties of the Avanti are listed in Table 16.4.

Table 16.4 Key characteristics of the Avanti UAS.

Parameter	Value	Units
Gross weight	70	lb
Empty weight	40	lb
Wing span	97.6	in.
Wing area	12.67	ft^2
Aspect ratio	5.2	—
Fuselage length	124	in.
Maximum thrust (take-off)	40	lb
Stall speed (clean)	60	kts
Parasitic drag coefficient (C_{D_0}) (predicted)	0.0260	—
Oswald efficiency factor (e) (predicted)	0.72	—

Figure 16.17 The Avanti UAS during the world record-setting flight (as seen from the chase plane). Note the extended nose gear, which resulted in a drag penalty of approximately 10 counts. Source: Courtesy of Ross Heidersbach, The Ohio State University.

16.6.2 Coast-Down Testing for the Drag Polar

During the record-setting flight of the Avanti UAS, there was an unfortunate mishap that led to the nose landing gear remaining extended throughout the flight (see Figure 16.17). The flight crew elected to continue the flight without attempting to cycle the landing gear, since that action would have invalidated the record attempt in the autonomous UAV category. However, the extended landing gear posed a significant drag penalty and ultimately lowered the maximum speed of the aircraft. Following the successful record flight, the team desired insight into how much the drag penalty ended up being due to the un-retracted nose landing gear. This need presented an opportunity to analyze the available flight data to extract the drag polar through coast-down testing.

The flight profile for the Avanti record flight involved an out-and-back course over Lake Erie, initiating and terminating at Kelleys Island airport (89D) in Ohio. The minimum required length of the speed course (set by the National Aeronautic Association, the record sanctioning body) was 15 km on both the outbound and return legs, with 5-km entry and exit segments on each leg. The flight profile involved a full-throttle dash in each of the speed segments, followed by idle-thrust coast-down segments to conserve fuel. These coast-down segments present an opportunity for evaluation of the drag polar, in a manner similar to level acceleration flight tests (with the simplification that thrust can be neglected).

We will start by considering the equations of motion for the aircraft during the coast-down maneuver. Level flight was maintained by the autopilot throughout the deceleration, and residual engine thrust is assumed negligible at idle conditions. This results in

$$\sum F_y = L - W = 0 \tag{16.3}$$

and

$$\sum F_x = -D = ma_x, \tag{16.4}$$

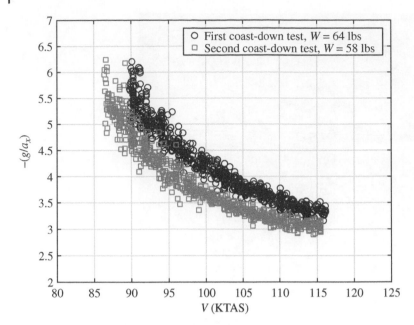

Figure 16.18 Measured deceleration as a function of airspeed during coast-down tests at different weight conditions.

where L is the lift, W is the aircraft weight, D is the drag, m is the aircraft mass, and a_x is the acceleration in the horizontal direction (Earth frame of reference). Substituting $W = mg$ and forming a ratio of Eqs. (16.3, 16.4), we have the simple relation for lift-to-drag ratio,

$$\frac{L}{D} = -\left(\frac{g}{a_x}\right). \tag{16.5}$$

Thus, in order to find L/D as a function of airspeed, we only need to measure longitudinal acceleration as a function of airspeed during the zero-thrust coast-down maneuver. Acceleration in the body axes is directly measured by the IMU on board the aircraft. Accounting for vehicle attitude throughout the deceleration maneuver, this acceleration in body axes can be transformed to the required acceleration (a_x) in Earth axes. Data for the two coast-down maneuvers are shown in Figure 16.18. Due to the significant fuel burn rate of the turbojet engine on the Avanti, the two coast-downs are at different weights (64 and 58 lb). Since L/D is also a function of vehicle weight, we can see a difference in measured acceleration between the two data records. The lower-weight coast-down maneuver shifts the L/D curve to the left.

In order to remove the effects of weight and density altitude on the measured accelerations, we can plot the data as a function of normalized velocity. In Chapter 7, we derived a "velocity independent of weight" parameter,

$$\mathrm{VIW} = \frac{V\sqrt{\sigma}}{(W/W_S)^{1/2}}, \tag{16.6}$$

where σ is the density ratio ρ/ρ_{SL}, W_S is a standardized weight (typically maximum take-off weight), and V is the true airspeed. Using measured values of weight and free stream density, along with the definition of VIW in Eq. (16.6), we can re-plot the data in Figure 16.19 (showing only 1 out of 20 recorded data points, for clarity). Here, we see that the curves collapse nicely, indicating the adequacy of the weight correction in the definition of VIW.

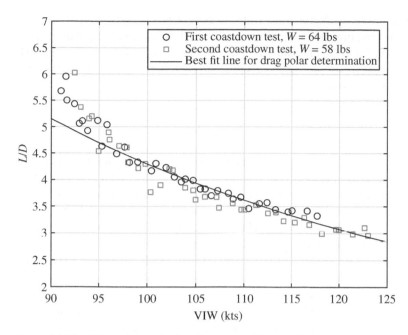

Figure 16.19 Lift-to-drag ratio data from coast-down testing, along with a model fit to determine the drag polar.

We now want to apply a curve fit to the data to find the drag polar for the aircraft with the nose gear extended. Expanding out Eqs. (16.3, 16.4) in terms of the definitions of lift coefficient, drag coefficient, and the drag polar, we have

$$L = W = \frac{1}{2}\rho V^2 S C_L \tag{16.7}$$

and

$$-\frac{W}{g}a_x = \frac{1}{2}\rho V^2 S \left(C_{D_0} + \frac{C_L^2}{\pi \, \text{AR} \, e} \right), \tag{16.8}$$

where the aspect ratio is $\text{AR} = b^2/S$ and b is the wing span. Solving Eq. (16.7) for lift coefficient, solving Eq. (16.6) for true airspeed, substituting both into Eq. (16.8) and equating the result to Eq. (16.5), we have

$$\frac{L}{D} = -\left(\frac{a_x}{g}\right)^{-1} = \left[\left(\frac{\rho_{\text{SL}} S}{2W_S} \right) (\text{VIW})^2 C_{D_0} + \left(\frac{2W_S}{\rho_{\text{SL}} \pi b^2} \right) (\text{VIW})^{-2} e^{-1} \right]^{-1}. \tag{16.9}$$

All parameters in Eq. (16.9) are known from the aircraft geometry or from in-flight measurements, such that we can find a best fit line based on this equation to determine the drag polar (see Figure 16.19). When we constrain the Oswald efficiency factor to the predicted value of $e = 0.72$, the parasitic drag coefficient resulting from the fit is $C_{D_0} = 0.0357$ for the Avanti when the nose gear is extended. This is nearly 10 drag counts higher than the clean value of $C_{D_0} = 0.0260$ predicted by analysis and prior flight testing work. A drag rise of 10 counts is commensurate with the drag increment predicted by methods such as Hoerner (1965).

16.6.3 Radio Range Testing

Ensuring robust command and control for operations BVLOS is a unique and critical aspect of UAS testing which requires additional flight testing to validate the models and predictive tools used to estimate maximum RF range. While these tests are generally less dynamic than aircraft performance testing, they do require the integration of additional data logging capabilities and precise trajectory information from the onboard guidance system. Radio range can be modeled using the Friis relation to approximate the received signal strength between two antennas (see Friis 1946 or Shaw 2013) and is given as

$$d = \left(\frac{\lambda}{4\pi}\right)\sqrt{\frac{P_t G_r G_t}{P_r}}, \tag{16.10}$$

where P_t and P_r are the transmit and receive power, respectively, G_t and G_r are the isotropic gain of the transmit and receive antennas, respectively, λ is the RF wavelength, and d is the distance between antennas. The gain parameters appearing in Eq. (16.10) were measured in a RF anechoic chamber (Figure 16.16) and are dependent on the aircraft's azimuth and attitude relative to the ground-based transceiver. The transmitted power level is provided by the radio manufacturer, and the received power level reported by the radio.

An example of one such data set for the primary 915 MHz command and control link is shown in Figure 16.20, where received power level is plotted on a log–log scale versus distance between the radio links. Using the log–log plot allows for extrapolation of the received power level to the point where it intersects the noise floor of the radio (minimum power the radio can receive to maintain the link), and in this case corresponds to approximately −95 dBm. In practice, the maximum range is generally taken to be 10 dBm above the noise floor to ensure adequate link integrity in the presence of unknown and randomly varying interference. Of note, the measured data in Figure 16.20 are shown shaded by the aircraft bank angle, and for this particular test, there is no strong coupling between vehicle attitude and received power. However, this trend is not expected to hold as the aircraft approaches its maximum range, and the change in orientation effectively changes the polarization of the antennas, with a corresponding drop in received power. This sensitivity presents a flight-testing challenge as the loss of signal occurs at great range, and with the aircraft generally

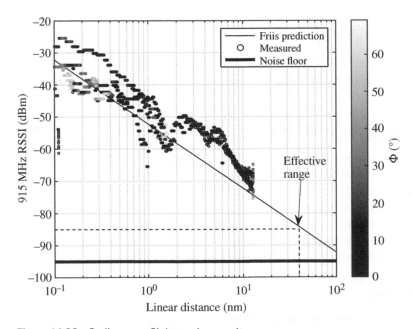

Figure 16.20 Radio range flight testing results.

not flying straight and level. A potential solution to this challenge includes intentional and controlled attenuation of the transmitted power such that the aircraft remains close to the operator. The aircraft can be safely maneuvered and flown in a controlled manner until loss of link occurs, and in that event, the attenuation can be rapidly removed from the transmitter in order to restore link integrity. The alternative approach is to rely upon extrapolation (as shown here), with a significant margin of safety built in to accommodate signal dropout during maneuvering.

16.6.4   Assessment of Autonomous System Performance

The Avanti UAS was designed to fly under the autonomous UAV record category established by the Fédération Aéronautique Internationale (FAI). (Here, "autonomous" was defined as autopilot-guided control of the flight vehicle that follows a series of waypoints.) Based on contemporary definitions of autonomy, this flight vehicle operated somewhere on the spectrum between automatic control (waypoint sequencing) and fully autonomous (high-level decision-making). For example, the flight vehicle was designed with real-time system identification, such that the flight computer would provide multi-sine inputs near the beginning of flight in order to determine or refine the flight control laws. The vehicle also had high-level autonomous capability such as automatic avoidance maneuvering in the event of a collision advisory, autonomous lost-link and ditching profiles, and automatic envelope protection.

This higher-level functioning can lead to unanticipated consequences, particularly when a single failure cascades through the complex autonomous logic in an unanticipated manner. A specific example of this occurred during the maneuvering of the flight vehicle at the turn-around point on the out-and-return course required for the speed record. In order to conserve fuel, the flight vehicle was programmed to execute a high-bank, low-speed turn. This condition can lead to what the aviation community terms "accelerated stall," where the higher load factor of the turn demands more lift, which elevates the lift coefficient at the condition and raises the stall speed (see Chapter 13 for the theory of turning flight). Flight data for this situation are presented in Figure 16.21.

The unanticipated failure in this case was an inadequate propulsion system model. When commanding low flight speeds, the propulsion model resorted to a "bang-bang" controller that modulated between the extremes of engine idle and full throttle. In a time-average sense, this results in adequate mean thrust to maintain the desired low flight speed. However, during the engine-idle portions of the flight profile, the airspeed decayed substantially. At the same time, the vehicle was maneuvering and sustaining a higher load factor. These two events nearly converged on a condition where the stall speed exceeded the flight speed of the vehicle, which could have resulted in loss of control. If the timing of the deceleration phase (engine idle) had been slightly different, the stall condition would have occurred. The flight vehicle's envelope protection system actually engaged near the 534-s mark, which is evidenced by throttle-up and oscillations of the airspeed (see Figure 16.21).

While the flight vehicle's envelope protection performed as designed, this was a hazardous condition that was not anticipated prior to flight. This situation is one example of an emergent property of an autonomous system. A robust V&V program that leads to certification will need to successfully identify these and other emergent failure modes.

16.7   Conclusion

The field of flight testing for UAS is rapidly changing as new regulations emerge. The basic principles of flight testing that have been detailed throughout this book are applicable to UAVs as well, with specific approaches and opportunities available due to the lack of a pilot on board. As system

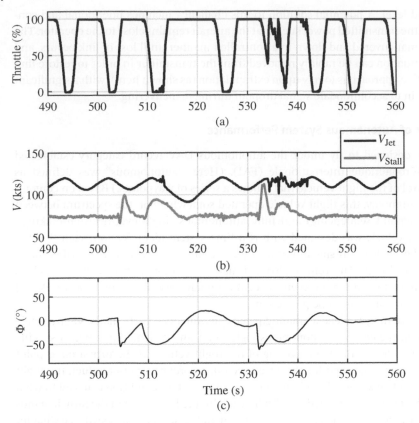

Figure 16.21 Example of an emergent property of an autonomous system, where stall was nearly reached in turning flight before the Avanti envelope protection system was activated at about 534 seconds. Shown are plots of (a) throttle setting, (b) airpseed, and (c) bank angle.

complexity increases, and new sensors are added and integrated as flight-critical design features, additional flight testing profiles will emerge over time.

One of the challenges of UAV flight testing is that the technical community still does not know all of the scenarios under which the flight vehicles must be tested for certification. Clear determination of flight control laws is critical, particularly for new and innovative concepts such as hybrid VTOL/fixed-wing aircraft with distributed electric propulsion and large pivoting wing (see Figure 16.6). Characteristics of the control link must be well established, such as the quality of the C2 link under various conditions. Maximum radio range must be defined, and lost-link protocols must be tested. For electric propulsion aircraft, battery life must be well characterized in order to define range and endurance limits of electric aircraft.

The impact of wind on the vehicle response is also critical. The gust tolerance of small UAS should be assessed under a wide variety of gusting wind conditions. This is particularly important for UAS with low wing loading operating in close proximity to buildings where wind gusts may lead to a stall (fixed wing), vortex ring state (VTOL), or some other upset condition.

Many of the performance capabilities and failure modes of UAS are still being identified. Over time, as clarity emerges on the key performance metrics that must be assessed in flight testing, protocols for assessing those metrics will be developed. This work will take the place in community-based standard organizations and ultimately inform policymaking on airworthiness certification standards for UAS.

Nomenclature

a_x	acceleration in the x-direction
AR	aspect ratio
b	wing span
C_{D_0}	parasitic drag coefficient
C_L	lift coefficient
d	distance from transmitter to receiver
D	drag
e	Oswald efficiency factor
F	force
g	gravitational acceleration
G_r	gain of a receiving antenna
G_t	gain of a transmitting antenna
L	lift
m	aircraft mass
P_r	receive power (RF)
P_S	specific excess power
P_t	transmit power (RF)
q	dynamic pressure
Re	Reynolds number
S	wing area
t	time
V	true airspeed
VIW	velocity independent of weight
W	aircraft weight
W_S	standard aircraft weight (maximum take-off weight)
x	horizontal direction
y	vertical direction
α	angle of attack
$\alpha_{L=0}$	angle of zero lift
λ	RF wavelength
Φ	bank angle
ρ	density
ρ_{SL}	density at standard sea level conditions
σ	density ratio, ρ/ρ_{SL}

Acronyms and Abbreviations

ACAS	airborne collision avoidance system
ADS-B	automatic dependent surveillance – broadcast
AGL	altitude above ground level
BEMT	blade element momentum theory
BVLOS	beyond visual line of sight
C2	command and control link
COA	certificate of authorization/waiver
DAA	detect and avoid

DAQ	data acquisition
EKF	extended Kalman filter
EMI	electromagnetic interference
EO	electrooptical
FAA	Federal Aviation Administration
FL	flight level
FMC	flight management computer
GA	general aviation
GBDAA	ground-based detect and avoid
GCS	ground control station
GNSS	global navigation satellite system
GPS	Global Positioning System
IMU	inertial measurement unit
INS	inertial navigation system
IR	infrared
ITU	International Telecommunication Union
KTAS	knots true airspeed
LiDAR	light detection and ranging
MSL	altitude above mean sea level
MFD	multifunction display
PFD	primary flight display
R/C	remote control
RF	radio frequency
RPA	remotely piloted aircraft
RPM	revolutions per minute
RTK	real-time kinematic
SATCOM	satellite communications
sUAS	small unmanned aircraft system
TCAS	traffic collision avoidance system
UAS	unmanned aircraft system
UAV	unmanned aerial vehicle
V&V	verification and validation
VHF	very high frequency
VLOS	visual line of sight
VMC	visual meteorological conditions
VTOL	vertical take-off and landing
WRC	World Radiocommunication Conference

References

Abdulrahim, M. (2019). Using simulated, unmanned, and manned aircraft in undergraduate flight test engineering education. *AIAA 2019-0065, AIAA Scitech 2019 Forum*, San Diego, CA. https://doi.org/10.2514/6.2019-0065.

Atkins, E., Ollero, A., Tsourdos, A. et al. (2016). *Unmanned Aircraft Systems*. Chechester, UK: Wiley.

Beard, R.W. and McLain, T.W. (2012). *Small Unmanned Aircraft: Theory and Practice*. Princeton, NJ: Princeton University Press.

Fahlstrom, P.G. and Gleason, T.J. (2012). *Introduction to UAV Systems*, 4e. Chichester, UK: Wiley.

Federal Aviation Administration (2016). Small Unmanned Aircraft Systems, 14 CFR §107 (28 June 2016).

Fédération Aéronautique Internationale (FAI) (2017a). Speed over a straight 15/25 km course, 236.89 km/h, world record, Class I (Experimental/New Technologies), Sub-class U-1 (Fixed wing), Category 7 (25 kg to less than 100 kg), Group 1 (Internal combustion and jet). Record ID 18242. https://fai.org/record/18242.

Fédération Aéronautique Internationale (FAI) (2017b). Distance over an out and return course, 45.23 km, world record, Class I (Experimental/New Technologies), Sub-class U-1 (Fixed wing), Category 7 (25 kg to less than 100 kg), Group 1 (Internal combustion and jet). Record ID 18243. https://fai.org/record/18243.

Fédération Aéronautique Internationale (FAI) (2017c). Speed, 236.89 km/h, world record, Class I (Experimental/New Technologies), Sub-class U-Absolute (Absolute Record of class U). Record ID 18244. https://fai.org/record/18244.

Friis, H.T. (1946). A note on a simple transmission formula. *Proceedings of the IRE* 34 (5): 254–256. https://doi.org/10.1109/JRPROC.1946.234568.

Gundlach, J. (2014). *Designing Unmanned Aircraft Systems: A Comprehensive Approach*, 2e. Reston, VA: American Institute of Aeronautics and Astronautics.

Hoerner, S.F. (1965). *Fluid Dynamic Drag*. Brick Town, NJ: Hoerner Fluid Dynamics. https://hoernerfluiddynamics.com.

Marqués, P. and Da Ronch, A. (2017). *Advanced UAV Aerodynamics, Flight Stability and Control*. Chichester, UK: Wiley.

Marshall, D.M., Barnhart, R.K., Shappee, E., and Most, M. (2016). *Introduction to Unmanned Aircraft Systems*, 2e. Boca Raton, FL: CRC Press, Taylor & Francis Group.

McCrink, M. H. (2015) Development of flight-test performance estimation techniques for small unmanned aerial systems. PhD Dissertation. Department of Mechanical and Aerospace Engineering, The Ohio State University, Columbus, OH.

McCrink, M.H. and Gregory, J.W. (2017). Blade element momentum modeling of low-Reynolds electric propulsion systems. *Journal of Aircraft* 54 (1) https://doi.org/10.2514/1.C033622.

McCrink, M.H. and Gregory, J.W. (2021). Design and development of a high-speed UAS for beyond visual line-of-sight operations. *Journal of Intelligent & Robotic Systems* 101: 31. https://doi.org/10.1007/s10846-020-01300-2.

National Research Council (2014). *Autonomy Research for Civil Aviation: Toward a New Era of Flight*. Washington, DC: The National Academies Press https://doi.org/10.17226/18815.

Shaw, J.A. (2013). Radiometry and the Friis transmission equation. *American Journal of Physics* 81 (1): 33–37. https://doi.org/10.1119/1.4755780.

Tennekes, H. (2009). *The Simple Science of Flight: From Insects to Jumbo Jets*, revised and expanded edition. Cambridge, MA: MIT Press.

UAS Task Force Airspace Integration Integrated Product Team (2011). *Department of Defense Unmanned Aircraft System Airspace Integration Plan*, 1-7ABA52E, version 2.0, March 2011, Appendix D.

Valasek, J. and Bowden, E. (2019). A UAS flight test engineering course modeled on professional methodologies and practices. *AIAA 2019-1078, AIAA Scitech 2019 Forum*, San Diego, CA. https://doi.org/10.2514/6.2019-1078.

Warwick, G. (2017). Ohio State pushes speed envelope in UAS research. In: *Aviation Week & Space Technology* (ed. J.C. Anselmo), (11 September), 179(36): 44. Shawnee, KS: Informa Plc.

Appendix A

Standard Atmosphere Tables

Below are tabulated values of the 1976 US Standard Atmosphere as a function of geometric altitude h_G. Chapter 2 provides complete information on the derivation and use of the standard atmosphere. Data are presented as a ratio with the standard sea-level value, with the temperature ratio $\theta = T/T_{SL}$, pressure ratio $\delta = P/P_{SL}$, and density ratio $\sigma = \rho/\rho_{SL}$. This format is selected for consistent representation of the data (without worrying about powers of 10), and because these ratios are often used in the flight testing environment (*e.g.* the density ratio, σ, is used to convert between true airspeed and equivalent airspeed, $V_e = V\sqrt{\sigma}$). The values of the standard atmosphere are provided here for ready reference when working quick or simple problems (these values are limited to lower altitudes, reflecting the emphasis in this book on light aircraft and UAVs). For engineering analysis, it may be more practical to rely upon computer codes based on the derived equations in Chapter 2, or one of the many freely available codes (*e.g.* the `stdatmo` function available on the MATLAB File Exchange). Data are presented in both English units (Tables A.1 and A.2) and SI units (Tables A.3 and A.4) up to altitudes of 15,000 ft or 5000 m, respectively.

Introduction to Flight Testing, First Edition. James W. Gregory and Tianshu Liu.
© 2021 John Wiley & Sons Ltd. Published 2021 by John Wiley & Sons Ltd.
Companion website: https://www.wiley.com/go/flighttesting

Table A.1 Standard sea-level properties in English units.

Property	Symbol	Value
Temperature	T_{SL}	518.67°R
Pressure	P_{SL}	2116.2 lb/ft^2
Density	ρ_{SL}	2.3769×10^{-3} slug/ft^3

Table A.2 Standard Atmosphere in English units.

Altitude, h_G (ft)	Temperature ratio, θ	Pressure ratio, δ	Density ratio, σ
−1,000	1.00688	1.03667	1.02959
−500	1.00344	1.01820	1.01471
0	1.00000	1.00000	1.00000
500	0.99656	0.98206	0.98545
1,000	0.99312	0.96439	0.97107
1,500	0.98969	0.94697	0.95684
2,000	0.98625	0.92982	0.94278
2,500	0.98281	0.91291	0.92887
3,000	0.97938	0.89626	0.91513
3,500	0.97594	0.87985	0.90154
4,000	0.97250	0.86369	0.88811
4,500	0.96907	0.84777	0.87483
5,000	0.96563	0.83209	0.86170
5,500	0.96219	0.81664	0.84873
6,000	0.95876	0.80143	0.83590
6,500	0.95532	0.78645	0.82323
7,000	0.95189	0.77170	0.81070
7,500	0.94845	0.75717	0.79832
8,000	0.94502	0.74287	0.78609
8,500	0.94158	0.72879	0.77400
9,000	0.93815	0.71492	0.76206
9,500	0.93471	0.70127	0.75025
10,000	0.93128	0.68783	0.73859
10,500	0.92784	0.67460	0.72707
11,000	0.92441	0.66158	0.71568
11,500	0.92097	0.64877	0.70444
12,000	0.91754	0.63615	0.69333
12,500	0.91411	0.62374	0.68235
13,000	0.91067	0.61152	0.67151
13,500	0.90724	0.59950	0.66080
14,000	0.90381	0.58768	0.65022

Table A.3 Standard sea-level properties in SI units.

Property	Symbol	Value
Temperature	T_{SL}	288.15 K
Pressure	P_{SL}	101,325 N/m^2
Density	ρ_{SL}	1.2250 kg/m^3

Table A.4 Standard Atmosphere in SI units.

Altitude, h_G (m)	Temperature ratio, θ	Pressure ratio, δ	Density ratio, σ
−400	1.00902	1.04835	1.03897
−200	1.00451	1.02394	1.01934
0	1.00000	1.00000	1.00000
200	0.99549	0.97652	0.98094
400	0.99098	0.95348	0.96216
600	0.98647	0.93089	0.94366
800	0.98196	0.90873	0.92543
1000	0.97745	0.88701	0.90748
1200	0.97294	0.86571	0.88979
1400	0.96843	0.84483	0.87237
1600	0.96392	0.82435	0.85521
1800	0.95941	0.80429	0.83832
2000	0.95490	0.78462	0.82168
2200	0.95039	0.76534	0.80529
2400	0.94588	0.74645	0.78916
2600	0.94137	0.72794	0.77328
2800	0.93687	0.70981	0.75764
3000	0.93236	0.69204	0.74225
3200	0.92785	0.67464	0.72710
3400	0.92334	0.65759	0.71219
3600	0.91884	0.64090	0.69751
3800	0.91433	0.62455	0.68307
4000	0.90983	0.60854	0.66885
4200	0.90532	0.59287	0.65487
4400	0.90081	0.57752	0.64111
4600	0.89631	0.56250	0.62758
4800	0.89180	0.54780	0.61426
5000	0.88730	0.53341	0.60117

Appendix B

Useful Constants and Unit Conversion Factors

The topic of units and unit conversions can be a confusing and frustrating experience for students and practitioners alike. From an engineering standpoint, and within a global context, the metric system offers distinct advantages for the analysis and presentation of data. Indeed, the International Civil Aviation Organization (ICAO) has called for standardization of units on the International System of Units (SI), the modern form of the metric system. However, the aviation community today uses a varied and diverse set of units, often mixing English (Imperial) with metric. The diversity of units can be downright maddening! Even in North America, where the Imperial unit system is used almost exclusively, there can be a mix of statute miles (what most North Americans use for terrestrial measurement of long distances), nautical miles (about 15% longer than a statute mile), and feet for measuring distance. Around the globe, a single aircraft may measure airspeed in kilometers per hour or knots (nautical miles per hour), rate of climb in feet per minute, altitude in feet, and altimeter setting in hectopascals! (see Figure 3.2 for an instrument panel that reports data in these units). Most altitudes and flight levels are defined in units of feet, although a few nations follow the ICAO standard of meters. For example, pilots on an international flight that enters Chinese airspace will often have to do a small climb or descent to adjust from an altitude based on feet to one based on meters. Mental math or lookup tables may be needed to follow air traffic control instructions (given in meters) if the aircraft's altimeter is calibrated in feet.

This may seem bewildering, but there are a few important principles that make the situation manageable for pilots and engineers alike. First, when performing engineering analysis, always convert data to standard consistent units. Second, unit conversion factors can be easily created by forming a ratio of the input and output units. Third, it is essential to carefully document and track units in the calculations in order to avoid errors in analysis based on faulty assumptions.

The consequences of unit conversion errors, or incorrect assumptions of units, can lead to dramatic and severe consequences. This is best illustrated by the specific example of a Boeing 767 that suffered fuel exhaustion and double-engine failure, turning the aircraft into a massive glider – what is now referred to as the "Gimli Glider" (Lockwood 1985).

On 23 July 1983, Air Canada Flight 143 was preparing for departure from Montreal, Quebec bound for Edmonton, Alberta with an intermediate stop in Ottawa, Ontario. On the day of the flight, however, the pilots noted a problem with the fuel quantity indicator system (FQIS) – the gas gauges were bad. The pilots instructed the ground crew to manually check the fuel levels with a dipstick, which required conversion between the measured fuel volume and fuel mass; the pilots then double-checked the calculations. In the stopover in Ottawa, the ground crew and flight crew again checked fuel levels and proceeded with the second leg of the flight. However, at an altitude of 41,000 ft over Red Lake, Ontario, the aircraft control panel gave warnings of low fuel pressure. Shortly thereafter, the left engine shut down, followed quickly by the right engine – it was not clear to the pilots at the time, but their aircraft had suffered from complete fuel exhaustion.

Introduction to Flight Testing, First Edition. James W. Gregory and Tianshu Liu.
© 2021 John Wiley & Sons Ltd. Published 2021 by John Wiley & Sons Ltd.
Companion website: https://www.wiley.com/go/flighttesting

The pilots declared an emergency and diverted the flight toward Winnipeg. However, it became evident that the glide distance of the aircraft was insufficient to make it to the airport, so the pilots again diverted to a disused military airbase in Gimli, Manitoba. Unbeknownst to the pilots, the selected runway had been converted into a drag racing strip, and there were a large number of drivers and their families camping in the immediate vicinity of the runway as the aircraft approached. Fortunately, the pilots brought the aircraft to a stop before reaching people on the ground, and there were no fatalities or significant injuries resulting from this incident.

The fuel exhaustion problem with the Gimli Glider was due to improper tracking and conversion of units. On the day of the flight, the FQIS on board the aircraft was inoperable, but the pilots decided to continue the flight and measure initial fuel quantity by having the ground crew manually measure the fuel level with dipsticks. Using this technique, the fuel quantity was measured as a volume (in units of liters, l). However, the flight computer on board the aircraft expects fuel quantity to be input with units of mass, where the density of fuel is used to calculate the mass of a given volume of fuel, $m_{fuel} = v_{fuel}\rho_{fuel}$. The ground crew used the standard conversion factor of 1.77, *without taking note of the units associated with this conversion*. Jet A-1 fuel has a density of 0.804 kg/l, and there are 2.2 lb in a kilogram, giving a conversion of 1.77 lb/l. However, the brand-new Boeing 767's flight computers expected metric data, such that the fuel weight should have been provided in kilograms instead of pounds. Thus, the ground crew should have used a conversion factor of 0.804 kg/l instead of 1.77 lbs/l. The result was that the aircraft was carrying a factor of 2.2 less fuel than the pilots thought, leading to the fuel exhaustion and double-engine failure event. If the ground crews in Montreal and Ottawa had tracked the units with the conversion factor, the error may have been caught in time for sufficient fuel to be loaded on board.

With this critical example in mind, let's turn our attention to the use of units in flight testing. In the preface to this text, we have addressed our approach and rationale for handling engineering units in this book. Nearly every equation in this book presumes standard, consistent units (defined in Table B.1), without any embedded unit conversion factors. This has the tremendous advantage of providing equations that illustrate the basic physics, without the confusion of embedded unit conversions. It also allows the equations to be used interchangeably with English or SI unit systems (as long as a consistent set is used throughout). Thus, when input parameters with a standard, consistent unit system are provided to an equation, the output of the equation is guaranteed to provide an answer in the same consistent unit system. When working engineering problems, the

Table B.1 Standard, consistent units for the English and SI unit systems.

	English units	SI units
Time	s	s
Pressure	lb/ft² (=psf)	N/m² (=Pa)
Temperature	°R	K
Density	slug/ft³	kg/m³
Velocity	ft/s	m/s
Force	lb	N (=kg m/s²)
Mass	slug	kg
Energy	ft lb	N m (=J)
Power	ft lb/s	J/s (=W)
Area	ft²	m²

Table B.2 Useful conversion factors.

Value		Value
	Time	
1 h	=	3600 s
	Distance	
1 ft	=	12 in.
1 m	=	3.28084 ft
1 mi	=	5280 ft
1 nm	=	6076.1 ft
	Velocity	
1 knot	=	1 nm/h
1 knot	=	1.15 mi/h (mph)
1 knot	=	1.6878 ft/s
1 m/s	=	1.9438 knots
	Mass	
1 slug	=	32.1740 lb$_m$ (pound mass)
1 slug	=	14.594 kg
1 kg	=	2.2046 lb$_m$
	Force	
1 lb	=	4.44822 N
	Force/mass equivalence[a)]	
1 lb$_m$	=	1 lb
1 kg	=	2.2046 lb
1 kg	=	9.80665 N
	Pressure	
1 atm	=	2116.2 lb/ft^2 (psf)
	=	14.69 lb/in.^{2} (psi)
	=	101 325 N/m^2 (Pa)
	=	101.325 kPa
	=	1013.25 hPa
	=	29.92 inHg
	=	407. 0 in. H$_2$O
	=	760.0 mm Hg
	=	760.0 Torr
	=	1.01325 bar
	=	1013.25 mbar
	Energy	
1 Btu	=	778 ft lb
1 ft lb	=	1.3558 J

(*Continued*)

Table B.2 (Continued)

Value		Value
Power		
1 hp	=	550 ft lb/s
1 hp	=	745.7 W
Temperature		
1 K	=	1.8°R
T (K)	=	T (°C) + 273.15
T (°R)	=	T (°F) + 459.67
Derived SI units		
1 N	=	1 kg m/s^2
1 J	=	1 N m
1 W	=	1 J/s

a) Assuming standard sea-level gravitational
    acceleration.

best practice is to take all input variables in the initial units (*e.g.*, knots), convert all to standard units (*e.g.*, from knots to ft/s), perform all desired calculations, and then convert the final results back to the desired output units.

Let's now briefly discuss best practices for handling units and unit conversions with engineering analysis of flight test data. First, it's critical to convert all measured parameters to a consistent set of units (see Table B.1). This can be either English or SI – it doesn't matter, as long as one is consistent. For example, if an engineer wishes to calculate dynamic pressure from flight test measurements, the measured airspeed must be converted from knots to feet per second, pressure altitude from the altimeter must be converted to a pressure in pounds per square foot through the standard atmosphere (Appendix A), and temperature must be converted from degrees Celsius to degrees Rankine. Only after these conversions have been made, the engineer can calculate density (from the ideal gas law) and dynamic pressure ($0.5\rho V^2$), which will be in units of slugs per cubic foot, and pounds per square foot, respectively. If dynamic pressure is desired in units of psi, an engineer may simply divide the answer in psf by 144 in^2/ft^2. It's critical to note that many common units used in aviation (knots, horsepower, degrees Celsius, etc.) are *nonstandard* and are often *inconsistent*. When writing data analysis computer codes, the best practice is to perform unit conversions as the first step and to clearly document units using comments throughout the code for clarity.

Convenient unit conversion factors are provided in Table B.2. If a specific conversion factor is not available in the table, an appropriate factor can be constructed by merging two or more other conversion factors. To illustrate this, let's consider the example of a pilot flying internationally in the US-manufactured aircraft with an altimeter Kollsman window calibrated in inches of mercury only. If the pilot receives an altimeter setting (QNH) of 1025 hPa from air traffic control, this must be converted to inches of mercury before being dialed into the altimeter. A suitable conversion factor may be constructed by recognizing that both 29.92 inHg and 1013.25 hPa are readings of standard sea-level pressure provided in Table B.2 (thus, the two indications are equal). We can form a ratio of these two equal quantities and multiply it by the provided altimeter reading, as follows:

$$1025\,\text{hPa} \times \frac{29.92\,\text{inHg}}{1013.25\,\text{hPa}} = 30.27\,\text{inHg}.$$

Table B.3 Useful constants in standard, consistent units.

Property	Symbol	English units	SI units
Standard sea-level temperature	T_{SL}	518.67 °R	288.15 K
Standard sea-level pressure	p_{SL}	2116.2 lb/ft^2	101325 N/m^2
Standard sea-level density	ρ_{SL}	2.3769×10^{-3} slug/ft^3	1.225 kg/m^3
Gravitational acceleration	g_0	32.1740 ft/s^2	9.80665 m/s^2
Earth's radius	r_{Earth}	2.085553×10^7 ft	6.356766×10^6 m
Gas constant for air	R	1716 ft lb/(slug °R)	287.05 J/(kg K)
Mean molecular mass of dry air	M	1.9847 slug/kmol	28.9644 kg/kmol
Universal gas constant	$\overline{R}$	3.4068×10^3 ft lb/(kmol °R)	8.31432×10^3 N m/(kmol K)
Ratio of specific heats	γ	1.4	1.4
Speed of sound	a	1116.4 ft/s	340.29 m/s
Dynamic viscosity	μ_{SL}	3.7372×10^{-7} slug/(ft s)	1.7894×10^{-5} kg/(m s)

Table B.4 SI prefix definitions.

Symbol	Prefix name	Factor		
M	mega	10^6	=	1,000,000
k	kilo	10^3	=	1,000
h	hecto	10^2	=	100
da	deka	10^1	=	10
—	—	1	=	1
d	deci	10^{-1}	=	0.1
c	centi	10^{-2}	=	0.01
m	milli	10^{-3}	=	0.001
μ	micro	10^{-6}	=	0.000001

Any of the equalities provided in Table B.2 can be combined, cancelling units as appropriate to yield the desired units at the output. Also, note that some units are derived, such as the watt, joule, newton, and others. Thus, it can be helpful to remember the definitions of these derived units when seeking to cancel units in the conversion process. Finally, we conclude by highlighting the availability of commonly used constants in aviation and engineering in Table B.3, along with standard SI prefixes in Table B.4 for ready reference.

Reference

Lockwood, G.H. (1985). Final Report of the Board of Inquiry into Air Canada Boeing 767 C-GAUN Accident – Gimli, Manitoba 23 July 1983. Ottawa, Canada: Canadian Government Publishing Centre. http://data2.collectionscanada.gc.ca/e/e444/e011083519.pdf.

Appendix C

Stability and Control Derivatives for a Notional GA Aircraft

Estimates of stability derivatives are for a notional general aviation aircraft, as defined by Example 8.2.1 on pp. 723–724 of Phillips (2004), are provided in Table C.1.

Table C.1 Stability derivatives for a notional GA aircraft.

S	$=$	$185\,\text{ft}^2$
b	$=$	$33\,\text{ft}$
W	$=$	$2800\,\text{lb}$
V	$=$	$180\,\text{ft/s}$
C_{D_0}	$=$	0.05
I_{xx}	$=$	$1000\,\text{slug}\cdot\text{ft}^2$
I_{yy}	$=$	$3000\,\text{slug}\cdot\text{ft}^2$
I_{zz}	$=$	$3500\,\text{slug}\cdot\text{ft}^2$
I_{xz}	$=$	$30\,\text{slug}\cdot\text{ft}^2$
C_{L_α}	$=$	4.40
C_{D_α}	$=$	0.35
C_{m_α}	$=$	-0.68
$C_{L_{\dot\alpha}}$	$=$	1.60
$C_{m_{\dot\alpha}}$	$=$	-4.35
C_{Y_β}	$=$	-0.56
C_{l_β}	$=$	-0.075
C_{n_β}	$=$	0.070
C_{D_q}	$\cong$	0.0
C_{L_q}	$=$	3.80
C_{m_q}	$=$	-9.95
C_{Y_p}	$\cong$	0.0

Introduction to Flight Testing, First Edition. James W. Gregory and Tianshu Liu.
© 2021 John Wiley & Sons Ltd. Published 2021 by John Wiley & Sons Ltd.
Companion website: https://www.wiley.com/go/flighttesting

Table C.1 (Continued)

C_{l_p}	=	-0.410
C_{n_p}	=	-0.0575
C_{Y_r}	=	0.240
C_{l_r}	=	0.105
C_{n_r}	=	-0.125

Source: Phillips (2004).

Reference

Phillips, W.F. (2004). *Mechanics of Flight*. Hoboken, NJ: Wiley.

Index

Introduction to Flight Testing, First Edition. James W. Gregory and Tianshu Liu.
© 2021 John Wiley & Sons Ltd. Published 2021 by John Wiley & Sons Ltd.
Companion website: https://www.wiley.com/go/flighttesting

Printed and bound by CPI Group (UK) Ltd, Croydon, CR0 4YY

27/10/2024

14580220-0001